THE
HUMAN
BODY

An Illustrated Guide
To Your Body and How it

General Editor:
Professor Peter Abrahams

amber
BOOKS

Published by
Amber Books Ltd
74–77 White Lion Street
London
N1 9PF
United Kingdom
www.amberbooks.co.uk
Appstore: itunes.com/apps/amberbooksltd
Facebook: www.facebook.com/amberbooks
Twitter: @amberbooks

ISBN: 978-1-78274-516-7

Project Editor: Sarah Uttridge
Picture Research: Terry Forshaw
Design: Ummagumma
Introduction by Conor Kilgallon

Printed in China

Contents

Introduction

Medicine and our understanding of human anatomy has developed enormously over the last few hundred years, often due to the ground-breaking discoveries of a few radical thinkers.

OUR FASCINATION WITH OUR BODIES and how they work, why they go wrong, and what to do to heal them is boundless. Throughout history, countless theories, mostly erroneous, explaining anatomy and physiology have been dreamed up by all manner of physicians, surgeons, quacks, witchdoctors, alchemists, faith-healers, astrologers and charlatans, who in their day, were often well-respected and highly paid professionals.

Despite this catalogue of bad practice, the history of medicine is punctuated by brilliant discoveries and truly visionary thinking that has, against all the odds, hauled us into the modern era of medical science. Hippocrates, 'the father of medicine', practised medicine on the Greek island of Cos in the fifth century BC, and is undoubtedly the most famous and recognizable figure of them all. His achievement was to establish a specialist body of physicians who were governed by a strict code of ethics, and who employed observable scientific methods in their research. This laid the foundation for modern medical practice.

The four 'humours'

Hippocrates' work had a profound influence on medicine, and his ideas were enthusiastically expanded by doctors in the centuries that followed. Unfortunately, his theories on anatomy and disease were factually inaccurate. He believed that four 'humours', (black bile, yellow bile, phlegm and blood) governed human health and that any illness

was a result of imbalances between them. With the exception of the monks, who grew herbs and plants with some genuine medicinal properties, factual inaccuracy was the trademark of medicine and anatomy during the Middle Ages. The 'humours' theory was still widely held to be true, and Christian and Islamic religious belief was highly influential on medical theory. All sorts of theories, such as blood letting, draining 'noxious fluids' from the body or encouraging 'excess fluids' to move about the body freely were commonly put into practice, often accompanied by

In this 17th century diagram by Anastasius Kircher, the human body represents the world in microcosm, which is described as a living organism with metabolic processes.

apothecaries' potions, which contained such bizarre and infamous ingredients as newt's tongues and worm's livers.

With the arrival of the Renaissance in Italy in the late fourteenth century, medical science moved forward. The rediscovery of classical learning encouraged physicians to re-apply scientific methods to medical research, and leave behind the influence of religion and superstition. Great names from the period, such as Leonardo Da Vinci, put forward new ideas. He believed that in order to treat disease it was necessary to first learn about the body and its processes, learning that could ultimately only come through the dissection of human cadavers. Dissection was not,

Surgeons can perform today what would have been a miracle only 200 years ago, with a patient survival rate that would have stunned early physicians.

however, a new idea. Claudius Galen, a highly influential second century physician, had dissected animals and had assumed that human anatomy followed the same patterns, an idea that became accepted wisdom for over 1500 years. But by the sixteenth century, the anatomist, Andreas Vasalius, showed Galen was wrong and revealed previously unknown anatomical structures in his book, *de Humani Corporis Fabrica* (*the Fabric of the Human Body*), in 1543. Procuring bodies for dissection, however, was neither easy or pleasant. The Church forbade human dissection, so anatomists across Europe infamously resorted to robbing graves and cutting down bodies from gallows in order to obtain fresh materials for their research. Other pioneering work recording what had been discovered was conducted by Da Vinci and Vasalius, who sought to accurately represent anatomical structure through detailed diagrams and illustrations.

Blood circulation
Still, these ideas and methods were controversial and often dismissed. In 1628, the English doctor, William Harvey, stunned the medical world when he published An Anatomical Disquisition on the Movement of the Heart and Blood. In this book, he showed that blood circulated around the body and further proposed that the heart pumped blood through arteries. He also realised the significance of the valves of the heart in controlling the flow of blood. Although his ideas were considered outlandish, this scientific method of research was again proved to be the way forward. His findings were confirmed by the invention of the microscope in the late seventeenth century: for the first time in history, scientists could observe more than the naked eye would allow.

By the end of the nineteenth century, many of the practices and procedures we now take for granted

were coming to the fore. Crude anaesthetics were developed by James Young Simpson, antiseptics were pioneered by Joseph Lister, and in 1896, Wilhelm Rontgen amazed the world with a new invention that allowed internal examination of the body without the need for surgery: the x-ray machine was born. Other ground-breaking work by figures such as Louis Pasteur, who established the link between germs and disease, and Karl Landsteiner, who discovered the four main blood groups, paved the way for much more complex surgery such as organ transplants. Surgeons can perform today what would have been a miracle only 200 years ago, with a patient survival rate that would have stunned early physicians.

Discovering human anatomy
So how much do we actually know about how our own body systems work and how can we better understand what the doctor or surgeon sees and does? *The Human Body* will show you what we are really made of through a thorough examination of human anatomy. The book is structured from the head to the toe, and is broken down into the head, neck, thorax, upper limbs, abdomen, reproductive system, pelvis, lower limbs and general body systems. In turn, each section examines the bones, muscles, nerves, soft tissue and organs and how they work and interact. This book is the beginning of a fascinating journey.

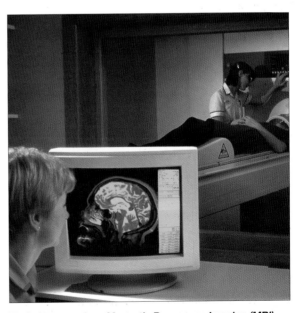

Techniques such as Magnetic Resonance Imaging (MRI) allow medical staff to gain a 'sliced' image through the body. This can be used to study tumours in soft tissue, such as the brain.

The skull

The skull is the head's natural crash helmet, protecting the brain and sense organs from damage. It is made up of 28 separate bones and is the most complex element of the human skeleton.

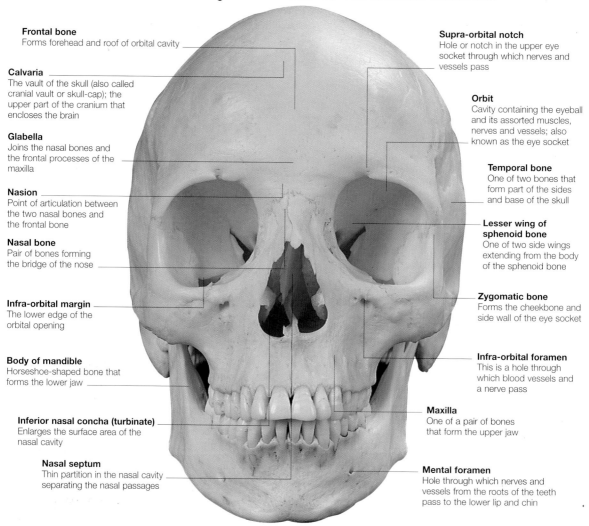

Frontal bone
Forms forehead and roof of orbital cavity

Calvaria
The vault of the skull (also called cranial vault or skull-cap); the upper part of the cranium that encloses the brain

Glabella
Joins the nasal bones and the frontal processes of the maxilla

Nasion
Point of articulation between the two nasal bones and the frontal bone

Nasal bone
Pair of bones forming the bridge of the nose

Infra-orbital margin
The lower edge of the orbital opening

Body of mandible
Horseshoe-shaped bone that forms the lower jaw

Inferior nasal concha (turbinate)
Enlarges the surface area of the nasal cavity

Nasal septum
Thin partition in the nasal cavity separating the nasal passages

Supra-orbital notch
Hole or notch in the upper eye socket through which nerves and vessels pass

Orbit
Cavity containing the eyeball and its assorted muscles, nerves and vessels; also known as the eye socket

Temporal bone
One of two bones that form part of the sides and base of the skull

Lesser wing of sphenoid bone
One of two side wings extending from the body of the sphenoid bone

Zygomatic bone
Forms the cheekbone and side wall of the eye socket

Infra-orbital foramen
This is a hole through which blood vessels and a nerve pass

Maxilla
One of a pair of bones that form the upper jaw

Mental foramen
Hole through which nerves and vessels from the roots of the teeth pass to the lower lip and chin

The skull is the skeleton of the face and the head. The basic role of the skull is protecting the brain, the organs of special sense such as the eyes, and the cranial parts of the breathing and digestive system. It also provides attachment for many of the muscles of the neck and head.

Although often thought of as a single bone, the skull is made up of 28 separate bones. For convenience, it is often divided into two main sections: the cranium and the mandible. The basis for this is that, whereas most of the bones of the skull articulate by relatively fixed joints, the mandible (jawbone) is easily detached.

The cranium is then subdivided into a number of smaller regions, including:
• cranial vault (upper dome part of the skull)
• cranial base
• facial skeleton
• upper jaw
• acoustic cavities (ears)
• cranial cavities (interior of skull housing the brain).

Illuminated skull

Most of the bones of the skull are connected by sutures –
immovable fibrous joints. These, and the bones inside the skull,
can be seen most clearly using a brightly illuminated skull.

The areas where skull bones meet are called 'sutures'. The coronal suture, for example, occurs between the frontal and parietal bones, and the sagittal suture connects the two parietal bones. It is important to learn the position of these joints, because they can be confused with fractures on X-rays.

In babies, there are relatively large gaps between skull bones, allowing the head to squeeze through the birth canal without fracturing. The gaps are covered in fibrous membranes called 'fontanelles'. In most 'head-first' births, the fontanelles can be palpated (examined using the fingertips) during vaginal examinations to determine the position of the head.

CHANGE OF FACE
Because children have only rudimentary teeth and sinuses, their faces are smaller proportionally to adults'. (The skull of a newborn, however, is one-quarter of its body size.) As we get older, the relative size of the face diminishes as our gums shrink and we lose our teeth and the bony sockets.

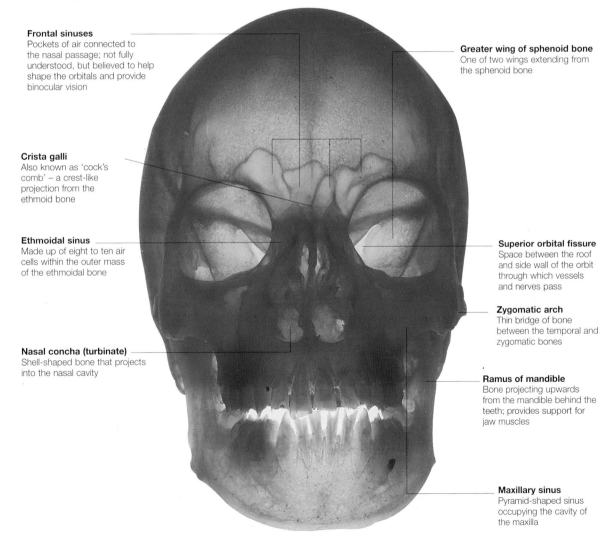

Frontal sinuses
Pockets of air connected to the nasal passage; not fully understood, but believed to help shape the orbitals and provide binocular vision

Crista galli
Also known as 'cock's comb' – a crest-like projection from the ethmoid bone

Ethmoidal sinus
Made up of eight to ten air cells within the outer mass of the ethmoidal bone

Nasal concha (turbinate)
Shell-shaped bone that projects into the nasal cavity

Greater wing of sphenoid bone
One of two wings extending from the sphenoid bone

Superior orbital fissure
Space between the roof and side wall of the orbit through which vessels and nerves pass

Zygomatic arch
Thin bridge of bone between the temporal and zygomatic bones

Ramus of mandible
Bone projecting upwards from the mandible behind the teeth; provides support for jaw muscles

Maxillary sinus
Pyramid-shaped sinus occupying the cavity of the maxilla

Side of the skull

A lateral or side view of the skull clearly reveals the complexity of the structure, with many separate bones and the joints between them.

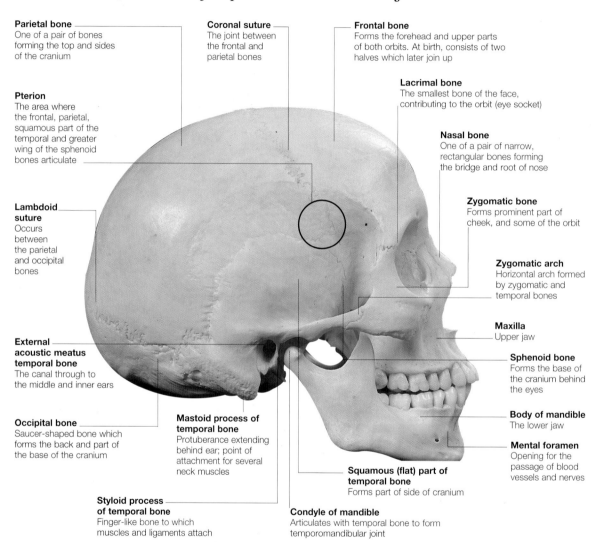

Parietal bone
One of a pair of bones forming the top and sides of the cranium

Pterion
The area where the frontal, parietal, squamous part of the temporal and greater wing of the sphenoid bones articulate

Lambdoid suture
Occurs between the parietal and occipital bones

External acoustic meatus temporal bone
The canal through to the middle and inner ears

Occipital bone
Saucer-shaped bone which forms the back and part of the base of the cranium

Styloid process of temporal bone
Finger-like bone to which muscles and ligaments attach

Coronal suture
The joint between the frontal and parietal bones

Mastoid process of temporal bone
Protuberance extending behind ear; point of attachment for several neck muscles

Squamous (flat) part of temporal bone
Forms part of side of cranium

Condyle of mandible
Articulates with temporal bone to form temporomandibular joint

Frontal bone
Forms the forehead and upper parts of both orbits. At birth, consists of two halves which later join up

Lacrimal bone
The smallest bone of the face, contributing to the orbit (eye socket)

Nasal bone
One of a pair of narrow, rectangular bones forming the bridge and root of nose

Zygomatic bone
Forms prominent part of cheek, and some of the orbit

Zygomatic arch
Horizontal arch formed by zygomatic and temporal bones

Maxilla
Upper jaw

Sphenoid bone
Forms the base of the cranium behind the eyes

Body of mandible
The lower jaw

Mental foramen
Opening for the passage of blood vessels and nerves

Several of the bones of the skull are paired, with one on either side of the midline of the head. The nasal, zygomatic, parietal and temporal bones all conform to this symmetry. Others, such as the ethmoid and sphenoid bones, occur singly along the midline. Some bones develop in two separate halves and then fuse at the midline, namely the frontal bone and the mandible (lower jaw).

The bones of the skull constantly undergo a process of remodelling: new bone develops on the outer surface of the skull, while the excess on the inside is reabsorbed into the bloodstream. This dynamic process is facilitated by the presence of numerous cells and also a good blood supply.

Occasionally, a deficiency in the cells responsible for reabsorption upsets the bone metabolism, which can result in severe thickening of the skull – osteopetrosis, or Paget's disease – and deafness or blindness may follow.

Inside the skull

The inside of the left half of the skull shows the large cranial vault (calvaria) and facial skeleton in section.

Comparing this photograph with the one of the skull's exterior, many of the same bones can be seen, as well as additional structures. The bony part of the nasal septum (the dividing wall of the nasal cavity) consists of the vomer and the perpendicular plate of the ethmoid bone.

In this skull, the sphenoidal air sinuses are large. The pituitary fossa, containing the pea-sized, hormone-producing pituitary gland, projects down into the sinus. The circle marks the pterion, corresponding to the position marked on the external photograph.

The skull covers the brain, and skull fractures can lead to potentially life-threatening situations. If the side of the skull is fractured in the region of the temporal bone, the blood vessel of the middle meningeal artery may be damaged (extra dural haemorrhage). This vessel supplies the skull bones and the meninges (outer coverings of the brain), and if ruptured, the escaping blood may cause pressure on vital centres in the brain. If not relieved, this can rapidly cause death. The artery is accessible to the surgeon if entry is made near the pterion.

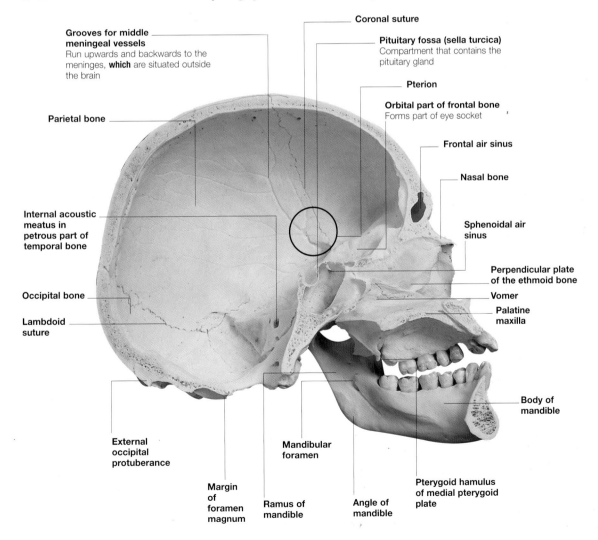

Coronal suture

Grooves for middle meningeal vessels
Run upwards and backwards to the meninges, **which** are situated outside the brain

Pituitary fossa (sella turcica)
Compartment that contains the pituitary gland

Pterion

Orbital part of frontal bone
Forms part of eye socket

Parietal bone

Frontal air sinus

Nasal bone

Internal acoustic meatus in petrous part of temporal bone

Sphenoidal air sinus

Perpendicular plate of the ethmoid bone

Vomer

Occipital bone

Palatine maxilla

Lambdoid suture

Body of mandible

External occipital protuberance

Mandibular foramen

Margin of foramen magnum

Ramus of mandible

Angle of mandible

Pterygoid hamulus of medial pterygoid plate

11

Top and base of the skull

The calvaria, or vault of the skull, is the upper section of the cranium, surrounding and protecting the brain.

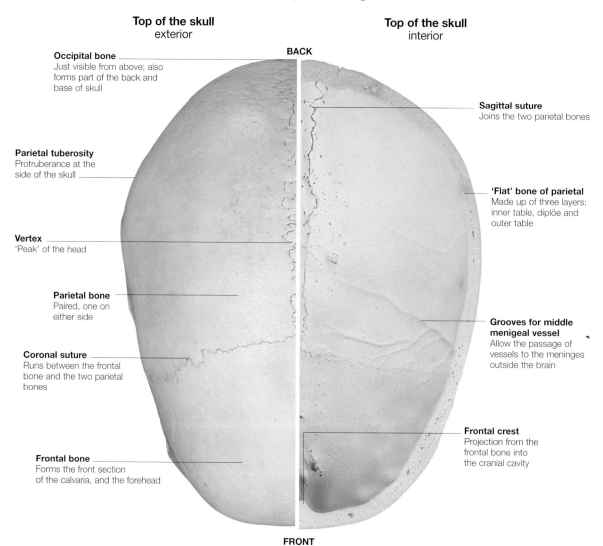

Top of the skull
exterior

Top of the skull
interior

BACK

Occipital bone
Just visible from above; also forms part of the back and base of skull

Sagittal suture
Joins the two parietal bones

Parietal tuberosity
Protruberance at the side of the skull

'Flat' bone of parietal
Made up of three layers: inner table, diplöe and outer table

Vertex
'Peak' of the head

Parietal bone
Paired, one on either side

Grooves for middle menigeal vessel
Allow the passage of vessels to the meninges outside the brain

Coronal suture
Runs between the frontal bone and the two parietal bones

Frontal crest
Projection from the frontal bone into the cranial cavity

Frontal bone
Forms the front section of the calvaria, and the forehead

FRONT

The four bones that make up the calvaria are the frontal bone, the two parietals and a portion of the occipital bone.

These bones are formed by a process in which the original soft connective tissue membrane ossifies (hardens) into bone substance, without going through the intermediate cartilage stage, as happens with some other bones of the skull.

Points of interest in the calvaria include:
• The sagittal suture running longitudinally from the lambdoid suture at the back of the head to the coronal suture.
• The vertex (highest point) of the skull; the central uppermost part, along the sagittal suture.
• The distance between the two parietal tuberosities is the widest part of the cranium.
• The complex, interlocking nature of the sutures which enable substantial skull growth in the formative years, and provide strength and stability in the adult skull.

Base of the skull

This unusual view of the skull is from below. The upper jaw and the hole through which the spinal cord goes can be seen.

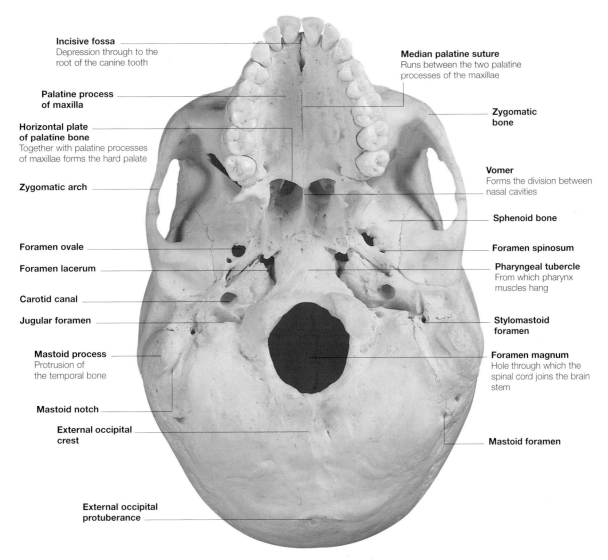

Incisive fossa
Depression through to the root of the canine tooth

Palatine process of maxilla

Horizontal plate of palatine bone
Together with palatine processes of maxillae forms the hard palate

Zygomatic arch

Foramen ovale

Foramen lacerum

Carotid canal

Jugular foramen

Mastoid process
Protrusion of the temporal bone

Mastoid notch

External occipital crest

External occipital protuberance

Median palatine suture
Runs between the two palatine processes of the maxillae

Zygomatic bone

Vomer
Forms the division between nasal cavities

Sphenoid bone

Foramen spinosum

Pharyngeal tubercle
From which pharynx muscles hang

Stylomastoid foramen

Foramen magnum
Hole through which the spinal cord joins the brain stem

Mastoid foramen

The bones found in the midline region of the base of the skull (the ethmoid, sphenoid and part of the occipital bone) develop in a different way from those of the vault of the skull. They are derived from an earlier cartilaginous structure in a process called endochondral ossification.

The maxillae are the two tooth-bearing bones of the upper jaw, one on each side. The palatine processes of the maxillae and the horizontal plates of the palatine bones form the hard palate.

PALATE DEFECTS
A cleft palate occurs when the structures of the palate do not fuse as normal before birth, creating a gap in the roof of the mouth. This links the oral and nasal cavities. If the gap extends through to the upper jaw, a harelip will become apparent on the upper lip. However, surgery can often improve the defect.

Children with narrow palates and crowded teeth can have an orthodontic appliance fitted which gradually increases tension across the longitudinally running midline palatine.

Over a period of months, the edges of the suture are forced apart, allowing for the growth of new bone, and extra space for the teeth.

13

Scalp

The scalp is composed of five layers of tissue that cover the bones of the skull. The skin is firmly attached to the muscles of the scalp by connective tissue which also carries numerous blood vessels.

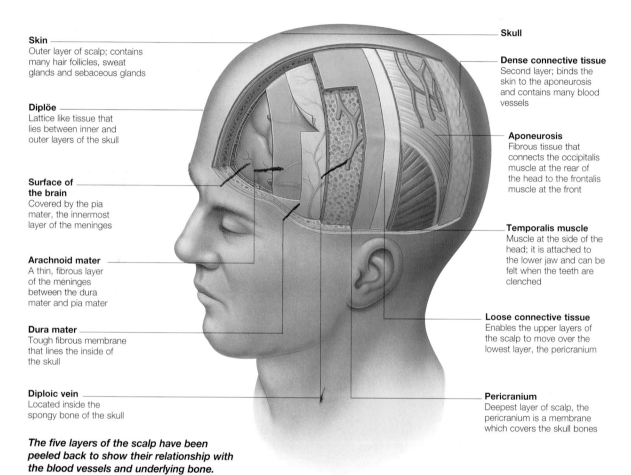

Skin
Outer layer of scalp; contains many hair follicles, sweat glands and sebaceous glands

Diplöe
Lattice like tissue that lies between inner and outer layers of the skull

Surface of the brain
Covered by the pia mater, the innermost layer of the meninges

Arachnoid mater
A thin, fibrous layer of the meninges between the dura mater and pia mater

Dura mater
Tough fibrous membrane that lines the inside of the skull

Diploic vein
Located inside the spongy bone of the skull

Skull

Dense connective tissue
Second layer; binds the skin to the aponeurosis and contains many blood vessels

Aponeurosis
Fibrous tissue that connects the occipitalis muscle at the rear of the head to the frontalis muscle at the front

Temporalis muscle
Muscle at the side of the head; it is attached to the lower jaw and can be felt when the teeth are clenched

Loose connective tissue
Enables the upper layers of the scalp to move over the lowest layer, the pericranium

Pericranium
Deepest layer of scalp, the pericranium is a membrane which covers the skull bones

The five layers of the scalp have been peeled back to show their relationship with the blood vessels and underlying bone.

The scalp is the covering of the top of the head which stretches from the hairline at the back of the skull to the eyebrows at the front. It is a thick, mobile, protective covering for the skull, and it has five distinct layers, the first three of which are bound tightly together.

The skin of the scalp is the thickest in the body and the hairiest. As well as its functions of hair-bearing and protection of the skull, the skin of the front of the scalp in particular has an important role in facial expression.

This is because many of the fibres of the scalp muscles are attached to the skin, allowing it to move backwards and forwards.

CONNECTIVE TISSUE
Under the skin, and attached firmly to it, is a layer of dense tissue which carries numerous arteries and veins. The arteries are branches of the external and internal carotid arteries, which interconnect to give a rich blood supply to all areas of the scalp. This layer of connective tissue is also attached firmly to the underlying layer of muscle.

The connective tissue binds the skin to the muscle in such a way that even if the scalp is torn from the head in an accident, these three layers will remain together.

Muscles of the scalp

The muscles of the scalp lie below the skin and a layer of connective tissue. They act to move the skin of the forehead and the jaw while chewing.

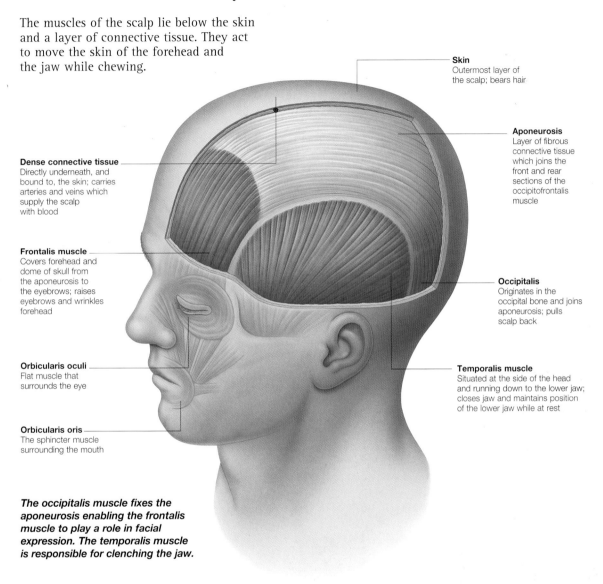

Skin
Outermost layer of the scalp; bears hair

Aponeurosis
Layer of fibrous connective tissue which joins the front and rear sections of the occipitofrontalis muscle

Dense connective tissue
Directly underneath, and bound to, the skin; carries arteries and veins which supply the scalp with blood

Frontalis muscle
Covers forehead and dome of skull from the aponeurosis to the eyebrows; raises eyebrows and wrinkles forehead

Occipitalis
Originates in the occipital bone and joins aponeurosis; pulls scalp back

Orbicularis oculi
Flat muscle that surrounds the eye

Temporalis muscle
Situated at the side of the head and running down to the lower jaw; closes jaw and maintains position of the lower jaw while at rest

Orbicularis oris
The sphincter muscle surrounding the mouth

The occipitalis muscle fixes the aponeurosis enabling the frontalis muscle to play a role in facial expression. The temporalis muscle is responsible for clenching the jaw.

The occipitofrontalis is a large muscle formed by two sections at the front and the back of the scalp, which are connected by a thin, tough, fibrous sheet (aponeurosis). The frontalis is the section of muscle over the forehead, arising from the skin overlying the eyebrow and passing back to become continuous with the aponeurosis. This muscle acts to raise the eyebrows, thus wrinkling the forehead or pulling the scalp forward, as when frowning.

The occipitalis is the section of muscle that arises from the top of the back of the neck and passes forward to the aponeurosis. It acts to pull the scalp backwards.

The temporalis muscle lies at the side of the scalp, above the ears, and runs from the skull down to the lower jaw. It is involved in the action of chewing.

LOOSE CONNECTIVE TISSUE
The fourth layer, underlying the muscle and aponeurosis, is a layer of loose connective tissue which allows the layers above to move relatively freely over the layer below. It is at this level that the scalp may be torn away during accidents, such as the head going forward through the windscreen of a car.

The pericranium, the fifth layer of the scalp, is the tough membrane covering the bone of the skull itself.

15

Brain

The brain is the part of the central nervous system that lies inside the skull. It controls many body functions including our heart rate, the ability to walk and run, and the creation of our thoughts.

Left cerebral hemisphere   **Right cerebral hemisphere**

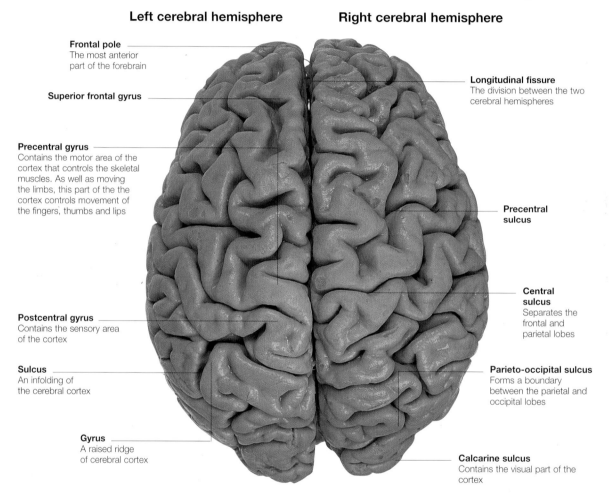

Frontal pole
The most anterior part of the forebrain

Superior frontal gyrus

Precentral gyrus
Contains the motor area of the cortex that controls the skeletal muscles. As well as moving the limbs, this part of the the cortex controls movement of the fingers, thumbs and lips

Postcentral gyrus
Contains the sensory area of the cortex

Sulcus
An infolding of the cerebral cortex

Gyrus
A raised ridge of cerebral cortex

Longitudinal fissure
The division between the two cerebral hemispheres

Precentral sulcus

Central sulcus
Separates the frontal and parietal lobes

Parieto-occipital sulcus
Forms a boundary between the parietal and occipital lobes

Calcarine sulcus
Contains the visual part of the cortex

The brain comprises three major parts: forebrain, midbrain and hindbrain. The forebrain is divided into two halves, forming the left and right cerebral hemispheres.

HEMISPHERES
The cerebral hemispheres form the largest part of the forebrain. Their outer surface is folded into a series of gyri (ridges) and sulci (furrows) that greatly increases its surface area. Most of the surface of each hemisphere is hidden in the depths of the sulci.

Each hemisphere is divided into frontal, parietal, occipital and temporal lobes, named after the closely related bones of the skull. Connecting the two hemispheres is the corpus callosum, a large bundle of fibres deep in the longitudinal fissure.

GREY AND WHITE MATTER
The hemispheres consist of an outer cortex of grey matter and an inner mass of white matter.
• Grey matter contains nerve cell bodies, and is found in the cortex of the cerebral and cerebellar hemispheres and in groups of sub-cortical nuclei.
• White matter comprises nerve fibres found below the cortex. They form the communication network of the brain, and can project to other areas of the cortex and spinal cord.

Inside the brain

A midline section between the two cerebral hemispheres reveals the main structures that control a vast number of activities in the body. While particular areas monitor sensory and motor information, others control speech and sleep.

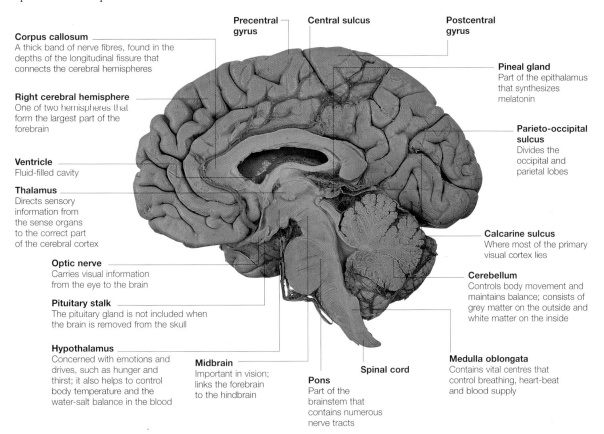

Precentral gyrus

Central sulcus

Postcentral gyrus

Corpus callosum
A thick band of nerve fibres, found in the depths of the longitudinal fissure that connects the cerebral hemispheres

Pineal gland
Part of the epithalamus that synthesizes melatonin

Right cerebral hemisphere
One of two hemispheres that form the largest part of the forebrain

Parieto-occipital sulcus
Divides the occipital and parietal lobes

Ventricle
Fluid-filled cavity

Thalamus
Directs sensory information from the sense organs to the correct part of the cerebral cortex

Calcarine sulcus
Where most of the primary visual cortex lies

Optic nerve
Carries visual information from the eye to the brain

Cerebellum
Controls body movement and maintains balance; consists of grey matter on the outside and white matter on the inside

Pituitary stalk
The pituitary gland is not included when the brain is removed from the skull

Hypothalamus
Concerned with emotions and drives, such as hunger and thirst; it also helps to control body temperature and the water-salt balance in the blood

Midbrain
Important in vision; links the forebrain to the hindbrain

Pons
Part of the brainstem that contains numerous nerve tracts

Spinal cord

Medulla oblongata
Contains vital centres that control breathing, heart-beat and blood supply

SPEECH, THOUGHT AND MOVEMENT
The receptive speech area (Wernicke's area) lies behind the primary auditory cortex and is essential for understanding speech. The prefrontal cortex has high-order cognitive functions, including abstract thinking, social behaviour and decision-making ability.

Within the white matter of the cerebral hemispheres are several masses of grey matter, known as the basal ganglia. This group of structures is involved in aspects of motor function,

including movement programming, planning and motor programme selection and motor memory retrieval.

DIENCEPHALON
The medial part of the forebrain comprises the structures surrounding the third ventricle. These form the diencephalon which includes the thalamus, hypothalamus, epithalamus and subthalamus of either side. The thalamus is the last relay station for information from the brainstem and spinal cord before it reaches the cortex.

The hypothalamus lies below the thalamus in the floor of the diencephalon. It is involved in a variety of homeostatic mechanisms, and controls the pituitary gland which descends from its base. The anterior (front) lobe of the pituitary secretes substances that influence the thyroid and adrenal glands, and the gonads and produces growth factors. The posterior lobe produces hormones that increase blood pressure, decrease urine production and cause uterine contraction. The hypothalamus also influences

the sympathetic and parasympathetic nervous systems and controls body temperature, appetite and wakefulness. The epithalamus is a relatively small part of the dorso-caudal diencephalon that includes the pineal gland, which synthesizes melatonin and is involved in the control of the sleep/wake cycle.

The subthalamus lies beneath the thalamus and next to the hypothalamus. It contains the subthalamic nucleus which controls movement.

17

Blood vessels of the brain

The arteries provide the brain with a rich supply of oxygenated blood.

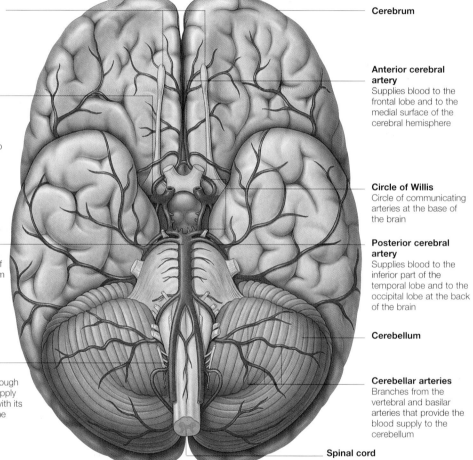

Inferior (from below) view of the brain

Right hemisphere · Left hemisphere

Olfactory bulbs
Organs of smell

Middle cerebral artery
This is the main branch of
the internal carotid artery,
supplying blood to two-
thirds of the cerebral
hemisphere and many deep
structures of the brain

Basilar artery
A large artery which lies
on the inferior surface of
the pons; divides to form
the two posterior
cerebral arteries

Vertebral artery
Arises from the subclavian
arteries, enters the skull through
the foramen magnum to supply
the brainstem, then fuses with its
opposite number to form the
basilar artery

Cerebrum

**Anterior cerebral
artery**
Supplies blood to the
frontal lobe and to the
medial surface of the
cerebral hemisphere

Circle of Willis
Circle of communicating
arteries at the base of
the brain

**Posterior cerebral
artery**
Supplies blood to the
inferior part of the
temporal lobe and to the
occipital lobe at the back
of the brain

Cerebellum

Cerebellar arteries
Branches from the
vertebral and basilar
arteries that provide the
blood supply to the
cerebellum

Spinal cord

The human brain weighs
about 1.4 kg and accounts
for two per cent of our total
body weight. However, it
requires 15–20 per cent of
the cardiac output to be able
to function properly. If the
blood supply to the the brain
is cut for as little as 10
seconds we lose
consciousness and, unless
blood flow is quickly
restored, it takes only a
matter of minutes before the
damage is irreversible.

THE ARTERIAL NETWORK
Blood reaches the brain via
two pairs of arteries. The
internal carotid arteries
originate from the common
carotid arteries in the neck,
enter the skull via the carotid
canal and then branch
to supply the cerebral cortex.
The two main branches of
the internal carotid are the
middle and anterior cerebral
arteries.

The vertebral arteries arise
from the subclavian arteries,
enter the skull via the
foramen magnum and supply
the brainstem and
cerebellum. They join,
forming the basilar artery
which then divides to
produce the two posterior
cerebral arteries that supply,
among other things, the
occipital or visual cortex at
the back of the brain.

These two sources of
blood to the brain are
linked by other arteries to
form a circuit at the base of
the brain called the 'circle
of Willis'.

Veins of the brain

Deep and superficial veins drain blood from the brain into a complex system of sinuses. These sinuses rely on gravity to return blood to the heart as, unlike other veins, they do not possess valves.

The veins of the brain can be divided into deep and superficial groups. These veins, none of which have valves, drain into the venous sinuses of the skull.

The sinuses are formed between layers of dura mater, the tough outer membrane covering the brain, and they are unlike the veins in the rest of the body in that they have no muscular tissue in their walls.

The superficial veins have a variable arrangement on the surface of the brain and many of them are highly interconnected. Most superficial veins drain into the superior sagittal sinus. By contrast, most of the deep veins, associated with structures within the body of the brain, drain into the straight sinus via the great cerebral vein (vein of Galen).

FUNCTIONS OF THE SINUSES

The straight sinus and the superior sagittal sinus converge. Blood flows through the transverse and sigmoid sinuses and exits the skull through the internal jugular vein before flowing back towards the heart.

Beneath the brain, on either side of the sphenoid bone, are the cavernous sinuses. These drain blood from the orbit (eye socket) and the deep parts of the face. This provides a potential route of infection into the skull.

Superior sagittal sinus
The largest of the venous sinuses; receives blood from many of the superficial cerebral veins and is also the site for reabsorption of cerebrospinal fluid into the circulation

Inferior sagittal sinus
Found at the lower margin of the falx cerebri (a large fold in the dura mater separating the two cerebral hemispheres); receives blood from superficial veins

Straight sinus
Drains blood from the inferior sagittal sinus and the deep cerebral veins via the great vein of Galen

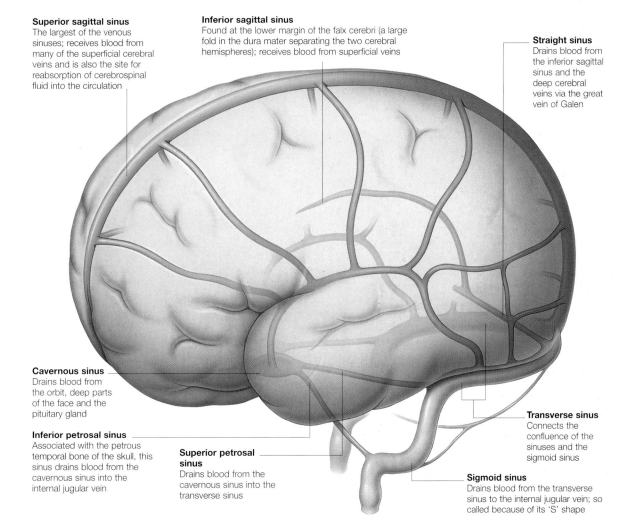

Cavernous sinus
Drains blood from the orbit, deep parts of the face and the pituitary gland

Inferior petrosal sinus
Associated with the petrous temporal bone of the skull, this sinus drains blood from the cavernous sinus into the internal jugular vein

Superior petrosal sinus
Drains blood from the cavernous sinus into the transverse sinus

Transverse sinus
Connects the confluence of the sinuses and the sigmoid sinus

Sigmoid sinus
Drains blood from the transverse sinus to the internal jugular vein; so called because of its 'S' shape

Ventricles of the brain

The brain 'floats' in a protective layer of cerebrospinal fluid – the watery
liquid produced in a system of cavities within the brain and brainstem.

The brain contains a system
of communicating
(connected) cavities known as
the ventricles. There are four
ventricles within the brain
and brainstem, each secreting
cerebrospinal fluid (CSF), the
fluid that surrounds and
permeates the brain and
spinal cord, protecting them
from injury and infection.

Three of the ventricles –
namely the two (paired)
lateral ventricles and the third
ventricle – lie within the
forebrain. The lateral
ventricles are the largest, and
lie within each cerebral
hemisphere. Each consists of
a 'body' and three 'horns' –
anterior (situated in the
frontal lobe), posterior
(occipital lobe) and inferior
(temporal lobe). The third
ventricle is a narrow cavity
between the thalamus and
hypothalamus.

HINDBRAIN VENTRICLE
The fourth ventricle is
situated in the hindbrain,
beneath the cerebellum.
When viewed from above, it
is diamond-shaped, but in
sagittal section (see right) it is
triangular. It is continuous
with the third ventricle via a
narrow channel called the
cerebral aqueduct of the
midbrain. The roof of the
fourth ventricle is incomplete,
allowing it to communicate
with the subarachnoid space
(see over).

*This sagittal section of the
brain and brainstem reveals
the four ventricles and the
foramina and aqueducts
that connect them.*

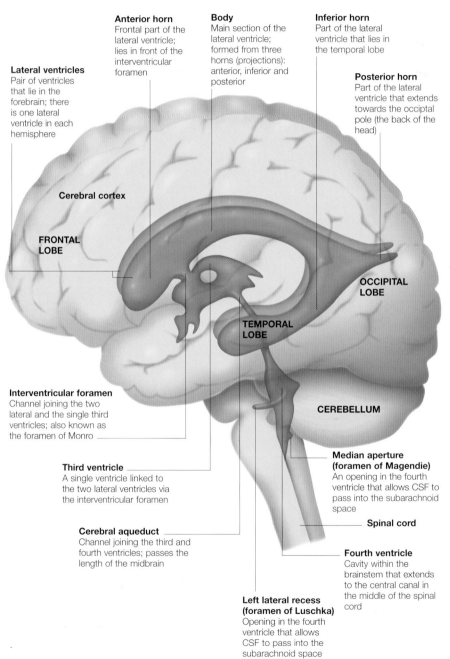

Anterior horn
Frontal part of the
lateral ventricle;
lies in front of the
interventricular
foramen

Body
Main section of the
lateral ventricle;
formed from three
horns (projections):
anterior, inferior and
posterior

Inferior horn
Part of the lateral
ventricle that lies in
the temporal lobe

Posterior horn
Part of the lateral
ventricle that extends
towards the occipital
pole (the back of the
head)

Lateral ventricles
Pair of ventricles
that lie in the
forebrain; there
is one lateral
ventricle in each
hemisphere

Cerebral cortex

FRONTAL LOBE

OCCIPITAL LOBE

TEMPORAL LOBE

CEREBELLUM

Interventricular foramen
Channel joining the two
lateral and the single third
ventricles; also known as
the foramen of Monro

Third ventricle
A single ventricle linked to
the two lateral ventricles via
the interventricular foramen

Cerebral aqueduct
Channel joining the third and
fourth ventricles; passes the
length of the midbrain

**Median aperture
(foramen of Magendie)**
An opening in the fourth
ventricle that allows CSF to
pass into the subarachnoid
space

Spinal cord

Fourth ventricle
Cavity within the
brainstem that extends
to the central canal in
the middle of the spinal
cord

**Left lateral recess
(foramen of Luschka)**
Opening in the fourth
ventricle that allows
CSF to pass into the
subarachnoid space

Circulation of cerebrospinal fluid

Cerebrospinal fluid (CSF) is produced by the choroid plexus within the lateral, third and fourth ventricles.

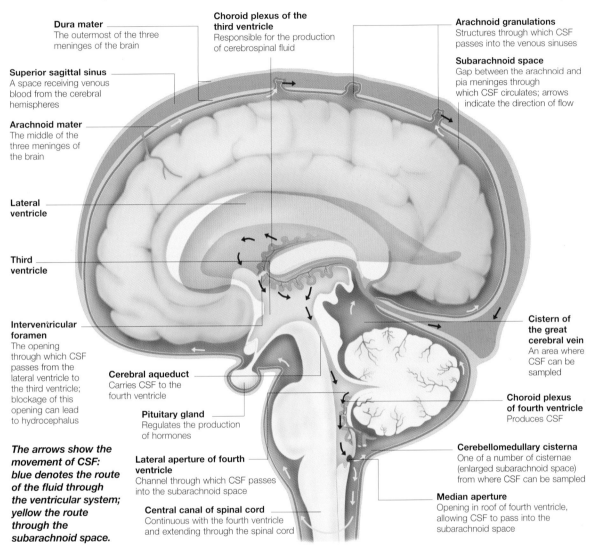

Dura mater
The outermost of the three meninges of the brain

Choroid plexus of the third ventricle
Responsible for the production of cerebrospinal fluid

Arachnoid granulations
Structures through which CSF passes into the venous sinuses

Superior sagittal sinus
A space receiving venous blood from the cerebral hemispheres

Subarachnoid space
Gap between the arachnoid and pia meninges through which CSF circulates; arrows indicate the direction of flow

Arachnoid mater
The middle of the three meninges of the brain

Lateral ventricle

Third ventricle

Interventricular foramen
The opening through which CSF passes from the lateral ventricle to the third ventricle; blockage of this opening can lead to hydrocephalus

Cerebral aqueduct
Carries CSF to the fourth ventricle

Pituitary gland
Regulates the production of hormones

Cistern of the great cerebral vein
An area where CSF can be sampled

Choroid plexus of fourth ventricle
Produces CSF

The arrows show the movement of CSF: blue denotes the route of the fluid through the ventricular system; yellow the route through the subarachnoid space.

Lateral aperture of fourth ventricle
Channel through which CSF passes into the subarachnoid space

Central canal of spinal cord
Continuous with the fourth ventricle and extending through the spinal cord

Cerebellomedullary cisterna
One of a number of cisternae (enlarged subarachnoid space) from where CSF can be sampled

Median aperture
Opening in roof of fourth ventricle, allowing CSF to pass into the subarachnoid space

The choroid plexuses are a rich system of blood vessels originating from the pia mater, the innermost tissue surrounding the brain. The plexuses contain numerous folds (villous processes) projecting into the ventricles, from which cerebrospinal fluid is produced.

From the choroid plexuses in the two lateral ventricles, CSF passes to the third ventricle via the interventricular foramen. Together with additional fluid produced by the choroid plexus in the third ventricle, CSF then passes through the cerebral aqueduct of the midbrain and into the fourth ventricle. Additional fluid is produced by the choroid plexus in the fourth ventricle.

SUBARACHNOID SPACE
From the fourth ventricle, CSF passes out into the subarachnoid space surrounding the brain. It does this through openings in the fourth ventricle – a median opening (foramen of Magendie) and two lateral ones (foramina of Luschka). Once in the subarachnoid space, the CSF circulates to surround the central nervous system. As CSF is produced constantly, it needs to be drained continuously to prevent build-up of pressure. This is achieved by passage of the CSF into the venous sinuses of the brain through protrusions (arachnoid granulations). These are particularly evident in the superior sagittal sinus.

Cerebral hemispheres

The cerebral hemispheres are the largest part of the brain. In humans, they have developed out of proportion to the other regions, distinguishing our brains from those of other animals.

Lobes of the cerebral hemispheres

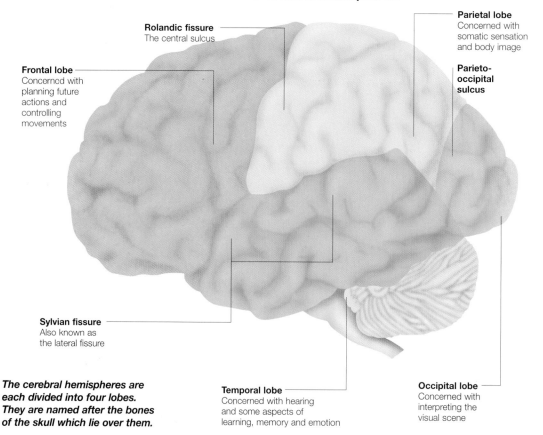

Rolandic fissure
The central sulcus

Parietal lobe
Concerned with somatic sensation and body image

Frontal lobe
Concerned with planning future actions and controlling movements

Parieto-occipital sulcus

Sylvian fissure
Also known as the lateral fissure

Temporal lobe
Concerned with hearing and some aspects of learning, memory and emotion

Occipital lobe
Concerned with interpreting the visual scene

The cerebral hemispheres are each divided into four lobes. They are named after the bones of the skull which lie over them.

The left and right cerebral hemispheres are separated by the longitudinal fissure which runs between them. Looking at the surface of the hemispheres from the top and side, there is a prominent groove running downwards, beginning about 1 cm behind the midpoint between the front and back of the brain.

This is the central sulcus or rolandic fissure. Further down on the side of the brain there is a second large groove, the lateral sulcus or sylvian fissure.

LOBES OF THE BRAIN

The cerebral hemispheres are divided into lobes, named after the bones of the skull which lie over them:

- The frontal lobe lies in front of the rolandic fissure and above the sylvian fissure
- The parietal lobe lies

behind the rolandic fissure and above the back part of the sylvian fissure; it extends back as far as the parieto-occipital sulcus, a groove separating it from the occipital lobe, which is at the back of the brain

- The temporal lobe is the area below the sylvian fissure and extends backward to meet the occipital lobe.

At the bottom of the sylvian fissure there is another distinct area known as the insula or island of Reil. This triangular region has been buried by the growth of the adjacent parts of the brain and is not normally visible unless the sylvian fissure is spread open.

Functions of the cerebral hemispheres

Different regions of the cortex have distinct
and highly specialized functions.

The cerebral cortex is divided
into:
• Motor areas, which initiate
and control movement. The
primary motor cortex
controls voluntary
movement of the opposite
side of the body. Just in
front of the primary motor
cortex is the pre-motor
cortex and a third area, the

supplementary motor area,
lies on the inner surface of
the frontal lobe. All of
these areas work with the
basal ganglia and
cerebellum to allow us to
perform complex
sequences of finely
controlled movements.
• Sensory areas, which
receive and integrate

information from sensory
receptors around the body.
The primary somatosensory
area receives information
from sensory receptors on
the opposite side of the
body about touch, pain,
temperature and the
position of joints and
muscles (proprioception).
• Association areas, which

are involved with the
integration of more
complex brain functions –
the higher mental processes
of learning, memory,
language, judgment and
reasoning, emotion and
personality.

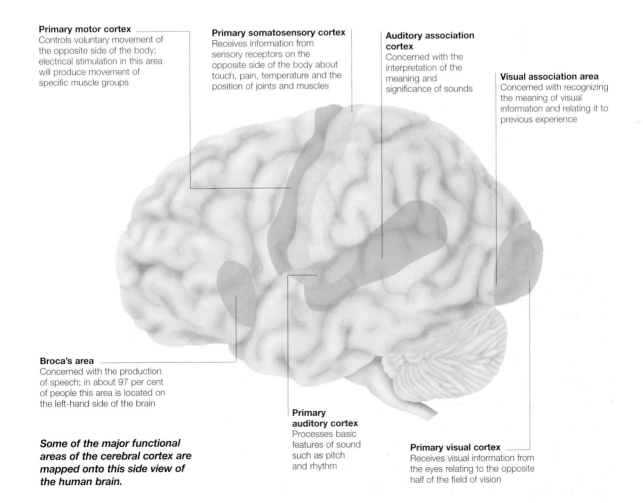

Primary motor cortex
Controls voluntary movement of
the opposite side of the body;
electrical stimulation in this area
will produce movement of
specific muscle groups

Primary somatosensory cortex
Receives information from
sensory receptors on the
opposite side of the body about
touch, pain, temperature and the
position of joints and muscles

**Auditory association
cortex**
Concerned with the
interpretation of the
meaning and
significance of sounds

Visual association area
Concerned with recognizing
the meaning of visual
information and relating it to
previous experience

Broca's area
Concerned with the production
of speech; in about 97 per cent
of people this area is located on
the left-hand side of the brain

*Some of the major functional
areas of the cerebral cortex are
mapped onto this side view of
the human brain.*

**Primary
auditory cortex**
Processes basic
features of sound
such as pitch
and rhythm

Primary visual cortex
Receives visual information from
the eyes relating to the opposite
half of the field of vision

Thalamus

The thalamus is a major sensory relay and integrating centre in the brain, lying deep within its central core. It consists of two halves, and receives sensory inputs of all types, except smell.

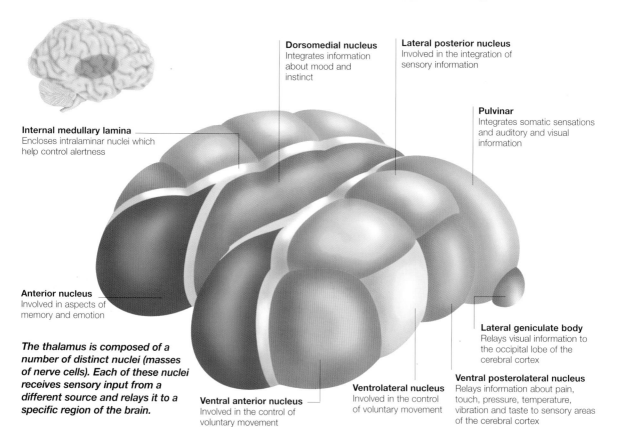

Dorsomedial nucleus
Integrates information about mood and instinct

Lateral posterior nucleus
Involved in the integration of sensory information

Pulvinar
Integrates somatic sensations and auditory and visual information

Internal medullary lamina
Encloses intralaminar nuclei which help control alertness

Anterior nucleus
Involved in aspects of memory and emotion

The thalamus is composed of a number of distinct nuclei (masses of nerve cells). Each of these nuclei receives sensory input from a different source and relays it to a specific region of the brain.

Ventral anterior nucleus
Involved in the control of voluntary movement

Ventrolateral nucleus
Involved in the control of voluntary movement

Lateral geniculate body
Relays visual information to the occipital lobe of the cerebral cortex

Ventral posterolateral nucleus
Relays information about pain, touch, pressure, temperature, vibration and taste to sensory areas of the cerebral cortex

The thalamus is made up of paired egg-shaped masses of grey matter (cell bodies of nerve cells) 3–4 cm long and 1.5 cm wide, located in the deep central core of the brain known as the diencephalon, or 'between brain'.

The thalamus makes up about 80 per cent of the diencephalon and lies on either side of the fluid-filled third ventricle.

The right and left parts of the thalamus are connected to each other by a bridge of grey matter – the massa intermedia, or interthalamic adhesion.

NEUROANATOMY
The front end of the thalamus is rounded and is narrower than the back, which is expanded into the pulvinar. The upper surface of the thalamus is covered with a thin layer of white matter – the stratum zonale. A second layer of white matter – the external medullary lamina – covers the lateral surface.

Its structure is very complex and it contains more than 25 distinct nuclei (collections of nerve cells with a common function).

These thalamic nuclear groups are separated by a vertical Y-shaped sheet of white matter – the internal medullary lamina. The anterior nucleus lies in the fork of the Y, and the tail divides the medial and lateral nuclei and splits to enclose the intralaminar nuclei.

These can be subdivided into several groups:
• Anterior nuclei
• Medial nuclei
• Lateral nuclei
Divided into a larger ventral group and a smaller dorsal group
• Intralaminar, midline and reticular nuclei
• Posterior nuclei
• pulvinar
• medial geniculate body
• lateral geniculate body

Hypothalamus

The hypothalamus is a complex structure located in the deep core of the brain. It regulates fundamental aspects of body function, and is critical for homeostasis – the maintenance of equilibrium in the body's internal environment.

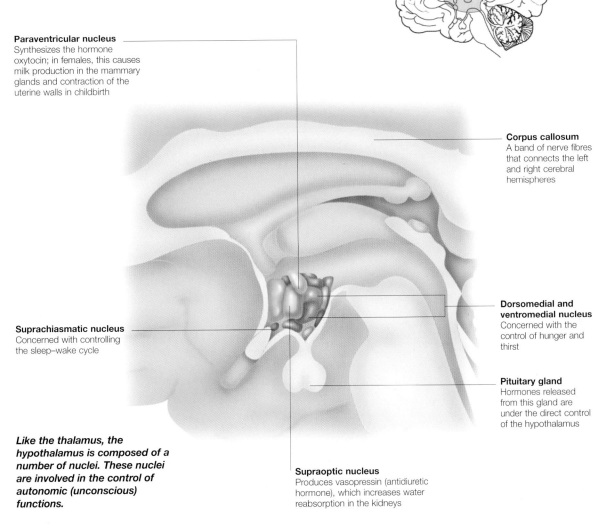

Paraventricular nucleus
Synthesizes the hormone oxytocin; in females, this causes milk production in the mammary glands and contraction of the uterine walls in childbirth

Corpus callosum
A band of nerve fibres that connects the left and right cerebral hemispheres

Suprachiasmatic nucleus
Concerned with controlling the sleep–wake cycle

Dorsomedial and ventromedial nucleus
Concerned with the control of hunger and thirst

Pituitary gland
Hormones released from this gland are under the direct control of the hypothalamus

Like the thalamus, the hypothalamus is composed of a number of nuclei. These nuclei are involved in the control of autonomic (unconscious) functions.

Supraoptic nucleus
Produces vasopressin (antidiuretic hormone), which increases water reabsorption in the kidneys

The hypothalamus is a small region of the diencephalon; it is the size of a thumbnail and weighs only about four grams. It lies below the thalamus and is separated from it by a shallow groove, the hypothalamic sulcus. The hypothalamus is just behind the optic chiasm, the point where the two optic nerves cross over as they travel from the eyes towards the visual area at the back of the brain.

Several distinct structures stand out on its undersurface:
• The mammillary bodies – two small, pea-like projections which are involved in the sense of smell
• The infundibulum or pituitary stalk – a hollow structure connecting the hypothalamus with the posterior part of the pituitary gland (neurohypophysis) which lies below it
• The tuber cinereum or median eminence – a greyish-blue, raised region surrounding the base of the infundibulum.

Limbic system

The limbic system is a ring of interconnected structures that lies deep within the brain. It makes connections with other parts of the brain, and is associated with mood and memory.

The limbic system is a collection of structures deep within the brain that is associated with the perception of emotions and the body's response to them.

The limbic system is not one, discrete part of the brain. Rather it is a ring of interconnected structures surrounding the top of the brainstem. The connections between these structures are complex, often forming loops or circuits and, as with much of the brain, their exact role is not fully understood.

STRUCTURE

The limbic system is made up from all or parts of the following brain structures:

- Amygdala – this almond-shaped nucleus appears to be linked to feelings of fear and aggression
- Hippocampus – this structure seems to play a part in learning and memory
- Anterior thalamic nuclei – these collections of nerve cells form part of the thalamus. One of their roles seems to lie in the control of instinctive drives
- Cingulate gyrus – this connects the limbic system to the cerebral cortex, the part of the brain that carries conscious thoughts
- Hypothalamus – this regulates the body's internal environment, including blood pressure, heart rate and hormone levels. The limbic system generates its effects on the body by sending messages to the hypothalamus.

Medial view of the limbic system within the brain

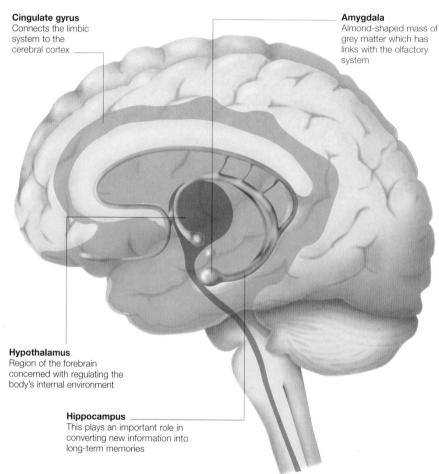

Cingulate gyrus
Connects the limbic system to the cerebral cortex

Amygdala
Almond-shaped mass of grey matter which has links with the olfactory system

Hypothalamus
Region of the forebrain concerned with regulating the body's internal environment

Hippocampus
This plays an important role in converting new information into long-term memories

The limbic system connections encircle the upper part of the brainstem. They link with other parts of the brain and are associated with emotion.

Connections of the limbic system

The limbic system has connections with the higher centres of the brain in the cortex, and with the more primitive brainstem. It not only allows the emotions to influence the body, but also enables the emotional response to be regulated.

The developing brain

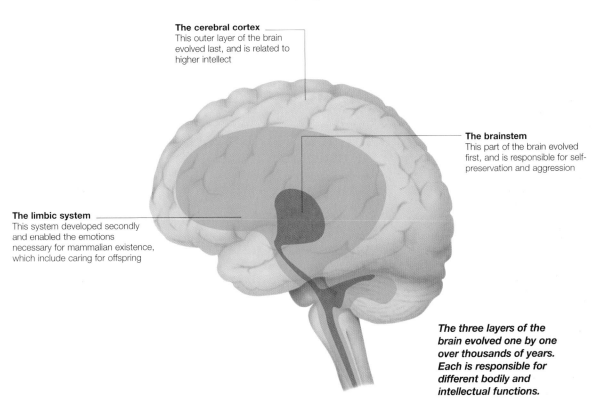

The cerebral cortex
This outer layer of the brain evolved last, and is related to higher intellect

The brainstem
This part of the brain evolved first, and is responsible for self-preservation and aggression

The limbic system
This system developed secondly and enabled the emotions necessary for mammalian existence, which include caring for offspring

The three layers of the brain evolved one by one over thousands of years. Each is responsible for different bodily and intellectual functions.

The human brain can be considered to be made up of three parts. These parts have evolved one after another over the millennia.

BRAINSTEM
The 'oldest' part of the brain, in evolutionary terms, is the brainstem, which is concerned largely with unconscious control of the internal state of the body. The brainstem can be seen as a sort of 'life support system'.

LIMBIC SYSTEM
With the evolution of mammals came another 'layer' of brain, the limbic system. The limbic system allowed the development of feelings and emotions in response to sensory information. It is also associated with the development of newer – in evolutionary terms – behaviours, such as closeness to offspring (maternal bonding).

CEREBRAL CORTEX
The final layer of the human brain is shared to some extent with higher mammals. It is the cerebral cortex, the part of the brain that allows humans to think and reason. With this part of the brain, individuals perceive the outside world and make conscious decisions about their behaviour and actions.

ROLE OF THE LIMBIC SYSTEM
The limbic system lies between the cortex and the brainstem and makes connections with both. Through its connections with the brainstem, the limbic system provides a way in which an individual's emotional state can influence the internal state of the body. This may prepare the body perhaps for an act of self-preservation such as running away in fear, or for a sexual encounter.

The extensive connections between the limbic system and the cerebral cortex allow human beings to use their knowledge of the outside world to regulate their response to emotions. The cerebral cortex can thus 'override' the more primitive limbic system when necessary.

Basal ganglia

The basal ganglia lie deep within the white matter of the cerebral hemispheres. They are collections of nerve cell bodies that are involved in the control of movement.

The common term basal ganglia is, in fact, a misnomer, as the term ganglion refers to a mass of nerve cells in the peripheral nervous system rather than the central nervous system, as here. The term basal nuclei is anatomically more appropriate.

COMPONENTS

There are a number of component parts to the basal nuclei which are all anatomically and functionally closely related to each other. The parts of the basal nuclei include:
- Putamen. Together with the caudate nucleus, the putamen receives input from the cortex
- Caudate nucleus. Named for its shape, as it has a long tail, this nucleus is continuous with the putamen at the anterior (front) end
- Globus pallidus. This nucleus relays information from the putamen to the pigmented area of the midbrain known as the substantia nigra, with which it bears many similarities.

GROUPING

Various names are associated with different groups of the basal nuclei. The term corpus striatum (striped body) refers to the whole group of basal nuclei, whereas the striatum includes only the putamen and caudate nuclei. Another term, the lentiform nucleus, refers to the putamen and the globus pallidus which, together, form a lens-shaped mass.

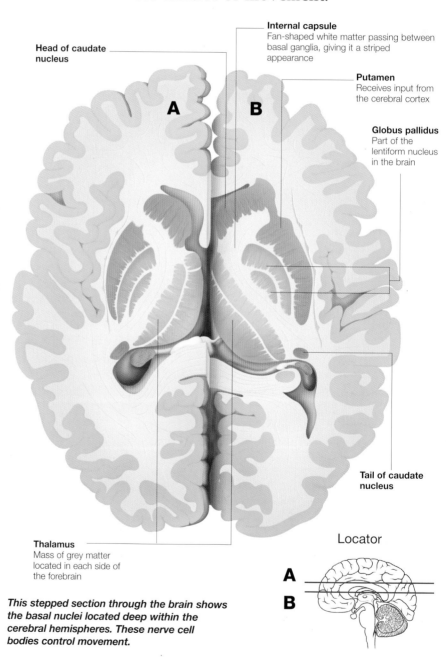

Head of caudate nucleus

Internal capsule
Fan-shaped white matter passing between basal ganglia, giving it a striped appearance

Putamen
Receives input from the cerebral cortex

Globus pallidus
Part of the lentiform nucleus in the brain

A   B

Tail of caudate nucleus

Thalamus
Mass of grey matter located in each side of the forebrain

Locator

A
B

This stepped section through the brain shows the basal nuclei located deep within the cerebral hemispheres. These nerve cell bodies control movement.

Structure and role of the basal ganglia

The overall shape of the basal ganglia (nuclei) is complex, and is hard to imagine by looking at two-dimensional cross-sections.

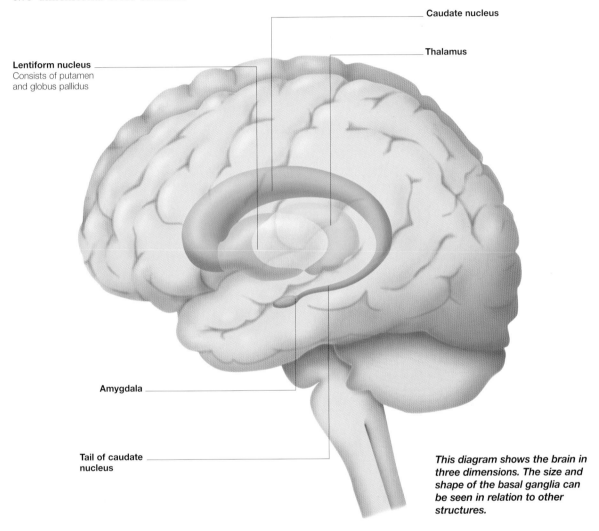

Caudate nucleus

Thalamus

Lentiform nucleus
Consists of putamen and globus pallidus

Amygdala

Tail of caudate nucleus

This diagram shows the brain in three dimensions. The size and shape of the basal ganglia can be seen in relation to other structures.

When seen in a three-dimensional view, the size and shape of the basal nuclei, together with their position within the brain as a whole, can be appreciated more easily.

In particular, the shape of the caudate nucleus can now be understood – it connects at its head with the putamen, then bends back to arch over the thalamus before turning forwards again. The tip of the tail of the caudate nucleus ends as it merges with the amygdala, part of the limbic system (concerned with unconscious, autonomic functions).

ROLE OF
THE BASAL NUCLEI
The functions of the basal nuclei have been difficult to study because they lie deep within the brain and are therefore relatively inaccessible. Much of what is known of their function derives from the study of those patients who have disorders of the basal nuclei that lead to particular disruptions of movement and posture, such as Parkinson's disease.

A summary of what is currently known about the function of the basal nuclei is that: they help to produce movements which are appropriate; and they inhibit unwanted or inappropriate movements.

Cerebellum

The cerebellum, which means 'little brain', lies under the occipital lobes of the cerebral cortex at the back of the brain. It is important to the subconscious control of movement.

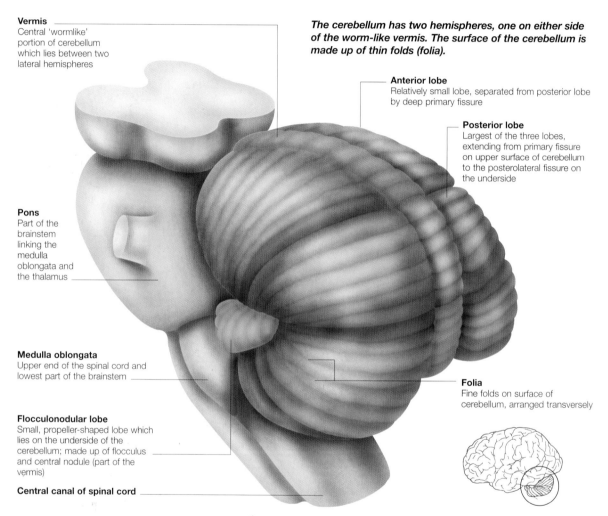

Vermis
Central 'wormlike' portion of cerebellum which lies between two lateral hemispheres

The cerebellum has two hemispheres, one on either side of the worm-like vermis. The surface of the cerebellum is made up of thin folds (folia).

Anterior lobe
Relatively small lobe, separated from posterior lobe by deep primary fissure

Posterior lobe
Largest of the three lobes, extending from primary fissure on upper surface of cerebellum to the posterolateral fissure on the underside

Pons
Part of the brainstem linking the medulla oblongata and the thalamus

Medulla oblongata
Upper end of the spinal cord and lowest part of the brainstem

Folia
Fine folds on surface of cerebellum, arranged transversely

Flocculonodular lobe
Small, propeller-shaped lobe which lies on the underside of the cerebellum; made up of flocculus and central nodule (part of the vermis)

Central canal of spinal cord

The part of the brain known as the cerebellum lies under the occipital lobes of the cerebral cortex at the back of the head. The vital roles of the cerebellum include the co-ordination of movement and the maintenance of balance and posture. The cerebellum works subconsciously and so an individual is not aware of its functioning.

STRUCTURE
The cerebellum is composed of two hemispheres which are bridged in the midline by the vermis. The hemispheres extend laterally (sideways) and posteriorly (backwards) from the midline to form the bulk of the cerebellum.

The surface of the cerebellum has a distinctive appearance. In contrast to the large folds of the cerebral hemispheres, the surface of the cerebellum is made up of numerous fine folds (folia).

LOBES
Between the folia of the cerebellar surface lie deep fissures which divide it into three lobes:
• Anterior lobe
• Posterior lobe
• Flocculonodular lobe.

30

Internal structure of the cerebellum

The cerebellum has an outer grey cortex and a core of nerve fibres, or white matter. Deep within the white matter lie four pairs of cerebellar nuclei: the fastigial, globose, emboliform and dentate nuclei.

Cross-section through cerebellum

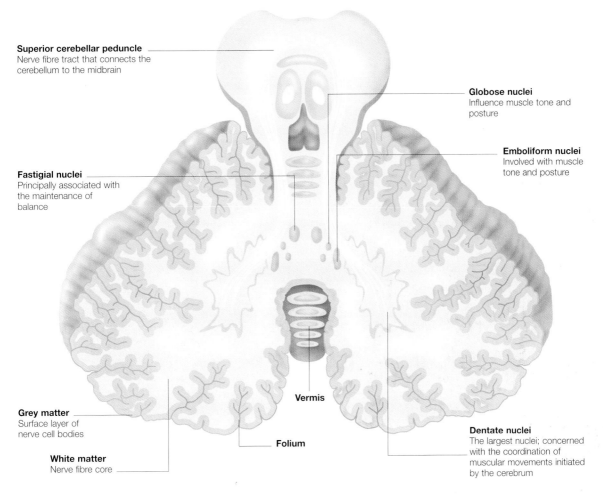

Superior cerebellar peduncle
Nerve fibre tract that connects the cerebellum to the midbrain

Globose nuclei
Influence muscle tone and posture

Emboliform nuclei
Involved with muscle tone and posture

Fastigial nuclei
Principally associated with the maintenance of balance

Grey matter
Surface layer of nerve cell bodies

White matter
Nerve fibre core

Vermis

Folium

Dentate nuclei
The largest nuclei; concerned with the coordination of muscular movements initiated by the cerebrum

The cerebellum is composed of a surface layer of nerve cell bodies, or grey matter, which overlies a core of nerve fibres, or white matter. Deep within the white matter lie the cerebellar nuclei.

CEREBELLAR CORTEX
Due to the presence of the numerous fine folia (folds) in the surface of the cerebellum, the cortex is very extensive. It is made up of the cell bodies and dendrites (cell processes) of the vast majority of cerebellar neurones.

The cells of the cortex receive information from outside the cerebellum via the cerebellar peduncles and make frequent connections between themselves within the cortex.

SIGNALS
In most cases, signals from the cerebellar cortex are conveyed in the fibres of the white matter down to the cerebellar nuclei. It is from here that information leaves the cerebellum to be carried to the rest of the central nervous system.

CEREBELLAR NUCLEI
There are four pairs of cerebellar nuclei which, from the midline outwards, are known as the:
- Fastigial nuclei
- Globose nuclei
- Emboliform nuclei
- Dentate nuclei.

Cranial nerves

There are twelve pairs of cranial nerves which leave the brain to supply structures mainly of the head and neck. The cranial nerves carry information to and from the brain.

Nerves are the routes by which information passes between the central nervous system (CNS) and the rest of the body. From the neck down, these nerves emerge from the spinal cord, passing out through openings in the bony spinal column. However, the cranial nerves emerge directly from the brain.

There are twelve pairs of cranial nerves, which are named and numbered with Roman numerals. The first two pairs attach to the forebrain while the rest come from the brainstem. The cranial nerves serve structures of the head and neck. To reach these they must pass through special openings, or foramina, in the bony skull.

CRANIAL NERVE FIBRES

Cranial nerves are made up of sensory and motor nerve fibres and so carry information to and from the CNS:

- Sensory nerve fibres bring information such as pain, touch and temperature sensations from the face as well as the senses of taste, vision and hearing
- Motor fibres send instructions to head, neck and face muscles, allowing various facial expressions and eye movements
- Autonomic nerve fibres allow the subconscious control of internal structures such as the salivary glands, the iris and some of the major organs of the chest and abdomen.

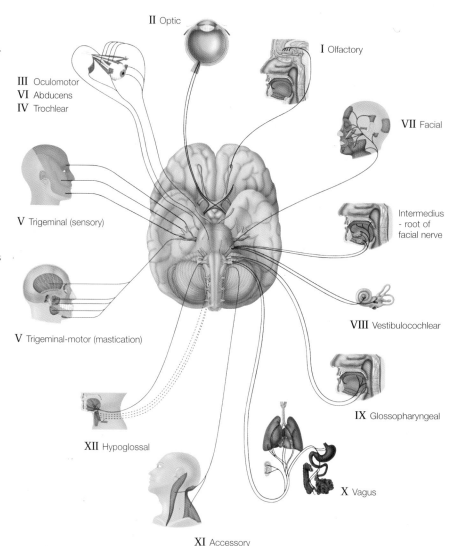

II Optic

I Olfactory

III Oculomotor
VI Abducens
IV Trochlear

VII Facial

V Trigeminal (sensory)

Intermedius - root of facial nerve

V Trigeminal-motor (mastication)

VIII Vestibulocochlear

XII Hypoglossal

IX Glossopharyngeal

X Vagus

XI Accessory

The twelve pairs of cranial nerves emerge directly from the brain. Each pair of nerves has responsibility for a specific sense or motor function.

The olfactory nerves

The olfactory nerves are the tiny sensory nerves of smell.
They run from the nasal mucosa to the olfactory bulbs.

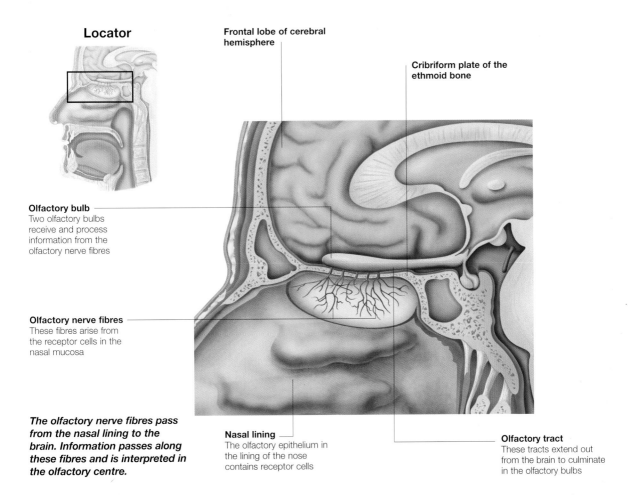

Locator

Frontal lobe of cerebral hemisphere

Cribriform plate of the ethmoid bone

Olfactory bulb
Two olfactory bulbs receive and process information from the olfactory nerve fibres

Olfactory nerve fibres
These fibres arise from the receptor cells in the nasal mucosa

The olfactory nerve fibres pass from the nasal lining to the brain. Information passes along these fibres and is interpreted in the olfactory centre.

Nasal lining
The olfactory epithelium in the lining of the nose contains receptor cells

Olfactory tract
These tracts extend out from the brain to culminate in the olfactory bulbs

The olfactory nerves carry the special sense of smell from the receptor cells in the nasal cavity to the brain above.

OLFACTORY EPITHELIUM
The olfactory epithelium is the part of the lining of the nasal cavity that carries special receptor cells for the sense of smell. It is found in the upper part of the nasal cavity and the septum, the partition between the two sides of the cavity.

The olfactory receptors are specialized neurones, or nerve cells, and are able to detect odorous substances which are present in the form of minute droplets in the air.

OLFACTORY NERVES
Information from the olfactory receptor neurones is passed up to the brain through their long processes, or axons, which group together to form about 20 bundles. These bundles

are the true olfactory nerves, which pass up through the thin perforated layer of bone, the cribriform plate of the ethmoid bone, to reach the olfactory bulbs in the cranial cavity.

The fibres of the olfactory nerves make connections (synapse) with the neurones within the olfactory bulb.

OLFACTORY BULBS
The paired olfactory bulbs are actually part of the brain, extended out on stalks, the

olfactory tracts, which contain fibres linking them to the cerebral hemispheres.

Large specialized neurones, known as mitral cells, connect with the olfactory nerves within the olfactory bulb. This connection permits information about smell to be passed on from the olfactory nerves.

The axons of these mitral cells then carry this information to the olfactory centre of the brain via the olfactory tracts.

Facial muscles

One of the features that distinguishes humans from animals is our ability to communicate using a wide range of facial expressions. The power behind this ability is a complex system of facial muscles.

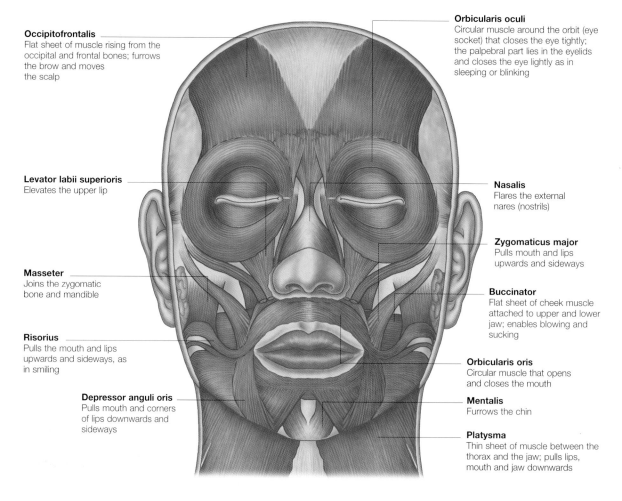

Occipitofrontalis
Flat sheet of muscle rising from the occipital and frontal bones; furrows the brow and moves the scalp

Levator labii superioris
Elevates the upper lip

Masseter
Joins the zygomatic bone and mandible

Risorius
Pulls the mouth and lips upwards and sideways, as in smiling

Depressor anguli oris
Pulls mouth and corners of lips downwards and sideways

Orbicularis oculi
Circular muscle around the orbit (eye socket) that closes the eye tightly; the palpebral part lies in the eyelids and closes the eye lightly as in sleeping or blinking

Nasalis
Flares the external nares (nostrils)

Zygomaticus major
Pulls mouth and lips upwards and sideways

Buccinator
Flat sheet of cheek muscle attached to upper and lower jaw; enables blowing and sucking

Orbicularis oris
Circular muscle that opens and closes the mouth

Mentalis
Furrows the chin

Platysma
Thin sheet of muscle between the thorax and the jaw; pulls lips, mouth and jaw downwards

Just under the skin of the scalp and face lies a group of very thin muscles, which are collectively known as the muscles of facial expression. These muscles play a vital role in a number of ways, in addition to their physiological function. They alter facial expression – providing a means of non-verbal communication by transmitting a range of emotional information – and are also one of the means of articulating speech.

Apart from this, the facial muscles also form sphincters that open and close the orifices of the face – the eyes and mouth.

SKIN AND BONE

The majority of facial muscles are attached to the skull bone at one end and to the deep layer of skin (dermis) at the other. From these attachments, it can be seen see how the numerous muscles alter facial expression, and also how they eventually cause creases and wrinkles in the overlying skin.

A number of small muscles called 'dilators' open the mouth. They radiate out from the corners of the mouth and lips, where they have an attachment to bone. The mouth and lips can be pulled up, pushed down and moved from side to side.

Opening and closing the eye

Whether fluttered alluringly or squeezed tightly shut for protection, the eyelids communicate a range of non-verbal signals. The eyelids are also vital for cleaning and lubricating the eyes.

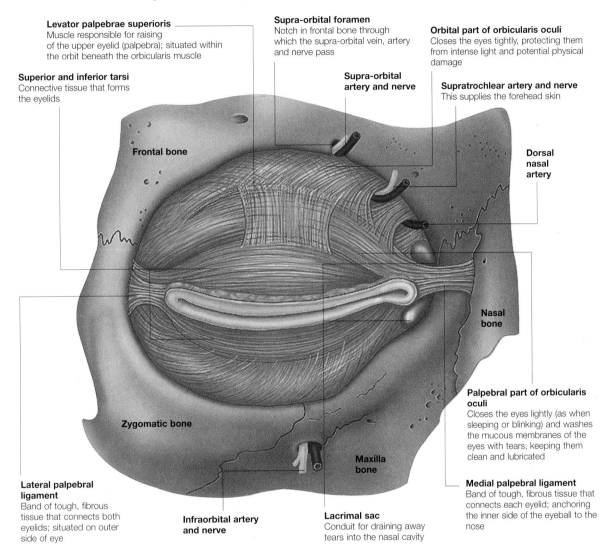

Levator palpebrae superioris
Muscle responsible for raising of the upper eyelid (palpebra); situated within the orbit beneath the orbicularis muscle

Superior and inferior tarsi
Connective tissue that forms the eyelids

Supra-orbital foramen
Notch in frontal bone through which the supra-orbital vein, artery and nerve pass

Orbital part of orbicularis oculi
Closes the eyes tightly, protecting them from intense light and potential physical damage

Supra-orbital artery and nerve

Supratrochlear artery and nerve
This supplies the forehead skin

Frontal bone

Dorsal nasal artery

Nasal bone

Zygomatic bone

Palpebral part of orbicularis oculi
Closes the eyes lightly (as when sleeping or blinking) and washes the mucous membranes of the eyes with tears, keeping them clean and lubricated

Maxilla bone

Lateral palpebral ligament
Band of tough, fibrous tissue that connects both eyelids; situated on outer side of eye

Infraorbital artery and nerve

Lacrimal sac
Conduit for draining away tears into the nasal cavity

Medial palpebral ligament
Band of tough, fibrous tissue that connects each eyelid; anchoring the inner side of the eyeball to the nose

The orbicularis oculi is the muscle responsible for the closing of the eye. This flat sphincter muscle lines the rim of the orbit (eye socket), and various sections of it can be manipulated individually.

Part of the orbicularis oculi lies in the eyelid (the palpebral part). This section of the muscle closes the eye lightly, as in sleeping or in routine blinking. This action also aids the flow of lacrimal secretion (tears) across the conjunctiva (the membrane covering the eyes) to keep it clean, free of foreign bodies and lubricated.

OPENING THE LIDS
A larger part of the orbicularis oculi consists of concentrically arranged fibres that cover the front of the eye socket. The role of this part of the muscle is to close the eye tightly (screw up the eyes) to protect against a blow or bright light.

The second orbital muscle is the levator palpebrae superioris. As its name suggests, this small muscle pulls on the upper lid to open the eye. Unlike the larger orbicularis, this muscle lies within the eye socket.

Arteries of the face and neck

The pulse you feel in your neck is blood being pumped to the head via the carotid artery.

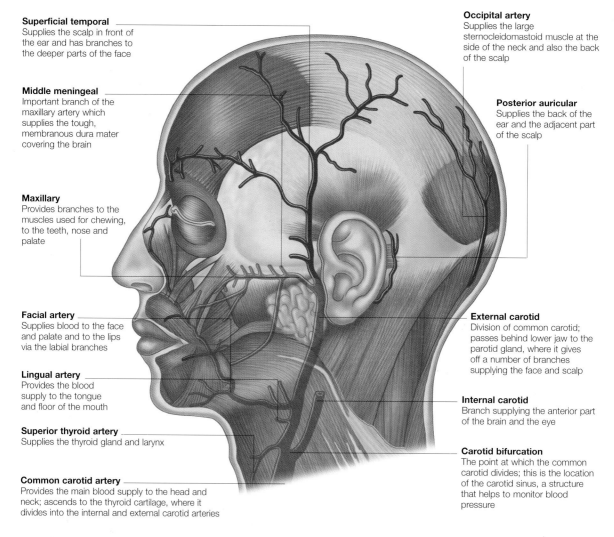

Superficial temporal
Supplies the scalp in front of the ear and has branches to the deeper parts of the face

Middle meningeal
Important branch of the maxillary artery which supplies the tough, membranous dura mater covering the brain

Maxillary
Provides branches to the muscles used for chewing, to the teeth, nose and palate

Facial artery
Supplies blood to the face and palate and to the lips via the labial branches

Lingual artery
Provides the blood supply to the tongue and floor of the mouth

Superior thyroid artery
Supplies the thyroid gland and larynx

Common carotid artery
Provides the main blood supply to the head and neck; ascends to the thyroid cartilage, where it divides into the internal and external carotid arteries

Occipital artery
Supplies the large sternocleidomastoid muscle at the side of the neck and also the back of the scalp

Posterior auricular
Supplies the back of the ear and the adjacent part of the scalp

External carotid
Division of common carotid; passes behind lower jaw to the parotid gland, where it gives off a number of branches supplying the face and scalp

Internal carotid
Branch supplying the anterior part of the brain and the eye

Carotid bifurcation
The point at which the common carotid divides; this is the location of the carotid sinus, a structure that helps to monitor blood pressure

The head and neck are supplied with blood from the two common carotid arteries that ascend either side of the neck. They are encased, along with the internal jugular vein and the vagus nerve, in a protective covering of connective tissue called the carotid sheath.

They have slightly different origins at the base of the neck with the left common carotid arising directly from the arch of the aorta while the right arises from the brachiocephalic trunk.

BRANCHING ARTERIES
The common carotid arteries divide at the level of the upper border of the thyroid cartilage (Adam's apple) to form the internal and external carotid arteries. The former enters the skull and supplies the brain and the latter provides branches that supply the face and scalp.

Many of the branches of the external carotid artery have a wavy or looped course. This flexibility ensures that when the mouth, larynx or pharynx are moved, during swallowing for example, the vessels are not stretched and damaged.

Veins of the face and neck

The veins have a similar distribution around the face and neck as the arteries. Many of the veins also share the same names.

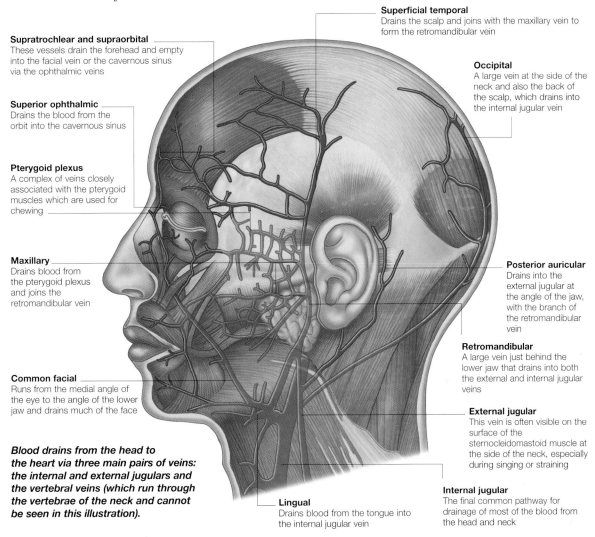

Superficial temporal
Drains the scalp and joins with the maxillary vein to form the retromandibular vein

Supratrochlear and supraorbital
These vessels drain the forehead and empty into the facial vein or the cavernous sinus via the ophthalmic veins

Occipital
A large vein at the side of the neck and also the back of the scalp, which drains into the internal jugular vein

Superior ophthalmic
Drains the blood from the orbit into the cavernous sinus

Pterygoid plexus
A complex of veins closely associated with the pterygoid muscles which are used for chewing

Maxillary
Drains blood from the pterygoid plexus and joins the retromandibular vein

Posterior auricular
Drains into the external jugular at the angle of the jaw, with the branch of the retromandibular vein

Retromandibular
A large vein just behind the lower jaw that drains into both the external and internal jugular veins

External jugular
This vein is often visible on the surface of the sternocleidomastoid muscle at the side of the neck, especially during singing or straining

Common facial
Runs from the medial angle of the eye to the angle of the lower jaw and drains much of the face

Blood drains from the head to the heart via three main pairs of veins: the internal and external jugulars and the vertebral veins (which run through the vertebrae of the neck and cannot be seen in this illustration).

Lingual
Drains blood from the tongue into the internal jugular vein

Internal jugular
The final common pathway for drainage of most of the blood from the head and neck

Blood drains from the head and neck back to the heart via the internal jugular veins that lie on either side of the neck. As with the common carotid arteries, the veins are protected by the carotid sheath.

Unlike in the rest of the body, the veins of this region generally lack valves, and the return of blood to the heart is by gravity and negative pressure in the thorax (chest).

The superficial (close to the surface) veins are often visible during exertion, and may be seen standing out on the necks of singers, for example.

JUGULAR VEIN
There is very little variation in the position of the internal jugular vein.

Because of this, the vein is used for monitoring central venous pressure (blood pressure within the right atrium of the heart). A cannula (hollow tube) is inserted into the vein and passed to the heart. The other end of the cannula is attached to a transducer, an instrument that records pressure. The blood volume may then be assessed.

As well as the veins draining the face, there is a series of emissary veins communicating between the venous sinuses (which drain blood from the brain) and the veins of the scalp. Along with the diploic veins (found in the bones of the skull), these provide a potential route for infection from the scalp into the brain.

37

Facial nerves

The facial muscles, and the involuntary functions such as tear formation, are served by the facial nerve, which transmits signals to and from the brain.

The facial nerve divides into five main branches: temporal, zygomatic, buccal, mandibular and cervical. These five branches fan out over the face and further divide, serving the muscles of facial expression.

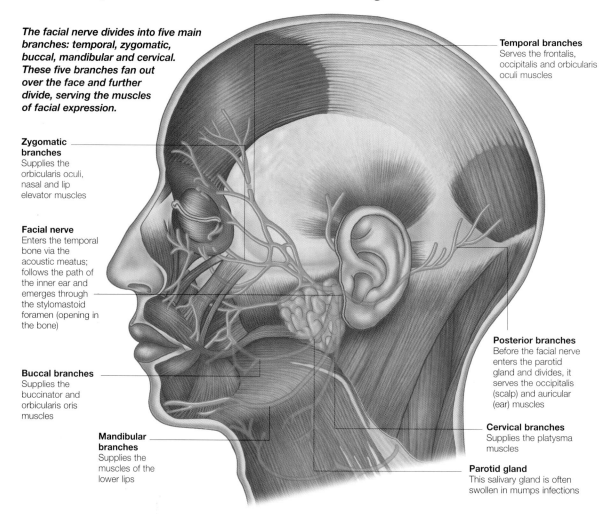

Temporal branches
Serves the frontalis, occipitalis and orbicularis oculi muscles

Zygomatic branches
Supplies the orbicularis oculi, nasal and lip elevator muscles

Facial nerve
Enters the temporal bone via the acoustic meatus; follows the path of the inner ear and emerges through the stylomastoid foramen (opening in the bone)

Buccal branches
Supplies the buccinator and orbicularis oris muscles

Mandibular branches
Supplies the muscles of the lower lips

Posterior branches
Before the facial nerve enters the parotid gland and divides, it serves the occipitalis (scalp) and auricular (ear) muscles

Cervical branches
Supplies the platysma muscles

Parotid gland
This salivary gland is often swollen in mumps infections

The muscles of facial expression are supplied by left and right facial nerves. Each nerve emerges through a hole in the skull (stylomastoid foramen) next to the lower part of the ear, and reaches the facial muscles by branching through a salivary gland (parotid gland) on the side of the face near the mouth and each ear.

Nerves are bundles of fibres that transmit electrical impulses from the brain or spinal column to the muscles, or from the sense organs to the brain or spinal column. Most nerves – including the facial nerves – are composed of a mixture of the two types, sending and receiving data to and from the brain.

NERVE DAMAGE
There are 12 pairs of cranial nerves serving various functions from moving the eyeballs to maintaining balance. The facial nerves are the seventh pair, and their principle task is providing motor impulses to the muscles of facial expression. (The muscles of mastication, used for chewing food, are served by the fifth cranial nerve, the trigeminal.)

As well as innervating the muscles (transmitting impulses to them), the facial nerves serve the autonomic functions, such as the production of tears and saliva. They also convey sensory impulses from the taste buds.

Muscles of mastication

The muscles that help us chew our food also play a part in speech, breathing and yawning.

The muscles of mastication are the muscles that move the mandible (jaw bone) up and down, and forwards and backwards, resulting in the opening and closing of the mouth.

This action is used in activities such as speaking, breathing through the mouth and in yawning. The closing action is also used very powerfully in the movements necessary for biting off and chewing up food (mastication), when side-to-side slewing of the jaw is also employed.

MOVING THE JAW

All jaw movements take place at the pair of temporo-mandibular joints, which lie in front of the ears.

The bones forming the joint are the head of the mandible (the rounded section at the top of the jaw bone) and the mandibular fossa of the temporal bone (the hollow in the skull in which the head of the mandible sits).

The hinge-like action allows up and down movements of the jaw. Additionally, the head of the mandible is covered with a closely fitting disc of cartilage, which allows forward and backward rocking movements. This latter movement enables the lower jaw to be slewed across the upper jaw on opening, and so provides the sideways forces necessary to grind up hard food on closing the mouth and chewing.

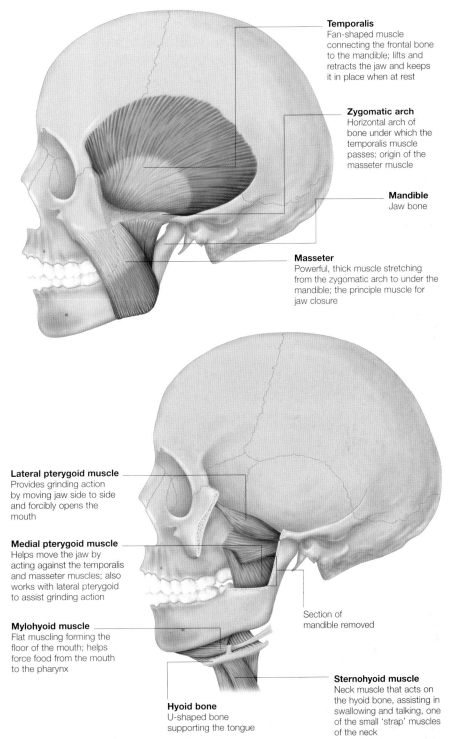

Temporalis
Fan-shaped muscle connecting the frontal bone to the mandible; lifts and retracts the jaw and keeps it in place when at rest

Zygomatic arch
Horizontal arch of bone under which the temporalis muscle passes; origin of the masseter muscle

Mandible
Jaw bone

Masseter
Powerful, thick muscle stretching from the zygomatic arch to under the mandible; the principle muscle for jaw closure

Lateral pterygoid muscle
Provides grinding action by moving jaw side to side and forcibly opens the mouth

Medial pterygoid muscle
Helps move the jaw by acting against the temporalis and masseter muscles; also works with lateral pterygoid to assist grinding action

Mylohyoid muscle
Flat muscling forming the floor of the mouth; helps force food from the mouth to the pharynx

Section of mandible removed

Hyoid bone
U-shaped bone supporting the tongue

Sternohyoid muscle
Neck muscle that acts on the hyoid bone, assisting in swallowing and talking, one of the small 'strap' muscles of the neck

Eyeball

The eyes are the specialized organs of sight, designed to respond to light.

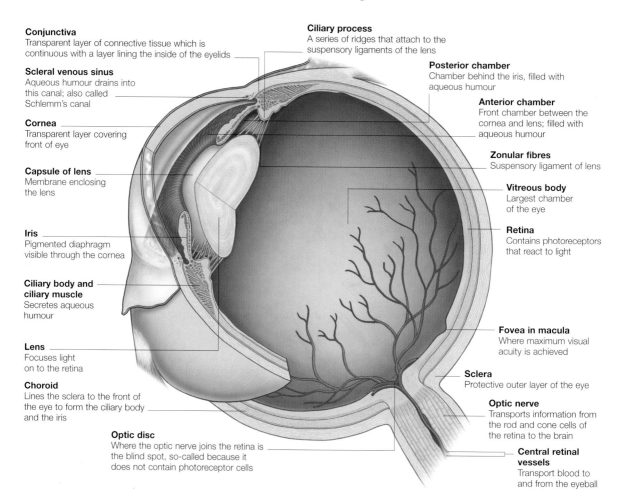

Conjunctiva
Transparent layer of connective tissue which is continuous with a layer lining the inside of the eyelids

Scleral venous sinus
Aqueous humour drains into this canal; also called Schlemm's canal

Cornea
Transparent layer covering front of eye

Capsule of lens
Membrane enclosing the lens

Iris
Pigmented diaphragm visible through the cornea

Ciliary body and ciliary muscle
Secretes aqueous humour

Lens
Focuses light on to the retina

Choroid
Lines the sclera to the front of the eye to form the ciliary body and the iris

Optic disc
Where the optic nerve joins the retina is the blind spot, so-called because it does not contain photoreceptor cells

Ciliary process
A series of ridges that attach to the suspensory ligaments of the lens

Posterior chamber
Chamber behind the iris, filled with aqueous humour

Anterior chamber
Front chamber between the cornea and lens; filled with aqueous humour

Zonular fibres
Suspensory ligament of lens

Vitreous body
Largest chamber of the eye

Retina
Contains photoreceptors that react to light

Fovea in macula
Where maximum visual acuity is achieved

Sclera
Protective outer layer of the eye

Optic nerve
Transports information from the rod and cone cells of the retina to the brain

Central retinal vessels
Transport blood to and from the eyeball

Our eyes allow us to receive information from our surroundings by detecting patterns of light. This information is sent to our brain, which processes it so that it can be perceived as different images.

Each eyeball is embedded in protective fatty tissue within a bony cavity (the orbit). The orbit has a large opening at the front to allow light to enter, and smaller openings at the back, allowing the optic nerve to pass to the brain, and blood vessels and nerves to enter the orbit.

CHAMBERS
The eyeball is divided into three internal chambers. The two aqueous chambers at the front of the eye are the anterior and posterior chambers, and are separated by the iris. These chambers are filled with clear, watery aqueous humour, which is secreted into the posterior chamber by a layer of cells covering the ciliary body.

This fluid passes into the anterior chamber through the pupil, then into the bloodstream via a number of small channels found where the base of the iris meets the margin of the cornea.

The largest of the chambers is the vitreous body, which lies behind the aqueous chambers, and is separated from them by the lens and the suspensory ligaments (zonular fibres), which connect the lens to the ciliary body. The vitreous body is filled with clear, jelly-like vitreous humour.

Layers of the eye

The eyeball is covered by three different layers, each of which has a special function.

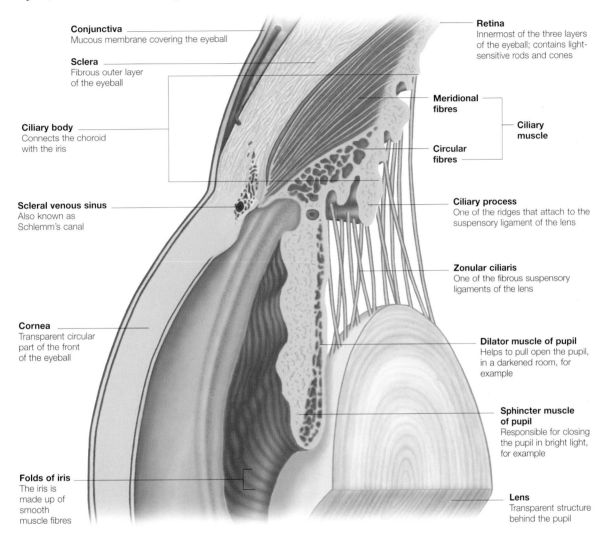

Conjunctiva
Mucous membrane covering the eyeball

Sclera
Fibrous outer layer
of the eyeball

Ciliary body
Connects the choroid
with the iris

Scleral venous sinus
Also known as
Schlemm's canal

Cornea
Transparent circular
part of the front
of the eyeball

Folds of iris
The iris is
made up of
smooth
muscle fibres

Retina
Innermost of the three layers
of the eyeball; contains light-
sensitive rods and cones

**Meridional
fibres**

**Circular
fibres**

**Ciliary
muscle**

Ciliary process
One of the ridges that attach to the
suspensory ligament of the lens

Zonular ciliaris
One of the fibrous suspensory
ligaments of the lens

Dilator muscle of pupil
Helps to pull open the pupil,
in a darkened room, for
example

**Sphincter muscle
of pupil**
Responsible for closing
the pupil in bright light,
for example

Lens
Transparent structure
behind the pupil

The outer layer of the eyeball is called the sclera, and is a tough, fibrous, protective layer. At the front of the eye, the sclera is visible as the 'white of the eye'. This is covered by the conjunctiva, a transparent layer of connective tissue. The transparent cornea covers the front of the eyeball, allowing light to enter the eye.

UVEA
The intermediate layer, the uvea, contains many blood vessels, nerves and pigmented cells. The uvea is divided into three main regions: the choroid, the ciliary body and the iris. The choroid extends from where the optic nerve meets the eyeball to the front of the eye, where it

forms both the ciliary body and the iris.

RETINA
The innermost layer of the eye is the retina, a layer of nerve tissue containing photosensitive (light-sensitive) cells called photoreceptors. It lines all but the most anterior (frontal) part of the vitreous

body. There are two types of photoreceptor cells: rods cells detect light intensity and are concentrated towards the periphery of the retina. Cone cells detect colour, and are most concentrated at the fovea at the most posterior part of the eyeball.

41

Muscles, blood vessels and nerves of the eye

The rotational movements of the eye are controlled by six rope-like extra-ocular muscles.

LEFT EYE (SIDE VIEW)

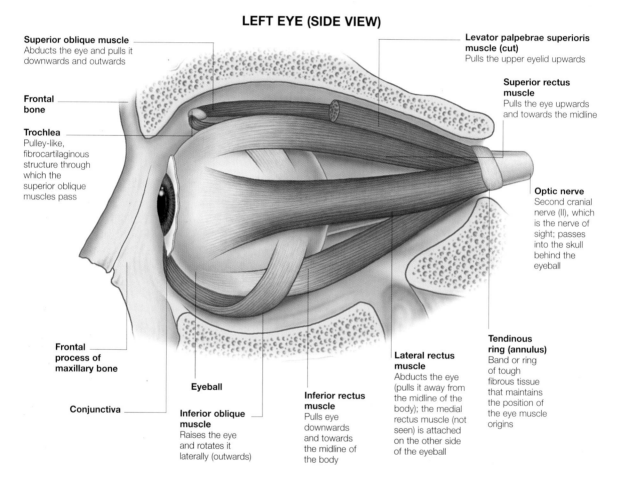

Superior oblique muscle
Abducts the eye and pulls it downwards and outwards

Frontal bone

Trochlea
Pulley-like, fibrocartilaginous structure through which the superior oblique muscles pass

Frontal process of maxillary bone

Conjunctiva

Eyeball

Inferior oblique muscle
Raises the eye and rotates it laterally (outwards)

Inferior rectus muscle
Pulls eye downwards and towards the midline of the body

Lateral rectus muscle
Abducts the eye (pulls it away from the midline of the body); the medial rectus muscle (not seen) is attached on the other side of the eyeball

Levator palpebrae superioris muscle (cut)
Pulls the upper eyelid upwards

Superior rectus muscle
Pulls the eye upwards and towards the midline

Optic nerve
Second cranial nerve (II), which is the nerve of sight; passes into the skull behind the eyeball

Tendinous ring (annulus)
Band or ring of tough fibrous tissue that maintains the position of the eye muscle origins

The muscles of the eye can be divided into three groups: the muscles inside the eyeball, the muscles of the eyelids and the extra-ocular muscles, which rotate the eyeball within its orbit.

The six extra-ocular muscles are rope-like, attaching directly to the sclera. Four of the muscles are rectus (straight) muscles – superior, inferior, lateral (the temple side of the eye) and medial (nasal side). Each rectus muscle arises from connective tissue, the common tendinous ring (annulus) at the back of the orbit that passes forward to insert just behind the junction of the sclera and cornea.

OBLIQUE MUSCLES
The two extra-ocular muscles are the oblique muscles. The superior oblique arises from bone near the back of the orbit, and extends to the front of the orbit. There, its tendon loops through the trochlea, a 'pulley' made of fibres and cartilage, and turns back to insert into the sclera.

The inferior oblique arises from the floor of the orbit, passing backwards and laterally under the eyeball to insert towards the back of the eye.

Nerves and blood vessels of the eye

The eye muscles are served by a series of nerves and blood vessels that help to make sight our dominant sense.

Nerves of the eye enter and leave the orbit through its openings posteriorly (at the back). Cranial nerve (CN) II – the optic nerve, which carries the visual signals from the retina to the brain – passes from the orbit to the cranial cavity through the optic canal. The other nerves – including branches of the ophthalmic nerve, the sensory nerve of the eye – enter the orbit through the orbital fissure.

Another nerve important for the eye is the facial nerve (CN VII). This supplies orbicularis oculi (a muscle of facial expression), causes blinking and also controls secretion from the lacrimal gland, which keeps the eye moist. They secrete fluid (tears) continuously, which is spread over the surface of the cornea by blinking. Irritation of the cornea can cause an increase in tear production.

ARTERIES OF THE EYE

The main artery of the eye is the ophthalmic artery, which is a branch of the internal carotid artery. The ophthalmic artery enters the orbit within the sheath of the optic nerve, and then branches to the extraocular muscles, the eyeball, the lacrimal gland and surrounding tissues.

The retinal artery remains within the optic nerve stalk until it reaches the optic disc, where it sends out branches supplying the retina. Veins drain the orbit to the cavernous sinus in the cranial cavity and to the facial vein, thus forming a connection between the blood vessels of the face and brain.

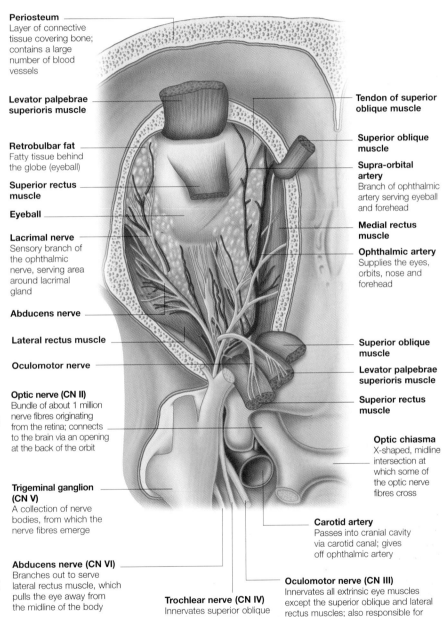

LEFT EYE (FROM ABOVE)

Periosteum
Layer of connective tissue covering bone; contains a large number of blood vessels

Levator palpebrae superioris muscle

Retrobulbar fat
Fatty tissue behind the globe (eyeball)

Superior rectus muscle

Eyeball

Lacrimal nerve
Sensory branch of the ophthalmic nerve, serving area around lacrimal gland

Abducens nerve

Lateral rectus muscle

Oculomotor nerve

Optic nerve (CN II)
Bundle of about 1 million nerve fibres originating from the retina; connects to the brain via an opening at the back of the orbit

Trigeminal ganglion (CN V)
A collection of nerve bodies, from which the nerve fibres emerge

Abducens nerve (CN VI)
Branches out to serve lateral rectus muscle, which pulls the eye away from the midline of the body

Trochlear nerve (CN IV)
Innervates superior oblique muscle, moving the eye down and outwards

Tendon of superior oblique muscle

Superior oblique muscle

Supra-orbital artery
Branch of ophthalmic artery serving eyeball and forehead

Medial rectus muscle

Ophthalmic artery
Supplies the eyes, orbits, nose and forehead

Superior oblique muscle

Levator palpebrae superioris muscle

Superior rectus muscle

Optic chiasma
X-shaped, midline intersection at which some of the optic nerve fibres cross

Carotid artery
Passes into cranial cavity via carotid canal; gives off ophthalmic artery

Oculomotor nerve (CN III)
Innervates all extrinsic eye muscles except the superior oblique and lateral rectus muscles; also responsible for elevating the upper eyelid and constricting the pupil

Eyelids and lacrimal apparatus

The eyelids are thin folds of skin that can close over the eye to protect it from injury and excessive light. The lacrimal apparatus is responsible for producing and draining lacrimal fluid.

Each lid is strengthened by a band of dense elastic connective tissue called a tarsal plate. These give the eyelids a curvature that matches that of the eye.

EYELID STRUCTURE

The tarsal plate of the upper eyelid is larger than that of the lower. The inner and outer ends of both tarsal plates are attached to the underlying bone by tiny ligaments. Between the front surface of the tarsal (meibomian) glands and the overlying skin, lie fibres of the orbicularis oculi muscle.

The eyelashes project from the free edge of the eyelids. The follicles of the eyelashes, from which the hairs emerge, have nerve endings which can sense any movement of the lashes.

The tarsal plates contain glands, called meibomian glands, that secrete an oily liquid which prevents the eyelids sticking together. There are also other tiny ciliary glands associated with the eyelash follicles.

EYELID MOVEMENT

The eye closes due to movement of the upper lid. The orbicularis oculi muscle contracts to close the eye, while the upper lid is opened by the levator palpebrae superioris muscle.

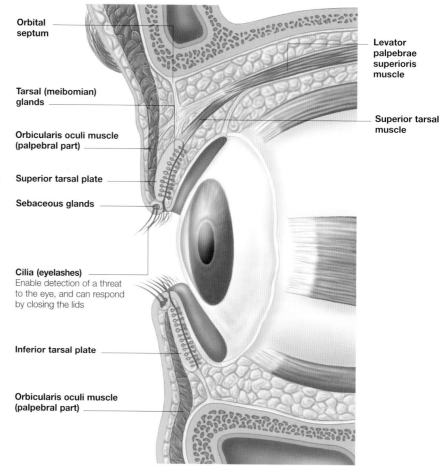

Orbital septum

Levator palpebrae superioris muscle

Tarsal (meibomian) glands

Superior tarsal muscle

Orbicularis oculi muscle (palpebral part)

Superior tarsal plate

Sebaceous glands

Cilia (eyelashes)
Enable detection of a threat to the eye, and can respond by closing the lids

Inferior tarsal plate

Orbicularis oculi muscle (palpebral part)

Movement of the two eyelids enable them to close in order to protect the eye. The upper eyelid is larger and more mobile than the lower lid.

Lacrimal apparatus

The eyes are protected and lubricated by lacrimal fluid, our tears. The lacrimal system produces this fluid and drains the excess to the nasal cavity.

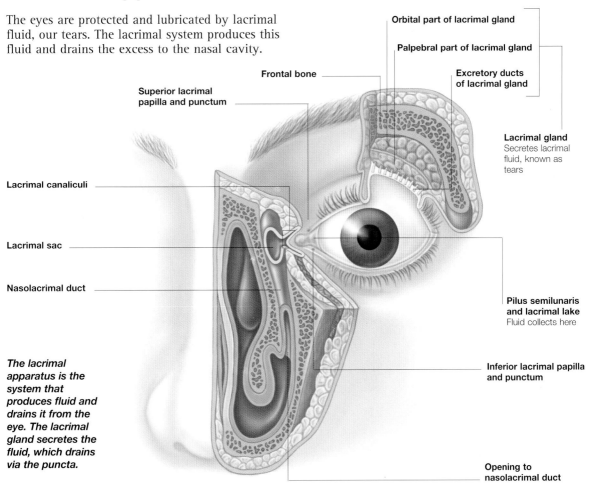

Orbital part of lacrimal gland

Palpebral part of lacrimal gland

Excretory ducts of lacrimal gland

Frontal bone

Superior lacrimal papilla and punctum

Lacrimal gland
Secretes lacrimal fluid, known as tears

Lacrimal canaliculi

Lacrimal sac

Nasolacrimal duct

Pilus semilunaris and lacrimal lake
Fluid collects here

The lacrimal apparatus is the system that produces fluid and drains it from the eye. The lacrimal gland secretes the fluid, which drains via the puncta.

Inferior lacrimal papilla and punctum

Opening to nasolacrimal duct

The eyes are kept moist by the continuous production of small amounts of lacrimal fluid by the lacrimal glands. This fluid also contains lysozyme, an antibacterial substance.

Most of the fluid, about 1 ml for each eye per day, is lost through evaporation. What is left is drained by the nasolacrimal ducts to the back of the nose.

LACRIMAL GLAND
The lacrimal gland, which produces the thin, watery lacrimal fluid, lies above the outer side of the eye within a recess in the bony eye socket. The lacrimal gland is about 2 cm long and roughly the shape of an almond.

The gland is divided into two parts; an upper orbital part and a lower palpebral part. There are also extra, accessory lacrimal glands which lie predominantly within the upper lid.

LACRIMAL DUCTS
The lacrimal glands each have up to 12 tiny ducts. These lacrimal ducts carry the secretions away from gland and release them into the conjunctival sac through openings under the upper lid in the superior fornix.

CANALICULI
After travelling across the eye during blinking, the lacrimal fluid collects in the lacrimal lake at the innermost corner.

The upper and lower eyelid both have a raised papilla at their inner ends, which has a tiny opening, the lacrimal punctum.

Excess lacrimal fluid enters these openings to be carried away by the underlying passages, the lacrimal canaliculi.

LACRIMAL SAC AND NASOLACRIMAL DUCT
Lacrimal fluid passes from the canaliculi into a collecting area known as the lacrimal sac.

From here the lacrimal fluid is carried down to the back of the nasal cavity by the nasolacrimal duct.

The fluid leaving the nasolacrimal duct normally evaporates within the nasal cavity, helping to humidify the air there.

Nose and nasal cavity

'Nose' commonly implies just the external structure, but anatomically it also includes the nasal cavity. The nose is the organ of smell and, as the opening of the respiratory tract, it serves to warm and filter air.

The external nose is a pyramid-shaped structure in the centre of the face, with the tip of the nose forming the apex of the pyramid. The underlying nasal cavity is a relatively large space and is the very first part of the respiratory tract (air passage).

The nasal cavity lies above the oral cavity (mouth) and is separated by a horizontal plate of bone called the hard palate. Both cavities open into the pharynx, a muscular, tube-like passageway.

EXTERNAL STRUCTURE

The external nose is made up of bone in its upper part and cartilage and fibrous tissue in its lower part. The upper part of the skeleton of the nose is mainly made up of a pair of plate-like bones called the nasal bones. These join, by their upper edges, with the frontal bone (forehead). Joining the outer edge of each nasal bone is the frontal process of the maxilla – a projection from the cheekbone between the nasal bone and the inner wall of the orbit (eye socket).

The bridge of the nose consists almost entirely of the two nasal bones, and adjoins the forehead between the two orbits. Because of their location and their relative fragility, the nasal bones are vulnerable to fracturing.

The lower half of the external nose is made up of plates of cartilage on each side. These join each other, and the cartilages of the other side along the midline of the nose.

Lateral view

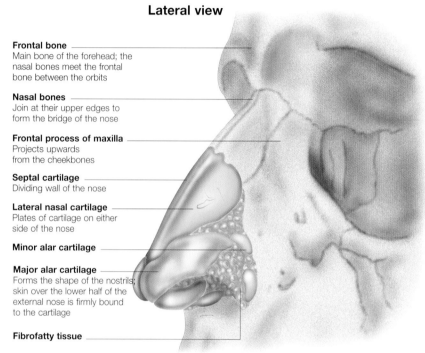

Frontal bone
Main bone of the forehead; the nasal bones meet the frontal bone between the orbits

Nasal bones
Join at their upper edges to form the bridge of the nose

Frontal process of maxilla
Projects upwards from the cheekbones

Septal cartilage
Dividing wall of the nose

Lateral nasal cartilage
Plates of cartilage on either side of the nose

Minor alar cartilage

Major alar cartilage
Forms the shape of the nostrils; skin over the lower half of the external nose is firmly bound to the cartilage

Fibrofatty tissue

Inferior view

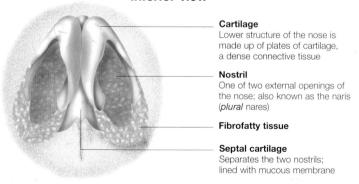

Cartilage
Lower structure of the nose is made up of plates of cartilage, a dense connective tissue

Nostril
One of two external openings of the nose; also known as the naris (*plural* nares)

Fibrofatty tissue

Septal cartilage
Separates the two nostrils; lined with mucous membrane

Inside the nasal cavity

The nasal cavity runs from the nostrils to the pharynx, and is divided in two by the septum. The roof forms part of the floor of the cranial cavity.

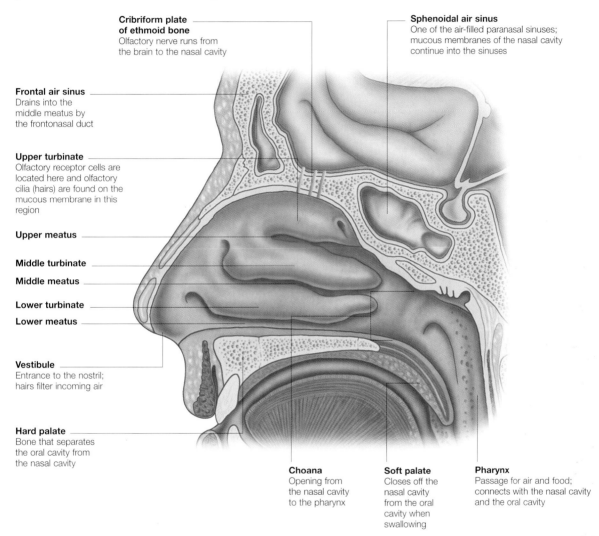

Cribriform plate of ethmoid bone
Olfactory nerve runs from the brain to the nasal cavity

Sphenoidal air sinus
One of the air-filled paranasal sinuses; mucous membranes of the nasal cavity continue into the sinuses

Frontal air sinus
Drains into the middle meatus by the frontonasal duct

Upper turbinate
Olfactory receptor cells are located here and olfactory cilia (hairs) are found on the mucous membrane in this region

Upper meatus

Middle turbinate

Middle meatus

Lower turbinate

Lower meatus

Vestibule
Entrance to the nostril; hairs filter incoming air

Hard palate
Bone that separates the oral cavity from the nasal cavity

Choana
Opening from the nasal cavity to the pharynx

Soft palate
Closes off the nasal cavity from the oral cavity when swallowing

Pharynx
Passage for air and food; connects with the nasal cavity and the oral cavity

The nasal cavity is partitioned into two halves by a vertical plate called the nasal septum, which is part bone and part cartilage. Each half of the nasal cavity is open in front at the nostril, and opens into the pharynx at the back through an opening called the choana.

NASAL CAVITY ROOF
The roof of the nasal cavity is arched from front to back. The central part of this roof is the cribriform plate of the ethmoid bone, a strip of bone perforated with a number of holes. This forms part of the floor of the cranial cavity, which contains the brain.

Running through the sieve-like cribriform plate from the nasal cavity to the brain is the olfactory nerve, which transmits the sensation of smell.

These anatomical features explain why head injuries involving fractures to the roof of the nasal cavity sometimes result in a leakage of cerebrospinal fluid (the clear fluid surrounding the brain) into the nose. If the head injury causes significant damage to the olfactory nerve, there may be a resultant loss of the ability to perceive smells. This is a condition called anosmia.

Paranasal sinuses

The term 'paranasal' means 'by the side of the nose'. The paranasal sinuses are air-filled cavities in the bones around the nasal cavity.

The paranasal sinuses are paired structures, and the two members of each pair are related to opposite halves of the nasal cavity.

The four pairs of paranasal sinuses are named according to the bones in which they are situated. These four pairs are:
• Maxillary sinuses
• Ethmoidal sinuses
• Frontal sinuses
• Sphenoidal sinuses.

Each member of a pair of paranasal sinuses opens into its half of the nasal cavity through a tiny opening called an ostium on the side of the nasal cavity.

The paranasal sinuses are very small, or even absent, at the time of birth, and remain small until puberty. Thereafter, the sinuses enlarge fairly rapidly; this enlargement accounts partially for the distinctive change in the size and shape of the face that occurs during adolescence.

FRONTAL SINUSES
The frontal sinuses are situated within the frontal bone (the bone of the forehead). Each is variable in size, corresponding to an area just above the inner part of the eyebrow.

The frontal sinuses are situated above the opening into the nasal cavity in the middle meatus. Drainage of mucous secretions is efficient and aided by gravity.

SPHENOIDAL SINUSES
The sphenoidal sinuses are behind the roof of the nasal cavity, within the sphenoid bone. The two sphenoidal

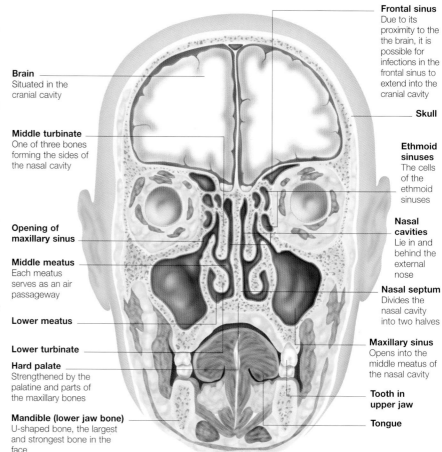

Brain
Situated in the cranial cavity

Middle turbinate
One of three bones forming the sides of the nasal cavity

Opening of maxillary sinus

Middle meatus
Each meatus serves as an air passageway

Lower meatus

Lower turbinate

Hard palate
Strengthened by the palatine and parts of the maxillary bones

Mandible (lower jaw bone)
U-shaped bone, the largest and strongest bone in the face

Frontal sinus
Due to its proximity to the the brain, it is possible for infections in the frontal sinus to extend into the cranial cavity

Skull

Ethmoid sinuses
The cells of the ethmoid sinuses

Nasal cavities
Lie in and behind the external nose

Nasal septum
Divides the nasal cavity into two halves

Maxillary sinus
Opens into the middle meatus of the nasal cavity

Tooth in upper jaw

Tongue

sinuses lie side by side, separated by a thin, vertical, bony partition. Each sphenoidal sinus opens into the uppermost part of the side wall of the nasal cavity (immediately above the upper turbinate) and also drains fairly efficiently into the nasal cavity.

ETHMOIDAL SINUSES
Each ethmoidal sinus is situated between the thin,

inner wall of the orbit (eye socket) and the side wall of the nasal cavity. Unlike the other paranasal sinuses, these sinuses are made up of multiple communicating cavities called ethmoid air cells. These cells are subdivided into front, middle and back groups. The front and middle groups of air cells open into the middle meatus, while the back group opens into the upper meatus.

The drainage into the nasal cavity is moderately efficient.

MAXILLARY SINUSES
The largest of the pairs of sinuses are the maxillary sinuses, situated within the maxillae (cheekbones). Infections are more common here than in any of the other paranasal sinuses because the drainage of mucous secretions from this sinus to the nasal cavity is not very efficient.

Inside the sinuses

The efficiency of mucous drainage from each of the pairs of sinuses depends on their location. Effective drainage lessens the risk of sinus infection.

Sphenoidal sinus
Tumours of the pituitary gland are commonly surgically removed by approach through the sphenoidal sinus

Pituitary gland
Immediately above the body of the sphenoid (in the cranial cavity) is the pituitary gland, which produces several important hormones

Brain

Frontal bone
Forms the forehead

Frontal sinus
Air-filled cavity in the frontal bone

Ethmoid cells
Multiple intercommunicating cavities divided into front, middle and back groups

Opening of maxillary sinus
The maxillary sinus is less efficient at draining mucus than the other sinus cavities

Middle turbinate (cut away)
Projection above the middle meatus

Hard palate
Separates the roof of the mouth from the nasal cavity

Lower turbinate
One of three projections of bone on the side wall of the nasal cavity

Opening of sphenoidal sinus
The two sphenoidal sinuses lie side by side within the sphenoid bone separated by a vertically placed thin bony partition

Pharynx
Common passage for air and food

SPHENOIDAL SINUSES
The sphenoidal sinuses are behind the roof of the nasal cavity, within the sphenoid bone. The two sphenoidal sinuses lie side by side, separated by a thin, vertical, bony partition. Each sphenoidal sinus opens into the uppermost part of the side wall of the nasal cavity (immediately above the upper turbinate) and also drains fairly efficiently into the nasal cavity.

ETHMOIDAL SINUSES
Each ethmoidal sinus is situated between the thin, inner wall of the orbit (eye socket) and the side wall of the nasal cavity. Unlike the other paranasal sinuses, these sinuses are made up of multiple communicating cavities called ethmoid air cells. These cells are subdivided into front, middle and back groups. The front and middle groups of air cells open into the middle meatus, while the back group opens into the upper meatus. The drainage into the nasal cavity is moderately efficient.

MAXILLARY SINUSES
The largest of the pairs of sinuses are the maxillary sinuses, situated within the maxillae (cheekbones). Infections and inflammation are more common here than in any of the other paranasal sinuses. This is because the drainage of mucous secretions from this sinus to the nasal cavity is not very efficient.

Oral cavity

Also known as the mouth, the oral cavity extends from the lips to the fauces, the opening leading to the pharynx.

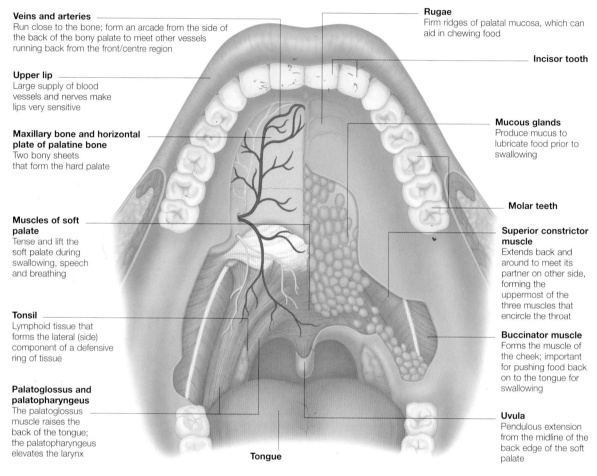

Veins and arteries
Run close to the bone; form an arcade from the side of the back of the bony palate to meet other vessels running back from the front/centre region

Upper lip
Large supply of blood vessels and nerves make lips very sensitive

Maxillary bone and horizontal plate of palatine bone
Two bony sheets that form the hard palate

Muscles of soft palate
Tense and lift the soft palate during swallowing, speech and breathing

Tonsil
Lymphoid tissue that forms the lateral (side) component of a defensive ring of tissue

Palatoglossus and palatopharyngeus
The palatoglossus muscle raises the back of the tongue; the palatopharyngeus elevates the larynx

Rugae
Firm ridges of palatal mucosa, which can aid in chewing food

Incisor tooth

Mucous glands
Produce mucus to lubricate food prior to swallowing

Molar teeth

Superior constrictor muscle
Extends back and around to meet its partner on other side, forming the uppermost of the three muscles that encircle the throat

Buccinator muscle
Forms the muscle of the cheek; important for pushing food back on to the tongue for swallowing

Uvula
Pendulous extension from the midline of the back edge of the soft palate

Tongue

This view of the mouth shows a dissection of the palate. The right side is stripped down to bone, revealing nerves and arteries; the left side has the mucosa intact at the front and side, but removed at the centre and rear to show underlying mucous glands.

The roof of the mouth, viewed from below, shows two distinct structures: the dental arch and the palate. The dental arch is the curved part of the maxilla bone at the front and sides of the roof, and the palate is a horizontal plate of tissue that separates the mouth from the nose.

The front two-thirds of the palate are bony and hard, and are formed by the maxillary bone. The hard palate is covered with a mucous membrane, beneath which run arteries, veins and nerves. These nourish and provide sensation to the palate and the overlying mucous glands, which often form fibrous ridges called rugae. The mucus secreted by these glands lubricates food to facilitate swallowing.

SOFT PALATE
The rear third of the palate is composed of glandula mucosa, muscle and tendon. Forming much of the soft palate are the tensor and levator palati muscles. These muscles close off the nasal cavity from the mouth during swallowing by respectively tensing and elevating the soft palate. They also act with other muscles to open the auditory (Eustachian) tube, which equalizes pressure on either side of the eardrum.

Floor of the mouth

The floor of the mouth acts as the foundation for a network of muscles and glands that are essential to its function.

The tongue is situated over the mylohyoid muscle, which forms the muscular floor of the mouth. It is the hyoglossus muscle that anchors the tongue to the hyoid bone and provides extra strength, while the genioglossus muscle stops the tongue from moving back into the throat.

The temporalis muscles are muscles of mastication (chewing). The lingula is a small bony projection of the mandible. The mandibular nerve passes below this, through the mandibular foramen and runs within the body of the mandible to supply the lower teeth and lower lip with sensation.

SALIVARY GLANDS
There are a pair each of the submandibular and sublingual salivary glands on either side of the oral floor and, with the paired parotid glands, they make up the six salivary glands. Saliva flows along the submandibular gland duct on the mylohyoid muscle, and emerges in the front of the oral cavity on either side of the tongue, behind the lower front teeth.

Saliva from the sublingual glands either runs into the submandibular duct, or flows out through openings in the mucosa to the side of the tongue. The lingual nerve provides taste and sensation to the front of the tongue.

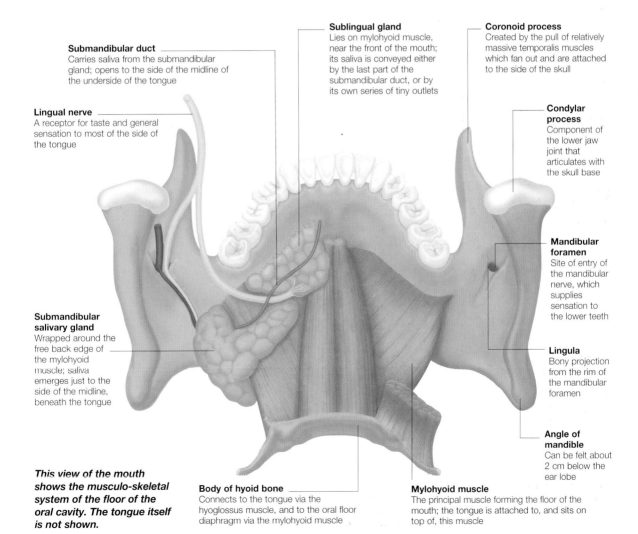

Submandibular duct
Carries saliva from the submandibular gland; opens to the side of the midline of the underside of the tongue

Lingual nerve
A receptor for taste and general sensation to most of the side of the tongue

Submandibular salivary gland
Wrapped around the free back edge of the mylohyoid muscle; saliva emerges just to the side of the midline, beneath the tongue

Sublingual gland
Lies on mylohyoid muscle, near the front of the mouth; its saliva is conveyed either by the last part of the submandibular duct, or by its own series of tiny outlets

Coronoid process
Created by the pull of relatively massive temporalis muscles which fan out and are attached to the side of the skull

Condylar process
Component of the lower jaw joint that articulates with the skull base

Mandibular foramen
Site of entry of the mandibular nerve, which supplies sensation to the lower teeth

Lingula
Bony projection from the rim of the mandibular foramen

Angle of mandible
Can be felt about 2 cm below the ear lobe

This view of the mouth shows the musculo-skeletal system of the floor of the oral cavity. The tongue itself is not shown.

Body of hyoid bone
Connects to the tongue via the hyoglossus muscle, and to the oral floor diaphragm via the mylohyoid muscle

Mylohyoid muscle
The principal muscle forming the floor of the mouth; the tongue is attached to, and sits on top of, this muscle

Teeth

Teeth are designed for biting and chewing up food, and each has a particular function.

The teeth are specialized hardened regions of gum tissue, partly embedded in the jaw bones. They break up solid foods by the actions of biting and chewing.

The visible part of a tooth is the crown. This is composed of a shell of hard, calcified material called dentine (similar to compact bone but without blood vessels), which is covered by a thin layer of even harder calcified material called enamel.

The hidden part (the root) is embedded in a socket of the jaw bone (the alveolus). It is also made of dentine, covered by a layer of cementum which, with dense, collagen-rich periodontal ligaments, anchors the root to the bone of the alveolar socket.

INSIDE THE TEETH

Inside inside an internal pulp cavity, containing soft connective tissue, blood vessels and nerves. The pulp is linked to the jaw via the root.

The layout of adult teeth is the same in the upper and lower jaws. Each side (quadrant) has eight teeth: two incisors, one canine, two premolars and three molars, making 32 in total. Children have 20 milk teeth, with only one molar in each quadrant.

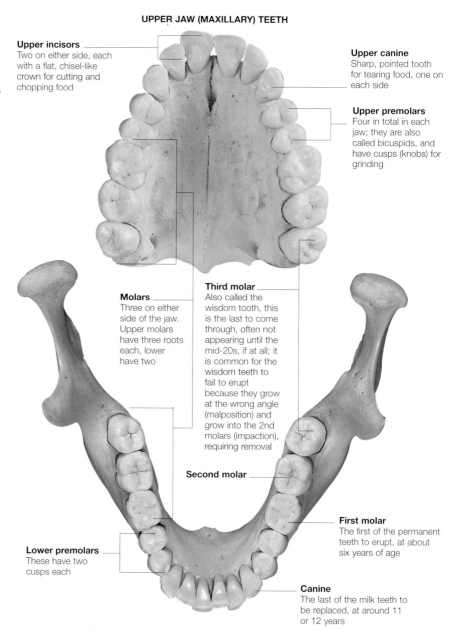

UPPER JAW (MAXILLARY) TEETH

Upper incisors
Two on either side, each with a flat, chisel-like crown for cutting and chopping food

Upper canine
Sharp, pointed tooth for tearing food, one on each side

Upper premolars
Four in total in each jaw; they are also called bicuspids, and have cusps (knobs) for grinding

Molars
Three on either side of the jaw. Upper molars have three roots each, lower have two

Third molar
Also called the wisdom tooth, this is the last to come through, often not appearing until the mid-20s, if at all; it is common for the wisdom teeth to fail to erupt because they grow at the wrong angle (malposition) and grow into the 2nd molars (impaction), requiring removal

Second molar

First molar
The first of the permanent teeth to erupt, at about six years of age

Lower premolars
These have two cusps each

Canine
The last of the milk teeth to be replaced, at around 11 or 12 years

LOWER JAW (MANDIBULAR) TEETH

Development of teeth

There are two major phases of tooth development during childhood. This is to allow the head to grow and adult teeth to develop.

The teeth begin to develop in the human embryo around the sixth week of pregnancy. Six to eight months after birth, root growth pushes the tooth crown through the gum in the process of eruption called teething.

This first set are the primary or deciduous teeth (milk teeth). These erupt in a specific order, usually the lower central incisors first, then the upper central incisors. The deciduous teeth do not include premolars.

ADULT TOOTH GROWTH

The tooth buds for the second wave of tooth production develop at the same time. These permanent teeth remain dormant until the ages of five to seven, when they begin to grow, causing the roots of the deciduous teeth to begin to break down.

This breakdown, together with the pressure of the underlying permanent teeth, results in the shedding of the deciduous teeth. The new teeth then start to appear and continue to do so until the ages of 10 to 12.

Eruption of the permanent set follows a similar pattern to the earlier growth (although premolars erupt between the canines and the molars). The permanent set has additional, third molars (wisdom teeth) that tend to appear after between 15 and 25 years.

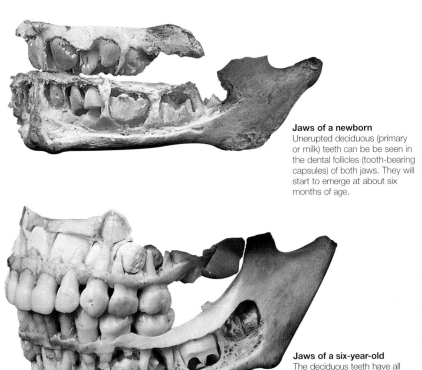

Jaws of a newborn
Unerupted deciduous (primary or milk) teeth can be be seen in the dental follicles (tooth-bearing capsules) of both jaws. They will start to emerge at about six months of age.

Jaws of a six-year-old
The deciduous teeth have all erupted. Beneath them in the alveoli (tooth sockets) are the permanent teeth ready to come through. This process continues until the early to mid teens.

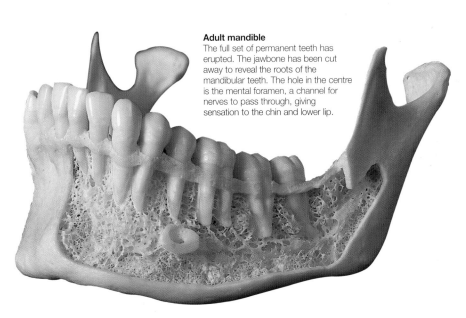

Adult mandible
The full set of permanent teeth has erupted. The jawbone has been cut away to reveal the roots of the mandibular teeth. The hole in the centre is the mental foramen, a channel for nerves to pass through, giving sensation to the chin and lower lip.

Tongue

The tongue is basically a mass of muscle, whose complex movement is essential for speech, mastication and swallowing. Its upper surface is lined with specialized tissue that contains taste buds.

The dorsal (upper) surface of the tongue is covered with an epithelium specialized for the sense of taste. The anterior two-thirds of the tongue at rest lies within the lower dental arcade. The posterior third slopes back and down to form part of the front wall of the oropharynx. Its musculature and movements are described in some detail overleaf.

DORSAL SURFACE

The tongue's upper surface is characterized by filiform papillae, tiny protuberances which give the surface a rough feel. The filiform papillae have tufts of keratin which, when elongated, may give the surface a 'hairy' appearance and feel. These 'hairs' can be stained by food, medicine and nicotine. Scattered among them are the larger fungiform papillae. Larger still are the 8–12 circumvallate papillae, which form an inverse V at the junction of the anterior two-thirds and posterior third. These papillae are the major site of taste buds, although they do occur in other papillae and are scattered over the tongue surface, as well as the cheek mucosa and the pharynx.

The posterior third of the dorsal surface has a cobbled appearance due to the presence of 40–100 nodules of lymphoid tissue, which form the lingual tonsil.

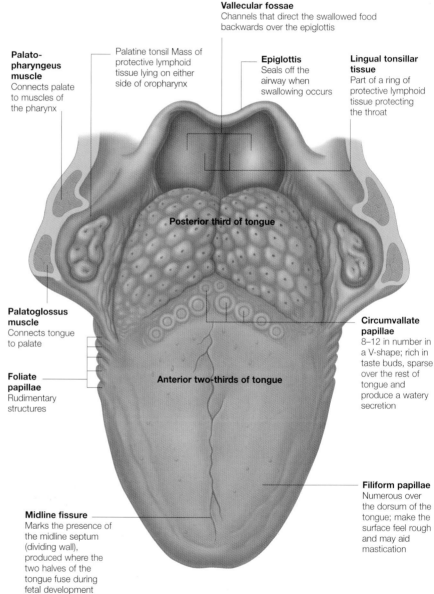

Vallecular fossae
Channels that direct the swallowed food backwards over the epiglottis

Palato-pharyngeus muscle
Connects palate to muscles of the pharynx

Palatine tonsil Mass of protective lymphoid tissue lying on either side of oropharynx

Epiglottis
Seals off the airway when swallowing occurs

Lingual tonsillar tissue
Part of a ring of protective lymphoid tissue protecting the throat

Posterior third of tongue

Palatoglossus muscle
Connects tongue to palate

Foliate papillae
Rudimentary structures

Circumvallate papillae
8–12 in number in a V-shape; rich in taste buds, sparse over the rest of tongue and produce a watery secretion

Anterior two-thirds of tongue

Filiform papillae
Numerous over the dorsum of the tongue; make the surface feel rough and may aid mastication

Midline fissure
Marks the presence of the midline septum (dividing wall), produced where the two halves of the tongue fuse during fetal development

Muscles of the tongue

The muscles within the tongue (intrinsic muscles) comprise three groups of fibre bundles running the length, breadth and depth of the organ.

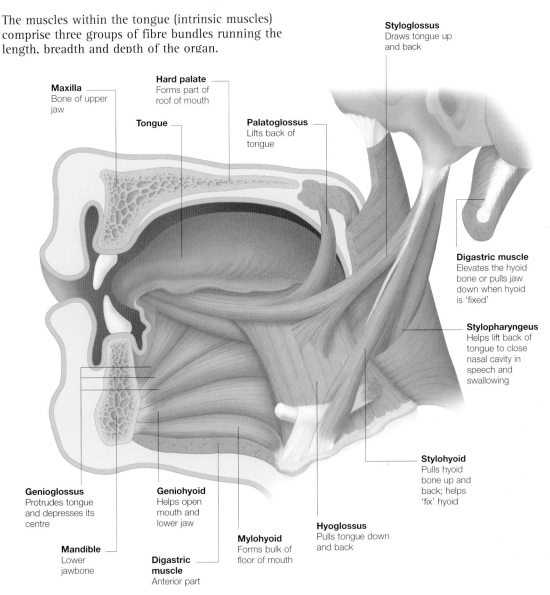

Styloglossus
Draws tongue up and back

Maxilla
Bone of upper jaw

Hard palate
Forms part of roof of mouth

Tongue

Palatoglossus
Lifts back of tongue

Digastric muscle
Elevates the hyoid bone or pulls jaw down when hyoid is 'fixed'

Stylopharyngeus
Helps lift back of tongue to close nasal cavity in speech and swallowing

Genioglossus
Protrudes tongue and depresses its centre

Geniohyoid
Helps open mouth and lower jaw

Mylohyoid
Forms bulk of floor of mouth

Hyoglossus
Pulls tongue down and back

Stylohyoid
Pulls hyoid bone up and back; helps 'fix' hyoid

Mandible
Lower jawbone

Digastric muscle
Anterior part

The intrinsic muscles of the tongue alter the shape of the tongue to facilitate speech, mastication (chewing) and swallowing. The other muscles attached to the tongue (extrinsic muscles), move the organ as a whole. The names of the extrinsic muscles denote their attachments and the general direction of the tongue's movement.

Protrusion of the tongue (sticking it out), elevation of its sides and depression of its centre are functions of the intrinsic muscles. They also, together with an intact palate, the lips and the teeth, allow the formation of specific sounds in speech.

SWALLOWING
When food has been chewed and mixed with lubricating saliva, it is forced up and back between the hard palate and the upper surface of the tongue by contraction of the styloglossus muscles which pull the tongue up and back. The palatoglossi then contract, squeezing the food bolus into the oral part of the pharynx. The levator palati muscles lift the soft palate to seal off the nasal passage, while the larynx and laryngopharynx are pulled up sealing the airway against the back of the epiglottis while the bolus passes over it.

Salivary glands

The salivary glands produce about three-quarters of a litre of saliva a day. Saliva plays a major role in lubricating and protecting the mouth and teeth, as well as aiding swallowing and mastication.

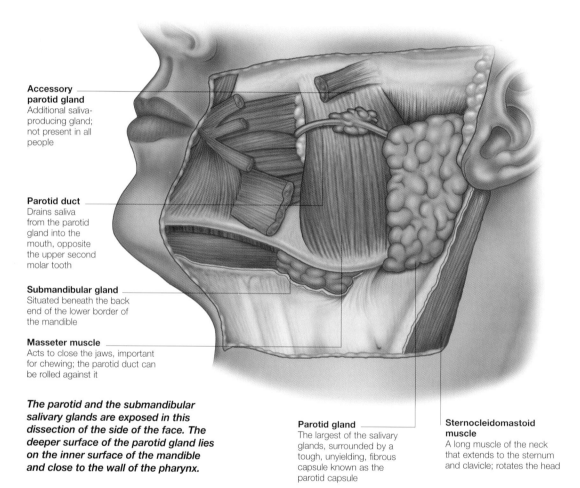

Accessory parotid gland
Additional saliva-producing gland; not present in all people

Parotid duct
Drains saliva from the parotid gland into the mouth, opposite the upper second molar tooth

Submandibular gland
Situated beneath the back end of the lower border of the mandible

Masseter muscle
Acts to close the jaws, important for chewing; the parotid duct can be rolled against it

The parotid and the submandibular salivary glands are exposed in this dissection of the side of the face. The deeper surface of the parotid gland lies on the inner surface of the mandible and close to the wall of the pharynx.

Parotid gland
The largest of the salivary glands, surrounded by a tough, unyielding, fibrous capsule known as the parotid capsule

Sternocleidomastoid muscle
A long muscle of the neck that extends to the sternum and clavicle; rotates the head

There are three pairs of major salivary glands, which produce about 90 per cent of saliva; the remaining 10 per cent is produced by minor salivary glands located in the cheeks, lips, tongue and palate. The major role of saliva is lubrication, allowing mastication, swallowing and speech. It also keeps the mouth and gums moist and limiting bacterial activity.

The cells producing saliva are located in clusters at the end of a branching series of ducts. Two different types of saliva are produced by two distinctive cell types, called mucous and serous cells. The secretory products of mucous cells form a viscous mucin-rich product; the serous cells produce a watery fluid containing the enzyme amylase.

PAROTID GLAND
The largest of salivary glands are the parotid glands, which secrete serous. Each parotid is superficial, lying just beneath the skin, situated between the mandible (lower jaw) and the ear.

Several important structures pass through the parotid gland. The deepest of these is the external carotid artery; the most superficial is the facial nerve, which supplies the muscles that allow facial expression.

Submandibular and sublingual glands

The two smaller pairs of salivary glands are the submandibular and the sublingual glands situated in the floor of the mouth.

The submandibular gland is situated beneath the lower border of the mandible towards the angle of the jaw. It is a mixed salivary gland containing serous cells (about 60 per cent) and mucous cells (about 40 per cent). It is about the size of a walnut, the gland has two parts: a large, superficial part and a smaller, deep part tucked behind the mylohyoid muscle which forms the floor of the mouth. The saliva produced by the submandibular gland is carried in the submandibular duct, which opens in the sublingual papilla (protuberance) underneath the tongue.

SUBLINGUAL GLANDS

The sublingual gland is the smallest of the three major salivary glands and is almond-shaped. It is composed of about 60 per cent mucous cells and 40 per cent serous cell and lies under the tongue in the sublingual fossa. The two sublingual glands almost meet in the midline, and lie on the mylohyoid muscle.

Behind, the sublingual gland sits close to the deep part of the submandibular gland. Unlike the other glands, the sublingual gland does not have a single major collecting duct, but many smaller ones opening separately into the floor of the mouth or into the submandibular duct.

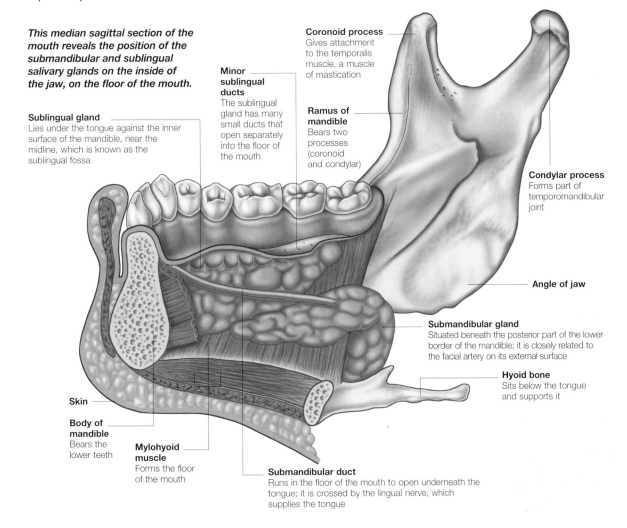

This median sagittal section of the mouth reveals the position of the submandibular and sublingual salivary glands on the inside of the jaw, on the floor of the mouth.

Coronoid process
Gives attachment to the temporalis muscle, a muscle of mastication

Minor sublingual ducts
The sublingual gland has many small ducts that open separately into the floor of the mouth

Ramus of mandible
Bears two processes (coronoid and condylar)

Sublingual gland
Lies under the tongue against the inner surface of the mandible, near the midline, which is known as the sublingual fossa

Condylar process
Forms part of temporomandibular joint

Angle of jaw

Submandibular gland
Situated beneath the posterior part of the lower border of the mandible; it is closely related to the facial artery on its external surface

Hyoid bone
Sits below the tongue and supports it

Skin

Body of mandible
Bears the lower teeth

Mylohyoid muscle
Forms the floor of the mouth

Submandibular duct
Runs in the floor of the mouth to open underneath the tongue; it is crossed by the lingual nerve, which supplies the tongue

Infratemporal fossa

The infratemporal fossa (a fossa is a depression or hollow) is a region at the side of the head which contains a number of important nerves, blood vessels and muscles involved in mastication (chewing).

The infratemporal fossa (fossa means a depression or hollow) is a region through which a number of important nerves, blood vessels and muscles pass.

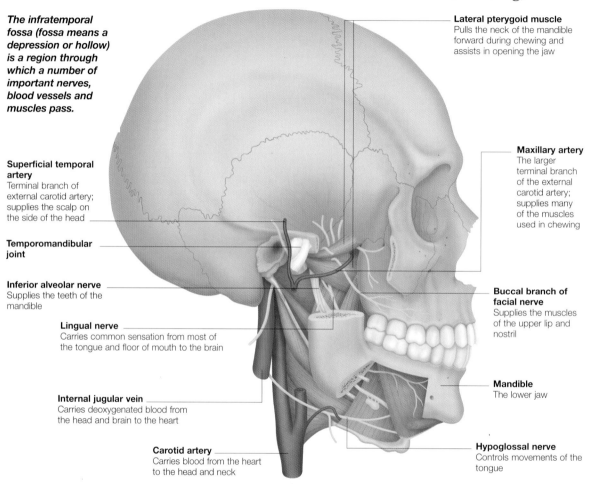

Lateral pterygoid muscle
Pulls the neck of the mandible forward during chewing and assists in opening the jaw

Superficial temporal artery
Terminal branch of external carotid artery; supplies the scalp on the side of the head

Temporomandibular joint

Inferior alveolar nerve
Supplies the teeth of the mandible

Lingual nerve
Carries common sensation from most of the tongue and floor of mouth to the brain

Internal jugular vein
Carries deoxygenated blood from the head and brain to the heart

Carotid artery
Carries blood from the heart to the head and neck

Maxillary artery
The larger terminal branch of the external carotid artery; supplies many of the muscles used in chewing

Buccal branch of facial nerve
Supplies the muscles of the upper lip and nostril

Mandible
The lower jaw

Hypoglossal nerve
Controls movements of the tongue

The infratemporal fossa is located below the base of the skull, between the pharynx and the ramus (side) of the mandible (lower jawbone). The region is important to dental surgeons, not only because many of its components are essential to the process of mastication, but also as many of the nerves and blood vessels supplying the mouth are transmitted through it.

ANATOMY OF THE FOSSA
The region is largely defined by the skeletal boundaries of the infratemporal fossa. The anterior boundary is the posterior surface of the maxillary bone, and the posterior boundary is the styloid process of the temporal bone and the carotid sheath. The midline boundary is formed by the lateral pterygoid plate of the sphenoid bone; the lateral boundary is the ramus of the mandible and the roof is the base of the greater wing of the sphenoid bone. The infratemporal fossa has no floor, and is continuous with the neck.

CONTENTS OF THE FOSSA
The fossa contains the pterygoid muscles, branches of the mandibular nerve, the chorda tympani branch of the facial nerve, the otic ganglion (part of the autonomic nervous system), the maxillary artery and the pterygoid venous plexus (vessels surrounding pterygoid muscles).

Mandibular nerve

The mandibular nerve leaves the skull (through the foramen ovale) to enter directly into the infratemporal fossa, where it divides into its many branches.

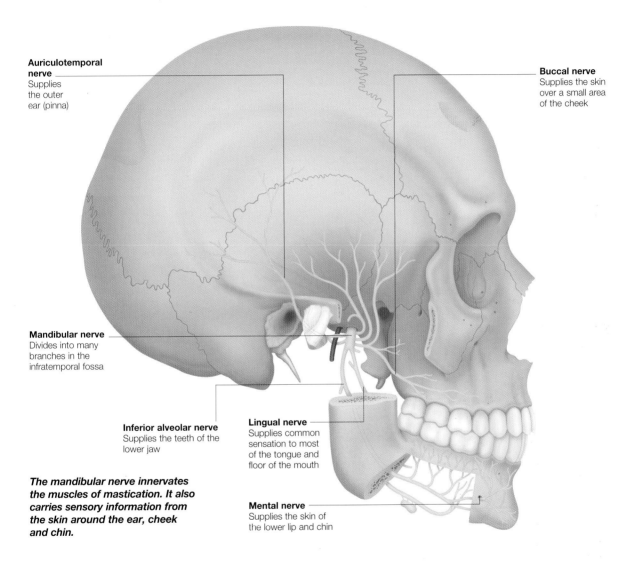

Auriculotemporal nerve
Supplies the outer ear (pinna)

Buccal nerve
Supplies the skin over a small area of the cheek

Mandibular nerve
Divides into many branches in the infratemporal fossa

Inferior alveolar nerve
Supplies the teeth of the lower jaw

Lingual nerve
Supplies common sensation to most of the tongue and floor of the mouth

Mental nerve
Supplies the skin of the lower lip and chin

The mandibular nerve innervates the muscles of mastication. It also carries sensory information from the skin around the ear, cheek and chin.

The mandibular nerve has motor branches which supply all of the muscles of mastication, allowing contractions of these muscles to take place. The nerve also supplies sensation to the skin of the temple and around the ear via the auriculotemporal nerve and to the skin on the outside of the cheek and the tissue lining the inside via the buccal nerve.

BRANCHES OF THE MANDIBULAR NERVE
The branch called the inferior alveolar nerve (or inferior dental nerve) travels downwards and then forwards to enter the body of the mandible. It supplies sensation to all of the lower teeth, but it also has a branch called the mental nerve which leaves the mandible through a foramen in the lower premolar region. This supplies sensation to the lower lip.

The lingual nerve branch supplies common sensation (for example touch, temperature and pain) to most of the tongue and the floor of the mouth.

Pterygopalatine fossa

The pterygopalatine fossa is a funnel-shaped space between the bones of the head. It contains important nerves and blood vessels that supply the eye, mouth, nose and face.

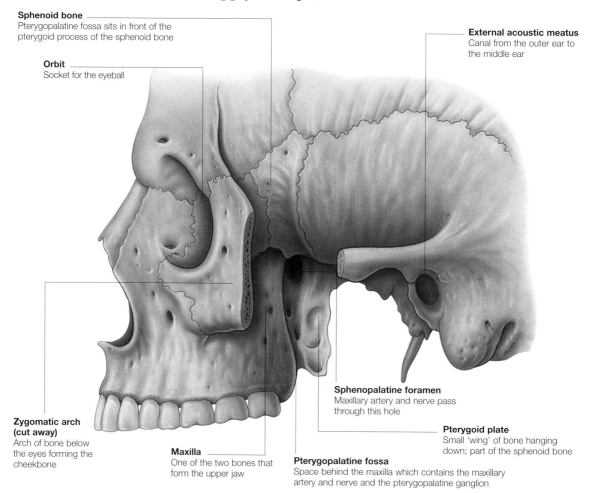

Sphenoid bone
Pterygopalatine fossa sits in front of the pterygoid process of the sphenoid bone

External acoustic meatus
Canal from the outer ear to the middle ear

Orbit
Socket for the eyeball

Sphenopalatine foramen
Maxillary artery and nerve pass through this hole

Pterygoid plate
Small 'wing' of bone hanging down; part of the sphenoid bone

Zygomatic arch (cut away)
Arch of bone below the eyes forming the cheekbone

Maxilla
One of the two bones that form the upper jaw

Pterygopalatine fossa
Space behind the maxilla which contains the maxillary artery and nerve and the pterygopalatine ganglion

The pterygopalatine fossa is an anatomical area that is difficult to find when the skull is intact, and furthermore disappears when the skull bones are separated. The easiest way of locating the fossa is via the pterygomaxillary fissure, which is a narrow triangular gap between the pterygoid plates of the sphenoid bone and the back of the upper jaw (maxilla). This leads to the lateral part of the fossa.

LOCATION OF THE FOSSA
The fossa is a small funnel-shaped space that tapers downwards and lies below the back of the orbit. It is located behind the maxilla and its back wall is formed by the pterygoid plates and the greater wing of the sphenoid bone. The palatine bone forms its midline and its floor. It is a very important distribution centre as it communicates with all of the important regions of the head including the mouth, nose, eye and face, infratemporal fossa and also with the brain.

The main components of the pterygopalatine fossa are the maxillary artery and nerve (branch of the trigeminal nerve) and the pterygopalatine ganglion. These enter and exit the region through the spheno-palatine foramen (hole).

Maxillary nerve

The maxillary nerve enters the pterygopalatine fossa before dividing into branches which supply sensation to large areas of the face.

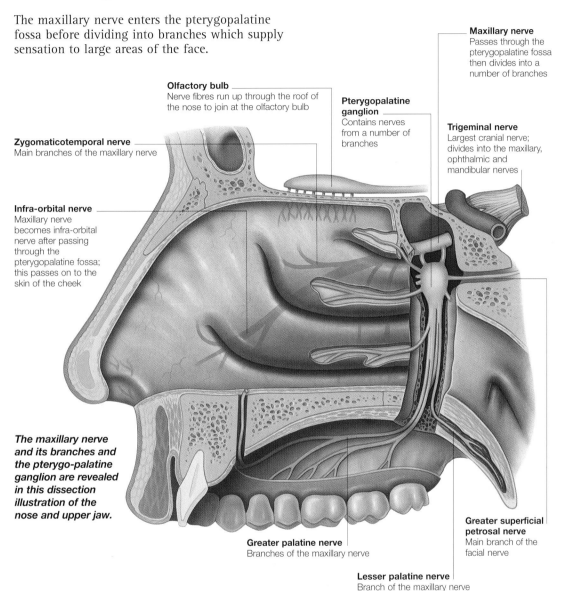

Maxillary nerve
Passes through the pterygopalatine fossa then divides into a number of branches

Olfactory bulb
Nerve fibres run up through the roof of the nose to join at the olfactory bulb

Pterygopalatine ganglion
Contains nerves from a number of branches

Trigeminal nerve
Largest cranial nerve; divides into the maxillary, ophthalmic and mandibular nerves

Zygomaticotemporal nerve
Main branches of the maxillary nerve

Infra-orbital nerve
Maxillary nerve becomes infra-orbital nerve after passing through the pterygopalatine fossa; this passes on to the skin of the cheek

The maxillary nerve and its branches and the pterygo-palatine ganglion are revealed in this dissection illustration of the nose and upper jaw.

Greater superficial petrosal nerve
Main branch of the facial nerve

Greater palatine nerve
Branches of the maxillary nerve

Lesser palatine nerve
Branch of the maxillary nerve

The maxillary nerve leaves the cranial part of the skull to enter directly into the pterygopalatine fossa by the foramen rotundum. On entering the fossa, the nerve contains only fibres for common sensation including touch, pain and temperature. It divides within the fossa to supply these sensations to the nose, palate, tonsils and gums, skin of the cheeks, upper lip and upper molar teeth.

BRANCHES OF THE MAXILLARY NERVE
The names of the main branches are the zygomaticotemporal and zygomaticofacial, greater and lesser palatine nerves, nasal nerves and posterior superior alveolar nerve. The main trunk of the maxillary nerve leaves the fossa through the inferior orbital fissure which is in the floor of the orbit.

As it leaves the pterygopalatine fossa, the maxillary nerve becomes the infra-orbital nerve. It travels in the floor of the orbit to emerge through a foramen in the maxilla, below the eye.

Branches from the infra-orbital nerve include the anterior superior alveolar nerve which supplies the front upper teeth.

61

Ear

The ears are vital sensory organs of hearing and balance. Each ear is divided into three parts – outer, middle and inner ear – each of which is designed to respond to sound or movement in a different way.

The ear can be divided anatomically into three different parts: the external, middle and inner ear. The external and middle ear are important in the gathering and transmitting of sound waves. The inner ear is the organ of hearing and is also vital in enabling us to maintain our balance.

TRANSMITTING INFORMATION

The external ear consists of the visible auricle or pinna (earlobe) and the canal that passes into the head – the external auditory meatus. At the inner end of the meatus is the tympanic membrane, or eardrum, which marks the border between the external and middle ear.

The middle ear is connected to the back of the throat via the auditory tube. Within the middle ear are three tiny bones called the ossicles. These bones are linked together in such a way that movements of the eardrum are transmitted via the footplate of the stapes to the oval window (the opening in between the middle and inner ear).

The inner ear contains the main organ of hearing, the cochlea, and the vestibular system that controls balance. Information from both these parts of the ear passes to specific areas within the brainstem via the vestibulocochlear nerve.

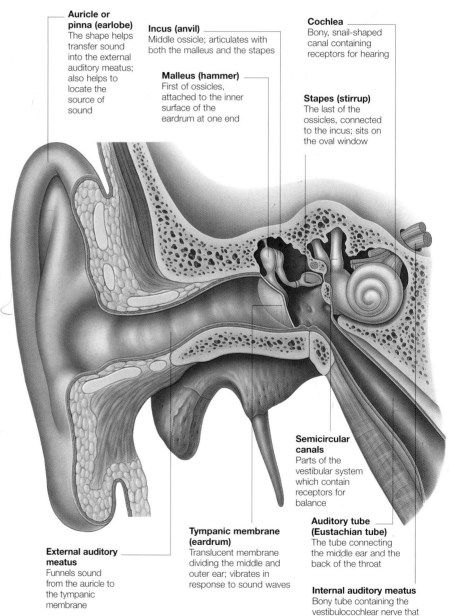

Auricle or pinna (earlobe)
The shape helps transfer sound into the external auditory meatus; also helps to locate the source of sound

Incus (anvil)
Middle ossicle; articulates with both the malleus and the stapes

Malleus (hammer)
First of ossicles, attached to the inner surface of the eardrum at one end

Cochlea
Bony, snail-shaped canal containing receptors for hearing

Stapes (stirrup)
The last of the ossicles, connected to the incus; sits on the oval window

Semicircular canals
Parts of the vestibular system which contain receptors for balance

External auditory meatus
Funnels sound from the auricle to the tympanic membrane

Tympanic membrane (eardrum)
Translucent membrane dividing the middle and outer ear; vibrates in response to sound waves

Auditory tube (Eustachian tube)
The tube connecting the middle ear and the back of the throat

Internal auditory meatus
Bony tube containing the vestibulocochlear nerve that conducts impulses to the brainstem

External ear

The pinna is the skin and cartilage that make up the external ear. It serves to channel sound into the middle ear.

The pinna, or auricle, collects sound from the environment and channels it into the external auditory meatus. It consists of a thin sheet of elastic cartilage and a lower portion called the lobule, consisting mainly of fatty tissue, with a tight covering of skin.

The auricle is attached to the head by a series of ligaments and muscles, and the external ear has a complex sensory nerve supply involving three of the cranial nerves.

PROTECTING THE EAR

The external auditory meatus is a tube extending from the lobule to the tympanic membrane and is about 2.5 cm long in adults. The outer third of the tube is made of cartilage (similar to that in the auricle), but the inner two thirds is bony (part of the temporal bone).

In the skin covering the cartilaginous part of the meatus, there are coarse hairs and ceruminous glands that secrete cerumen (earwax). Usually wax dries up and falls out of the ear, but it can build up and interfere with hearing. The combination of wax and hairs prevents dust and foreign objects from entering the ear.

The boundary between the outer and middle ear is the tympanic membrane, or eardrum. This is a translucent membrane which can be viewed using an auriscope. The tympanic membrane can sometimes be perforated due to middle ear infection or high-pressure sound waves.

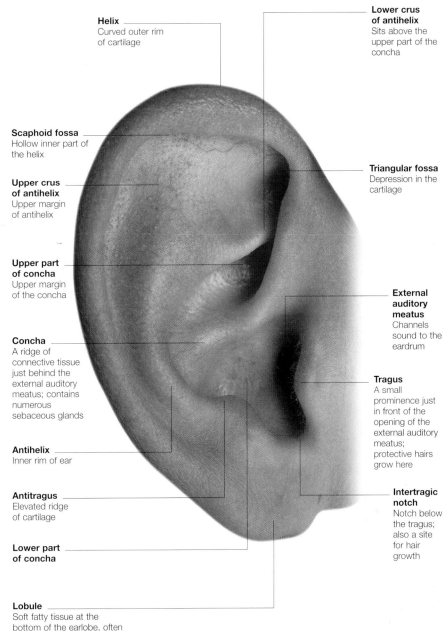

Helix
Curved outer rim of cartilage

Lower crus of antihelix
Sits above the upper part of the concha

Scaphoid fossa
Hollow inner part of the helix

Triangular fossa
Depression in the cartilage

Upper crus of antihelix
Upper margin of antihelix

Upper part of concha
Upper margin of the concha

External auditory meatus
Channels sound to the eardrum

Concha
A ridge of connective tissue just behind the external auditory meatus; contains numerous sebaceous glands

Tragus
A small prominence just in front of the opening of the external auditory meatus; protective hairs grow here

Antihelix
Inner rim of ear

Antitragus
Elevated ridge of cartilage

Intertragic notch
Notch below the tragus; also a site for hair growth

Lower part of concha

Lobule
Soft fatty tissue at the bottom of the earlobe, often the site for ear-piercing; contains no cartilage

Inside the ear

The middle ear is an air-filled cavity that contains the eardrum and three small bones that help transmit sound to the inner ear. It is also connected to the throat via the auditory tube.

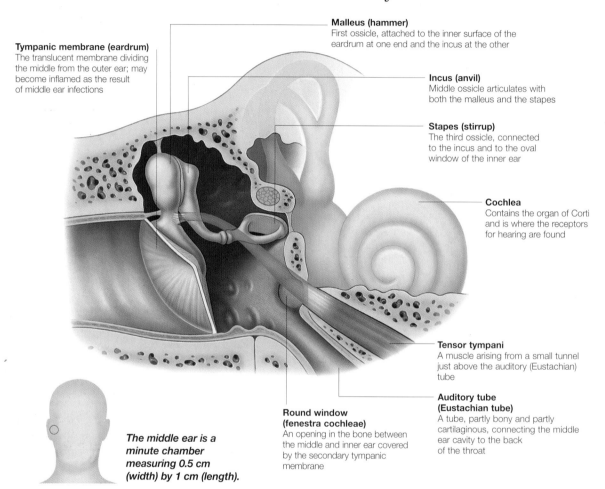

Malleus (hammer)
First ossicle, attached to the inner surface of the eardrum at one end and the incus at the other

Tympanic membrane (eardrum)
The translucent membrane dividing the middle from the outer ear; may become inflamed as the result of middle ear infections

Incus (anvil)
Middle ossicle articulates with both the malleus and the stapes

Stapes (stirrup)
The third ossicle, connected to the incus and to the oval window of the inner ear

Cochlea
Contains the organ of Corti and is where the receptors for hearing are found

Tensor tympani
A muscle arising from a small tunnel just above the auditory (Eustachian) tube

Auditory tube (Eustachian tube)
A tube, partly bony and partly cartilaginous, connecting the middle ear cavity to the back of the throat

Round window (fenestra cochleae)
An opening in the bone between the middle and inner ear covered by the secondary tympanic membrane

The middle ear is a minute chamber measuring 0.5 cm (width) by 1 cm (length).

The middle ear is an air-filled, box-shaped cavity within the temporal bone of the skull. It contains small bones or ossicles – the malleus, incus and stapes – that span the space between the tympanic membrane (the eardrum) and the medial wall of the cavity.

Two small muscles are also present: the tensor tympani, attached to the handle of the malleus; and the stapedius, attached to the stapes. Both help to modulate the movements of the ossicles. The medial wall divides the middle ear from the inner ear and contains two membrane-covered openings; the oval and round windows.

AUDITORY TUBE
The middle ear is connected to the throat by the auditory (Eustachian) tube. This tube is a possible route of infection into the middle ear. If left untreated, infections can spread into the mastoid air cells that lie just behind the middle ear cavity, and may breach the roof of the temporal bone and infect the membranous covering of the brain (the meninges).

Just below the floor of the middle ear cavity is the bulb of the internal jugular vein, and just in front is the internal carotid artery.

The inner ear

This part of the ear contains the organs of balance and hearing. It contains the labyrinth that helps us to orientate ourselves, and the cochlea, the organ of hearing.

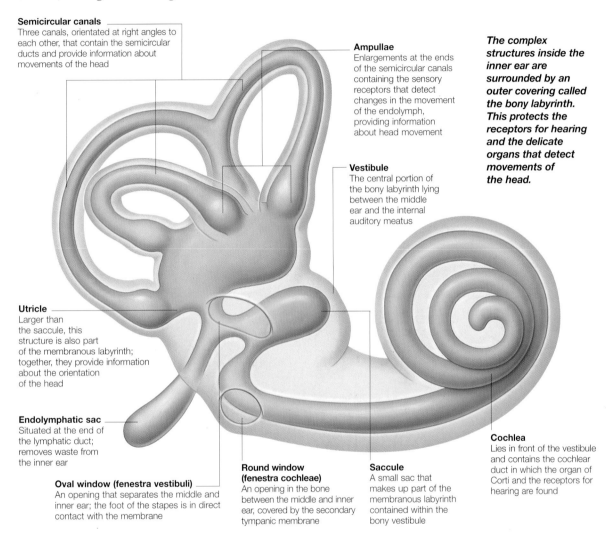

Semicircular canals
Three canals, orientated at right angles to each other, that contain the semicircular ducts and provide information about movements of the head

Ampullae
Enlargements at the ends of the semicircular canals containing the sensory receptors that detect changes in the movement of the endolymph, providing information about head movement

Vestibule
The central portion of the bony labyrinth lying between the middle ear and the internal auditory meatus

The complex structures inside the inner ear are surrounded by an outer covering called the bony labyrinth. This protects the receptors for hearing and the delicate organs that detect movements of the head.

Utricle
Larger than the saccule, this structure is also part of the membranous labyrinth; together, they provide information about the orientation of the head

Endolymphatic sac
Situated at the end of the lymphatic duct; removes waste from the inner ear

Oval window (fenestra vestibuli)
An opening that separates the middle and inner ear; the foot of the stapes is in direct contact with the membrane

Round window (fenestra cochleae)
An opening in the bone between the middle and inner ear, covered by the secondary tympanic membrane

Saccule
A small sac that makes up part of the membranous labyrinth contained within the bony vestibule

Cochlea
Lies in front of the vestibule and contains the cochlear duct in which the organ of Corti and the receptors for hearing are found

The inner ear, also known as the labyrinth because of its contorted shape, contains the organ of balance (the vestibule) and of hearing (the cochlea). It can be divided into an outer bony labyrinth and an inner membranous labyrinth. The bony labyrinth is filled with perilymph and the membranous labyrinth contains a fluid called endolymph, with a different chemical composition.

ORIENTATION
The membranous labyrinth contains the utricle and saccule – two linked, sac-like structures within the bony vestibule. They help detect orientation within the environment.

Related to these are the semicircular ducts lying within the bony semicircular canals. Where they are connected to the utricle, the semicircular canals enlarge to form ampullae, containing sensory receptors. Changes in the movement of the fluid in the ducts provides information about

acceleration and deceleration of the head.

The cochlea is a bony spiral canal, wound around a central pillar – the modiolus. Within the cochlea are hair cells, the hearing receptors, that react to vibrations in the endolymph caused by the movement of the stapes on the oval window. They lie within the organ of Corti.

Inside the neck

The neck is one of the most anatomically complex areas of the body. Many vital structures, including the spinal cord and thyroid gland, are closely packed together within layers of connecting tissue and muscle.

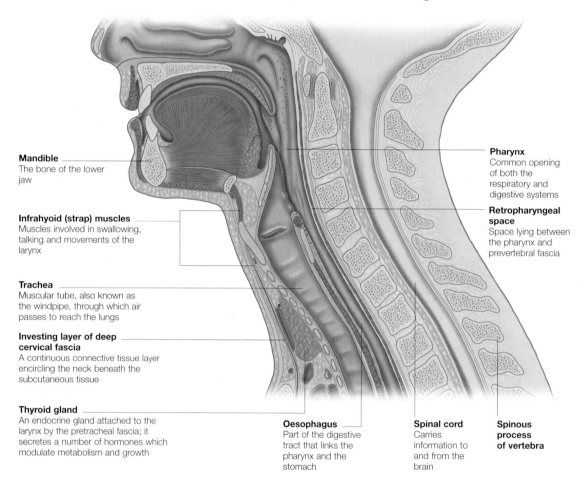

Mandible
The bone of the lower jaw

Infrahyoid (strap) muscles
Muscles involved in swallowing, talking and movements of the larynx

Trachea
Muscular tube, also known as the windpipe, through which air passes to reach the lungs

Investing layer of deep cervical fascia
A continuous connective tissue layer encircling the neck beneath the subcutaneous tissue

Thyroid gland
An endocrine gland attached to the larynx by the pretracheal fascia; it secretes a number of hormones which modulate metabolism and growth

Pharynx
Common opening of both the respiratory and digestive systems

Retropharyngeal space
Space lying between the pharynx and prevertebral fascia

Oesophagus
Part of the digestive tract that links the pharynx and the stomach

Spinal cord
Carries information to and from brain

Spinous process of vertebra

The neck is defined as the region lying between the bottom of the lower jaw and the top of the clavicle (collar bone). Within this relatively small area there are numerous vital structures that are closely packed together between layers of connective tissue.

The outermost layer of the neck is the skin. The skin of the neck contains sensory nerve endings from the second, third and fourth cervical nerves. A number of natural stress lines can be seen on the skin that run around the neck horizontally. When the skin is cut during surgical procedures, the incisions are made along, rather than across, these stress lines in order to minimize scarring.

EXTERNAL JUGULAR VEIN
Just beneath the skin is the thin layer of subcutaneous fat and connective tissue called the superficial fascia. Embedded in this layer are blood vessels, such as the external jugular vein and its tributaries. These veins drain blood from the face, scalp and neck. Closely associated with the external jugular vein are the superficial lymph nodes.

One other important structure that can be found in this layer, at the front of the neck, is the very thin platysma muscle that helps to depress the lower jaw.

Cross-section of the neck

Deeper layers of the neck reveal interconnected sheets of
tissue. These bind to and protect a variety of structures.

*A cross-section of the neck
exposes the muscles essential
for movement and facial
expression. Among these are
connecting tissues and vessels.*

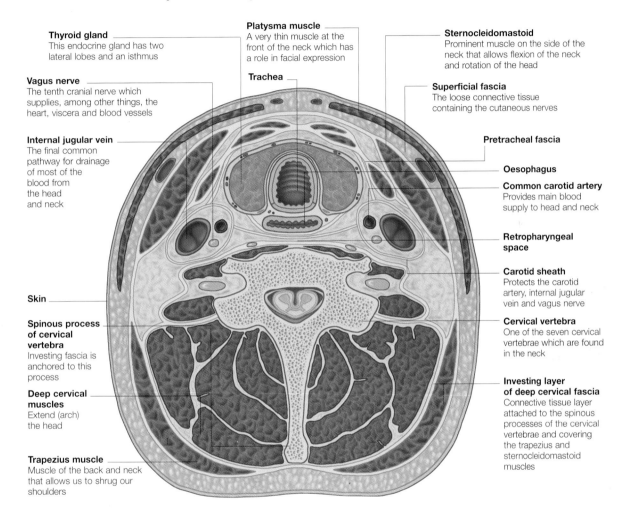

Thyroid gland
This endocrine gland has two
lateral lobes and an isthmus

Vagus nerve
The tenth cranial nerve which
supplies, among other things, the
heart, viscera and blood vessels

Internal jugular vein
The final common
pathway for drainage
of most of the
blood from
the head
and neck

Skin

**Spinous process
of cervical
vertebra**
Investing fascia is
anchored to this
process

**Deep cervical
muscles**
Extend (arch)
the head

Trapezius muscle
Muscle of the back and neck
that allows us to shrug our
shoulders

Platysma muscle
A very thin muscle at the
front of the neck which has
a role in facial expression

Trachea

Sternocleidomastoid
Prominent muscle on the side of the
neck that allows flexion of the neck
and rotation of the head

Superficial fascia
The loose connective tissue
containing the cutaneous nerves

Pretracheal fascia

Oesophagus

Common carotid artery
Provides main blood
supply to head and neck

**Retropharyngeal
space**

Carotid sheath
Protects the carotid
artery, internal jugular
vein and vagus nerve

Cervical vertebra
One of the seven cervical
vertebrae which are found
in the neck

**Investing layer
of deep cervical fascia**
Connective tissue layer
attached to the spinous
processes of the cervical
vertebrae and covering
the trapezius and
sternocleidomastoid
muscles

Moving deeper into the neck, the connective tissue of the deep cervical fascia is arranged into a number of fibrous sheets. These fasciae surround different groups of muscles, blood vessels and nerves, allowing them to move relative to each other with minimal friction.

The first of these is the investing fascia. This encircles the neck and is anchored to the spinous processes of the cervical vertebrae. It encloses the large sternocleidomastoid muscle at the front and side of the neck, and the trapezius muscle at the back, both of which are important in the movements of the head and neck.

LARYNX AND TRACHEA
The thin pretracheal fascia binds the thyroid gland to the larynx and trachea at the front of the neck. It is anchored to the cricoid cartilage, allowing movement during swallowing.

The pretracheal fascia is continuous with a sheet of tissue (the carotid sheath) which provides protection for the carotid artery, the internal jugular vein and the vagus nerve. Behind the trachea is the oesophagus (gullet), and behind the larynx is the pharynx, the muscular tube connecting the mouth and the oesophagus.

The last deep connective tissue layer is the prevertebral fascia, enclosing the remaining muscles of the neck, the vertebral column and the spinal cord, positioned in the centre of the neck for maximum protection.

Vertebral column

The vertebral column gives our bodies flexibility and keeps us upright. It also protects the delicate spinal cord.

The vertebral column forms the part of the skeleton commonly known as the backbone or spine. The spine supports the skull and gives attachment to the pelvic girdle, supporting the lower limbs. As well as its obvious role in posture and locomotion, the vertebral column surrounds and protects the spinal cord. Like all bones, its marrow is a source of blood cells, and it acts as a reservoir for calcium ions.

The spine exhibits four curvatures when viewed from the side. The cervical and lumbar curvatures are convex anteriorly (forward). The thoracic and sacral curvatures are convex posteriorly (backwards). The cervical curvature develops in infancy as the baby learns to hold its head upright; similarly, the lumbar curvature forms as the baby learns to walk.

In the fetus, the backbone has a single curvature in the thoracic and sacral regions. Other curvatures develop as the baby sits, stands and walks.

Frontal aspect

Lateral (side) view

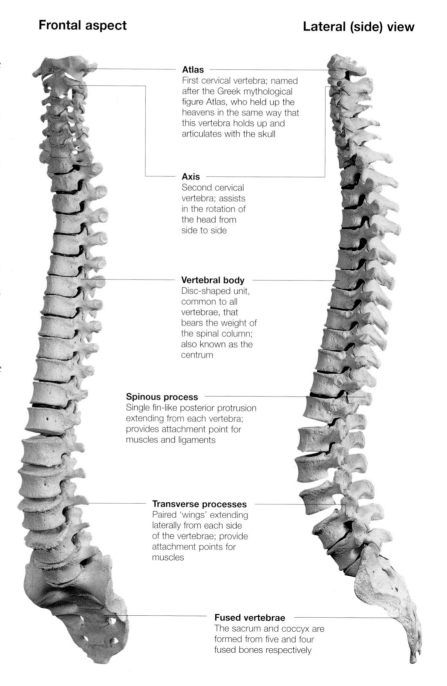

Atlas
First cervical vertebra; named after the Greek mythological figure Atlas, who held up the heavens in the same way that this vertebra holds up and articulates with the skull

Axis
Second cervical vertebra; assists in the rotation of the head from side to side

Vertebral body
Disc-shaped unit, common to all vertebrae, that bears the weight of the spinal column; also known as the centrum

Spinous process
Single fin-like posterior protrusion extending from each vertebra; provides attachment point for muscles and ligaments

Transverse processes
Paired 'wings' extending laterally from each side of the vertebrae; provide attachment points for muscles

Fused vertebrae
The sacrum and coccyx are formed from five and four fused bones respectively

Vertebral connections

The spinal column is divided into five main sections. Each section has a specific function and together they maintain the stability of the skeleton as a whole.

The vertebral column consists of 33 bones, known as the vertebrae. There are seven cervical vertebrae, twelve thoracic, five lumbar, five sacral and four coccygeal. Whereas the cervical, thoracic and lumbar vertebrae are separate bones, the sacral and coccygeal vertebrae are fused (inflexible). The vertebrae show differences along the column, and these will be featured on other sheets.

STRUCTURE

Each vertebra conforms to the same basic plan, consisting of a body in front and a neural arch at the back, which surrounds and protects the spinal cord. From the neural arch arise the transverse processes and spines, which give attachment to muscles and ligaments. Adjacent vertebrae articulate at joints, allowing movement. Movements between adjacent vertebrae are relatively small but, when taken over the whole length of the vertebral column, give the trunk considerable mobility.

The nerves leaving and entering the spinal cord do so through gaps, the intervertebral foramina, between adjacent vertebrae.

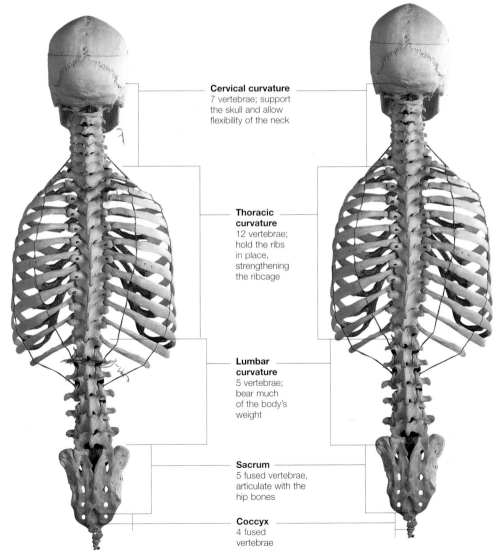

Cervical curvature
7 vertebrae; support the skull and allow flexibility of the neck

Thoracic curvature
12 vertebrae; hold the ribs in place, strengthening the ribcage

Lumbar curvature
5 vertebrae; bear much of the body's weight

Sacrum
5 fused vertebrae, articulate with the hip bones

Coccyx
4 fused vertebrae

Cervical vertebrae

There are seven cervical vertebrae, which together make up the skeletal structure of the neck. These vertebrae protect the spinal cord, support the skull and allow a range of movement.

Front view　　　　　　　　　　**Side view**

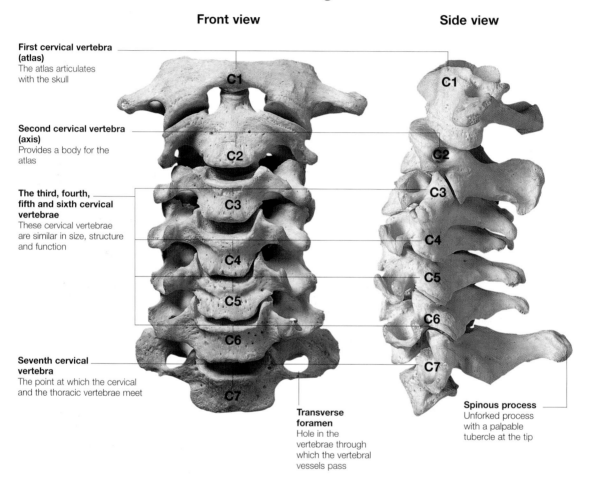

First cervical vertebra (atlas)
The atlas articulates with the skull

Second cervical vertebra (axis)
Provides a body for the atlas

The third, fourth, fifth and sixth cervical vertebrae
These cervical vertebrae are similar in size, structure and function

Seventh cervical vertebra
The point at which the cervical and the thoracic vertebrae meet

Transverse foramen
Hole in the vertebrae through which the vertebral vessels pass

Spinous process
Unforked process with a palpable tubercle at the tip

Of the seven cervical vertebrae, the lower five appear similar, although the seventh has some distinctive features. The first cervical vertebra (atlas) and the second cervical vertebra (axis) show specializations related to the articulation of the vertebral column with the skull.

TYPICAL CERVICAL VERTEBRA
The third to the sixth cervical vertebrae are comprised of two main components, a body towards the front and a vertebral arch at the rear. These surround the vertebral foramen (hole) which, as part of the vertebral column, forms the vertebral

canal. The body is small compared with vertebrae in other regions, and is nearly cylindrical.

The vertebral arch can be subdivided in to two main elements. The pedicles, by which it is attached to the body, contain notches that allow the passage of spinal nerves. The laminae are thin plates of bone which are

directed backwards and fuse in the midline, forming a bifid (divided) spine.

Associated with each vertebral arch are a pair of transverse processes. These are sites for muscle attachment, allowing movement and containing foramina through which blood vessels run.

Examining the cervical vertebrae

The first, second and seventh cervical vertebrae differ structurally
from the others, in relation to their unique functions.

First cervical vertebra (atlas)

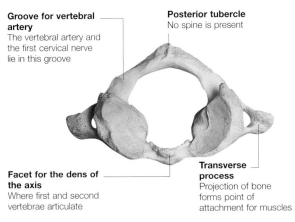

Groove for vertebral artery
The vertebral artery and the first cervical nerve lie in this groove

Posterior tubercle
No spine is present

Facet for the dens of the axis
Where first and second vertebrae articulate

Transverse process
Projection of bone forms point of attachment for muscles

Second cervical vertebra (axis)

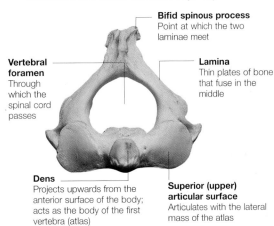

Bifid spinous process
Point at which the two laminae meet

Vertebral foramen
Through which the spinal cord passes

Lamina
Thin plates of bone that fuse in the middle

Dens
Projects upwards from the anterior surface of the body; acts as the body of the first vertebra (atlas)

Superior (upper) articular surface
Articulates with the lateral mass of the atlas

Fifth (typical) cervical vertebra

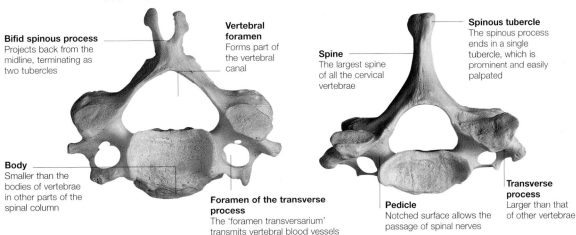

Bifid spinous process
Projects back from the midline, terminating as two tubercles

Vertebral foramen
Forms part of the vertebral canal

Body
Smaller than the bodies of vertebrae in other parts of the spinal column

Foramen of the transverse process
The 'foramen transversarium' transmits vertebral blood vessels

Seventh cervical vertebra

Spine
The largest spine of all the cervical vertebrae

Spinous tubercle
The spinous process ends in a single tubercle, which is prominent and easily palpated

Pedicle
Notched surface allows the passage of spinal nerves

Transverse process
Larger than that of other vertebrae

FIRST CERVICAL VERTEBRA

The first cervical vertebra, the atlas, is the vertebra that articulates with the skull. Unlike the other vertebrae, it does not have a body, this being incorporated into the second cervical vertebra as the dens. It also has no spine. Instead, the atlas takes the form of a thin ring of bone with anterior and posterior arches, the surface of which show grooves related to the vertebral arteries before they enter the skull through the foramen magnum.

SECOND CERVICAL VERTEBRA

The second cervical vertebra, the axis, can be distinguished from other cervical vertebrae by the presence of a tooth-like process called the dens (odontoid process). The dens articulates with the facet on the bottom surface of the anterior (front) arch of the atlas. Rotation of the head occurs at this joint.

The body of the axis resembles the bodies of the other cervical vertebrae.

SEVENTH CERVICAL VERTEBRA

This vertebra has the largest spine of any cervical vertebra and, as the first to be readily palpated, it has been termed the vertebra prominens.

The transverse processes are also larger than those of the other cervical vertebrae and the oval foramen transversarium transmits an accessory vertebral vein.

Muscles of the neck

The muscles running up the front of the neck are divided into the suprahyoid and infrahyoid muscle groups. They attach to the hyoid bone and act to raise and lower it and the larynx during swallowing.

Infrahyoid and suprahyoid muscles

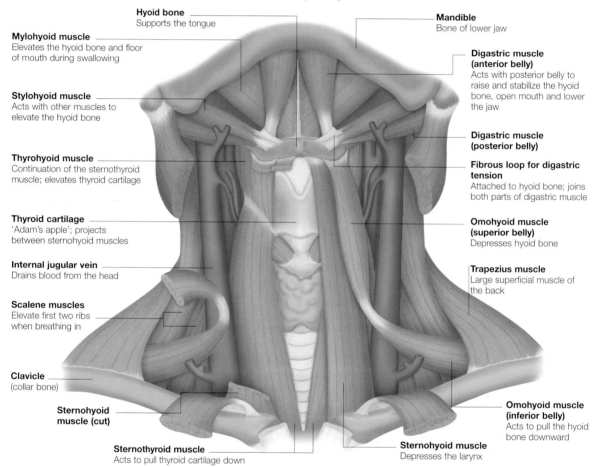

Hyoid bone
Supports the tongue

Mandible
Bone of lower jaw

Mylohyoid muscle
Elevates the hyoid bone and floor of mouth during swallowing

Digastric muscle (anterior belly)
Acts with posterior belly to raise and stabilize the hyoid bone, open mouth and lower the jaw

Stylohyoid muscle
Acts with other muscles to elevate the hyoid bone

Digastric muscle (posterior belly)

Thyrohyoid muscle
Continuation of the sternothyroid muscle; elevates thyroid cartilage

Fibrous loop for digastric tension
Attached to hyoid bone; joins both parts of digastric muscle

Thyroid cartilage
'Adam's apple'; projects between sternohyoid muscles

Omohyoid muscle (superior belly)
Depresses hyoid bone

Internal jugular vein
Drains blood from the head

Trapezius muscle
Large superficial muscle of the back

Scalene muscles
Elevate first two ribs when breathing in

Clavicle
(collar bone)

Omohyoid muscle (inferior belly)
Acts to pull the hyoid bone downward

Sternohyoid muscle (cut)

Sternothyroid muscle
Acts to pull thyroid cartilage down

Sternohyoid muscle
Depresses the larynx

Two groups of muscles run longitudinally within the front of the neck from the mandible (jaw) to the sternum (breastbone). These muscles control movements of the jaw, hyoid bone and larynx. The hyoid bone divides these two groups into the suprahyoid (above) and the infrahyoid (below).

THE SUPRAHYOID MUSCLES
This group of paired muscles, lie between the jaw and the hyoid bone. The digastric muscle has two spindle-shaped bellies connected by a tendon in the middle. The anterior belly is attached to the mandible near the mid-line, while the posterior belly arises from the base of the

skull. The connecting tendon slides freely through a fibrous 'sling' which is attached to the hyoid bone.

The stylohyoid is a small muscle passing from the styloid process, a bony projection of the base of the skull, forward and downward to the hyoid bone. Arising from the back of the

mandible, the mylohyoid muscles of each side unite in the mid-line to form the floor of the mouth. Posteriorly, they attach to the hyoid bone.

The geniohyoid is a narrow muscle that runs along the floor of the mouth from the back of the mandible in the mid-line to the hyoid bone below.

Action of the neck muscles

The suprahyoid and infrahyoid groups of muscles have opposing actions on the larynx and hyoid bone. This enables us to swallow.

Action of the infrahyoid and suprahyoid muscles

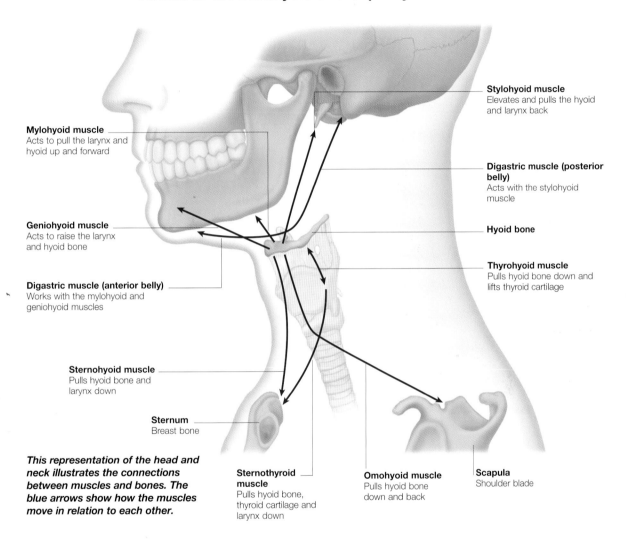

Stylohyoid muscle
Elevates and pulls the hyoid and larynx back

Mylohyoid muscle
Acts to pull the larynx and hyoid up and forward

Digastric muscle (posterior belly)
Acts with the stylohyoid muscle

Geniohyoid muscle
Acts to raise the larynx and hyoid bone

Hyoid bone

Digastric muscle (anterior belly)
Works with the mylohyoid and geniohyoid muscles

Thyrohyoid muscle
Pulls hyoid bone down and lifts thyroid cartilage

Sternohyoid muscle
Pulls hyoid bone and larynx down

Sternum
Breast bone

This representation of the head and neck illustrates the connections between muscles and bones. The blue arrows show how the muscles move in relation to each other.

Sternothyroid muscle
Pulls hyoid bone, thyroid cartilage and larynx down

Omohyoid muscle
Pulls hyoid bone down and back

Scapula
Shoulder blade

The mylohyoid, geniohyoid and the anterior belly of the digastric muscle act together to pull the hyoid and the larynx forward and up during swallowing. They also enable the mouth to be opened against resistance.

The stylohyoid and the posterior belly of the digastric muscle together lift and pull back the hyoid bone and the larynx. The suprahyoid muscles can be tested by asking the patient to open their mouth widely against resistance.

OPPOSING ACTION
The infrahyoid group of muscles act together to pull the hyoid and the larynx

back down to their normal positions, as at the end of the act of swallowing.

When contracted, the infrahyoid muscles lower and fix the hyoid bone so that the suprahyoid muscles can pull against it to open the mouth.

The infrahyoid muscles can be tested by asking a

patient to open their mouth against resistance while the doctor lightly holds the hyoid bone. The hyoid should move down as it is lowered and fixed by the muscles below. If there is weakness of the infrahyoid muscles the hyoid bone will rise up due to the unopposed action of the muscles above.

Brainstem

The brainstem lies at the junction of the brain and spinal cord. It helps to regulate breathing and blood circulation as well as having an effect upon a person's level of consciousness.

The brainstem is made up of three distinct parts: the midbrain, the pons, and the medulla oblongata. The midbrain connects with the higher brain above; the medulla is continuous with the spinal cord below.

BRAINSTEM APPEARANCE
The three parts of the brainstem can be viewed from underneath:
- The medulla oblongata – a bulge at the top of the spinal column. Pyramids, or columns, lie at either side of the midline. Nerve fibres within these columns carry messages from the cerebral cortex to the body. Raised areas known as the olives lie either side of the pyramids
- The pons – contains a system of nerve fibres which originate in the nerve cell bodies deep within the substance of the pons
- The midbrain – appears as two large columns, the cerebral crura, separated in the midline by a depression.

CRANIAL NERVES
Also present in the brainstem are some of the cranial nerves which supply much of the head. These nerves carry fibres which are associated with the cranial nerve nuclei, collections of grey matter, that lie inside the brainstem.

The brainstem connects the cerebral hemispheres with the spinal cord. There are three parts to the brain stem; the pons, medulla and the midbrain.

Ventral surface of the brainstem

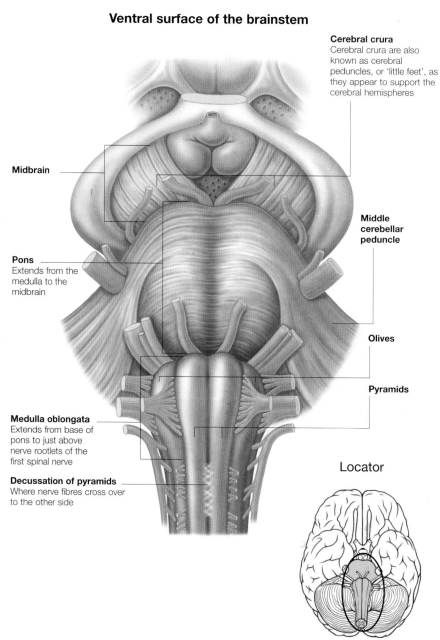

Cerebral crura
Cerebral crura are also known as cerebral peduncles, or 'little feet', as they appear to support the cerebral hemispheres

Midbrain

Middle cerebellar peduncle

Pons
Extends from the medulla to the midbrain

Olives

Pyramids

Medulla oblongata
Extends from base of pons to just above nerve rootlets of the first spinal nerve

Decussation of pyramids
Where nerve fibres cross over to the other side

Locator

Internal structure of the brainstem

The brainstem contains many areas of neural tissue which have a variety of functions vital to life and health. Responses to visual and auditory stimuli that influence head movement are also controlled here.

Cross sections through the brainstem reveal its internal structure, the arrangement of white and grey matter, which differs according to the level at which the section is taken.

MEDULLA
The features of a section through the medulla are:
- The inferior olivary nucleus – a bag-like collection of grey matter which lies just under the olives. Other nuclei lying within the medulla include some belonging to the cranial nerves, such as the hypoglossal and the vagus nerves
- The vestibular nuclear complex – an area that receives information from the ear and is concerned with balance and equilibrium
- The reticular formation – a complex network of neurones, which is seen here and throughout the brainstem. It has a number of functions vital to life such as the control of respiration and circulation. The reticular formation is present in the midbrain as are several of the cranial nerve nuclei.

MIDBRAIN
A section through the midbrain shows the presence of:
- The cerebral aqueduct – the channel which connects the third and fourth ventricles. Above the aqueduct lies an area called the tectum, while below it lie the large cerebral peduncles
- The cerebral peduncles – within these lie two

structures on each side; the red nucleus and the substantia nigra.
The red nucleus is involved in control of movement, while damage to the substantia nigra is associated with Parkinson's disease.

Cross sections of the brainstem

Midbrain (A)

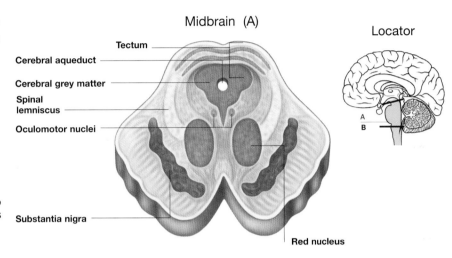

- Tectum
- Cerebral aqueduct
- Cerebral grey matter
- Spinal lemniscus
- Oculomotor nuclei
- Substantia nigra
- Red nucleus

Locator

Medulla (B)

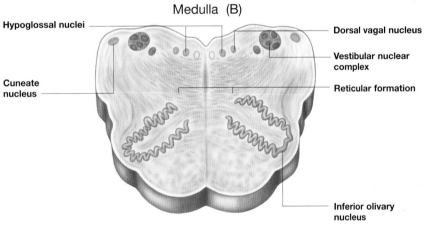

- Hypoglossal nuclei
- Cuneate nucleus
- Dorsal vagal nucleus
- Vestibular nuclear complex
- Reticular formation
- Inferior olivary nucleus

The numerous nuclei and tracts that are within the brainstem can be seen in these cross sections. They are involved in most functions of the brain.

PONS
The pons (not illustrated) is divided into upper and lower parts:
- The lower part – mostly made up of transverse nerve fibres, running across from the nuclei of the pons

to the cerebellum
- The upper portion – contains a number of cranial nerve nuclei. The pons also contains part of the reticular formation.

Brachial plexus

Lying within the root of the neck and extending into the axilla, the brachial plexus is a complicated network of nerves from which arise the major nerves supplying the upper limb.

Origins of the brachial plexus

The brachial plexus is a complex of nerves originating in the neck. They branch and rejoin as the nerves pass over the first rib to enter the axilla.

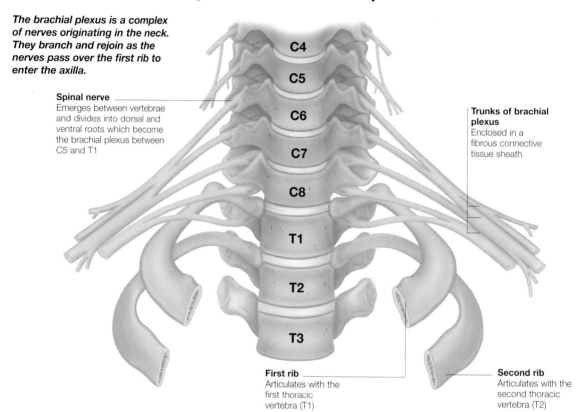

Spinal nerve
Emerges between vertebrae and divides into dorsal and ventral roots which become the brachial plexus between C5 and T1

Trunks of brachial plexus
Enclosed in a fibrous connective tissue sheath

C4
C5
C6
C7
C8
T1
T2
T3

First rib
Articulates with the first thoracic vertebra (T1)

Second rib
Articulates with the second thoracic vertebra (T2)

At the level of each vertebra of the spine there emerges a 'spinal nerve' which divides into dorsal and ventral parts, called 'rami'.

The brachial plexus is formed by the joining and intermixing of the ventral rami at the level of the fifth to the eighth cervical vertebrae and most of the ventral rami from the level of the first thoracic vertebra. These ventral rami are known as the 'roots' of the brachial plexus.

STRUCTURE
The roots of the brachial plexus join to form three 'trunks': superior, middle and inferior.

As the complexity of the brachial plexus increases, each of the three trunks then divides into an anterior (front) and a posterior (back) 'division'. In general, the nerve fibres within the anterior divisions are those which will go on to supply the anterior structures of the upper limb, while the fibres of the posterior divisions will supply posterior upper limb structures.

From the six divisions, three 'cords' are formed, which are named for their positions in relation to the axillary artery to which they lie adjacent: lateral, medial and posterior.

The final part of the brachial plexus consists of the branches given off by the three cords, although other branches also arise at higher levels.

Dermatomes

A dermatome is an area
of skin which receives its sensory nerve supply from a single spinal nerve (and
therefore a single segment of the spinal cord); however, that nerve supply may
actually be taken to the skin in two or more cutaneous branches.

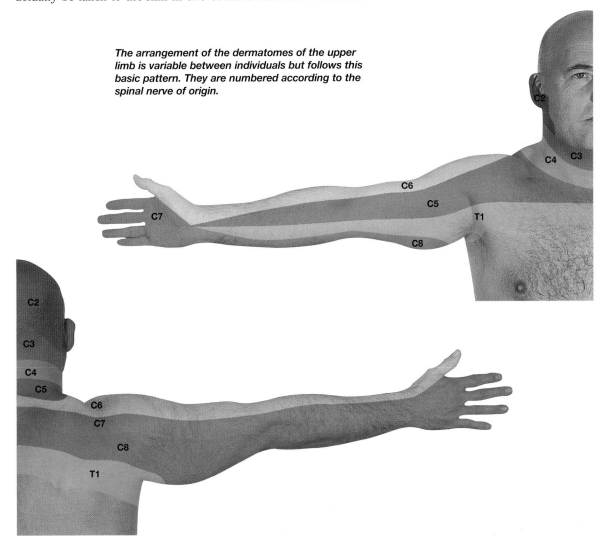

*The arrangement of the dermatomes of the upper
limb is variable between individuals but follows this
basic pattern. They are numbered according to the
spinal nerve of origin.*

The nerve supply to the skin, and so the dermatomes, of the upper limb comes via the brachial plexus from the ventral rami of the spinal nerves of C5 to T1.

The segmental pattern of the dermatomes can most readily be seen by visualizing the arm held lifted to the side with the thumb uppermost. This pattern has arisen because, during fetal development, the limbs begin as 'buds' from the side of a segmentally arranged embryo which becomes elongated, stretching the dermatome bands as they grow outwards.

ARRANGEMENT OF DERMATOMES

The exact arrangement of the dermatomes of the upper limb may vary but, in general:

T1 supplies a strip along the front of the arm but not extending to the hand.

This segmental pattern is continued down the length of the body, and is most striking over the thorax and abdomen.

77

Pharynx

The pharynx, situated at the back of the throat, is a passage both for food to the alimentary system and air to the lungs. It can be divided into three major parts, and the entrance is guarded by the tonsils.

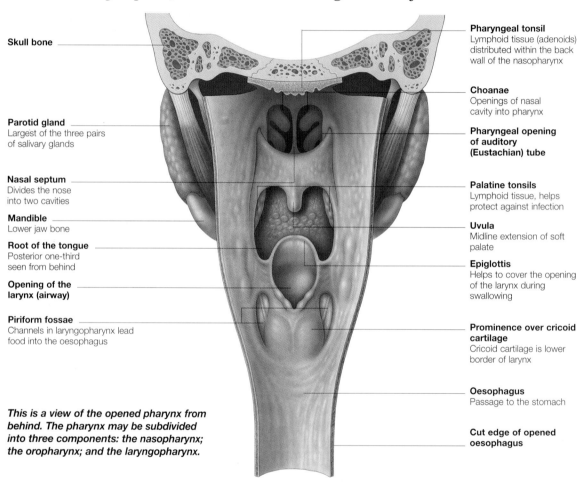

Skull bone

Parotid gland
Largest of the three pairs
of salivary glands

Nasal septum
Divides the nose
into two cavities

Mandible
Lower jaw bone

Root of the tongue
Posterior one-third
seen from behind

**Opening of the
larynx (airway)**

Piriform fossae
Channels in laryngopharynx lead
food into the oesophagus

*This is a view of the opened pharynx from
behind. The pharynx may be subdivided
into three components: the nasopharynx;
the oropharynx; and the laryngopharynx.*

Pharyngeal tonsil
Lymphoid tissue (adenoids)
distributed within the back
wall of the nasopharynx

Choanae
Openings of nasal
cavity into pharynx

**Pharyngeal opening
of auditory
(Eustachian) tube**

Palatine tonsils
Lymphoid tissue, helps
protect against infection

Uvula
Midline extension of soft
palate

Epiglottis
Helps to cover the opening
of the larynx during
swallowing

**Prominence over cricoid
cartilage**
Cricoid cartilage is lower
border of larynx

Oesophagus
Passage to the stomach

**Cut edge of opened
oesophagus**

The pharynx, a
fibromuscular, 15 cm long
tube at the back of the throat,
is a passage for food and air.
The constrictor muscles of
the pharynx allow food to be
squeezed into the
oesophagus.

NASOPHARYNX
The uppermost part of the
pharynx, lying above the

soft palate, is the
nasopharynx. The most
prominent feature on each
side is the tubal elevation,
the end of the auditory
(Eustachian) tube that
enables air pressure to be
equalized between the
nasopharynx and the middle
ear cavity. Lymphoid
(adenoid) tissue is found on
the back wall.

OROPHARYNX
The oropharynx lies at the
back of the throat. Its roof is
the undersurface of the soft
palate; the floor is the back
of the tongue. The palatine
tonsil lies in the side wall,
and is bounded by the
palatoglossal fold in front
and the palatopharyngeal
fold behind.

LARYNGOPHARYNX
The laryngopharynx extends
from the upper border of
the epiglottic cartilage
(which covers the opening
of the airway during
swallowing) to the lower
border of the cricoid
cartilage, where it continues
into the oesophagus. The
inlet of the airway lies in
the front section.

Muscles of the pharynx

There are six pairs of muscles which make up the pharynx. These muscles can be divided into two groups.

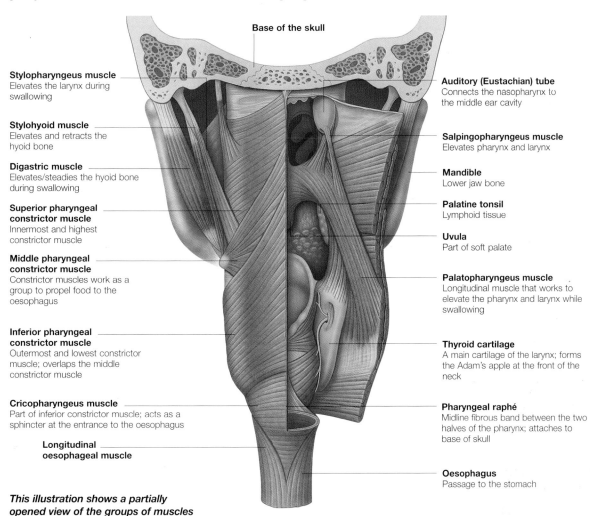

Base of the skull

Stylopharyngeus muscle
Elevates the larynx during swallowing

Stylohyoid muscle
Elevates and retracts the hyoid bone

Digastric muscle
Elevates/steadies the hyoid bone during swallowing

Superior pharyngeal constrictor muscle
Innermost and highest constrictor muscle

Middle pharyngeal constrictor muscle
Constrictor muscles work as a group to propel food to the oesophagus

Inferior pharyngeal constrictor muscle
Outermost and lowest constrictor muscle; overlaps the middle constrictor muscle

Cricopharyngeus muscle
Part of inferior constrictor muscle; acts as a sphincter at the entrance to the oesophagus

Longitudinal oesophageal muscle

Auditory (Eustachian) tube
Connects the nasopharynx to the middle ear cavity

Salpingopharyngeus muscle
Elevates pharynx and larynx

Mandible
Lower jaw bone

Palatine tonsil
Lymphoid tissue

Uvula
Part of soft palate

Palatopharyngeus muscle
Longitudinal muscle that works to elevate the pharynx and larynx while swallowing

Thyroid cartilage
A main cartilage of the larynx; forms the Adam's apple at the front of the neck

Pharyngeal raphé
Midline fibrous band between the two halves of the pharynx; attaches to base of skull

Oesophagus
Passage to the stomach

This illustration shows a partially opened view of the groups of muscles of the pharynx, viewed from behind.

One group of pharyngeal muscles comprises three pairs of constrictor muscles that run across the pharynx: the superior, middle and inferior constrictors. These constrict the pharynx, squeezing food downward into the oesophagus.

The other group comprises three pairs of muscles running from above down into the pharynx: the salpingo-pharyngeus, stylopharyngeus and palatopharyngeus. These muscles raise the pharynx during swallowing, elevating the larynx and protecting the airway.

The constrictor muscles overlap each other from below upwards (like three stacked plastic cups inside each other). Important structures enter the pharynx in the intervals between these muscles. The constrictor muscle fibres sweep backwards into a longitudinally running fibrous band in the midline, the pharyngeal raphé, which is attached to the base of the skull.

NERVE SUPPLY
Most of the pharynx derives its sensory nerve supply from the glossopharyngeal (ninth cranial) nerve. Stimulation of the oropharynx at the back of the throat triggers the swallowing and gagging reflexes. The pharynx muscles are supplied chiefly by the 11th cranial (accessory) nerve.

Larynx

The larynx is situated in the neck below and in front of the pharynx. It is the inlet protecting the lungs, and contains the vocal cords. In men, part of the larynx is visible as the Adam's apple.

Midline sagittal view   **Front view**

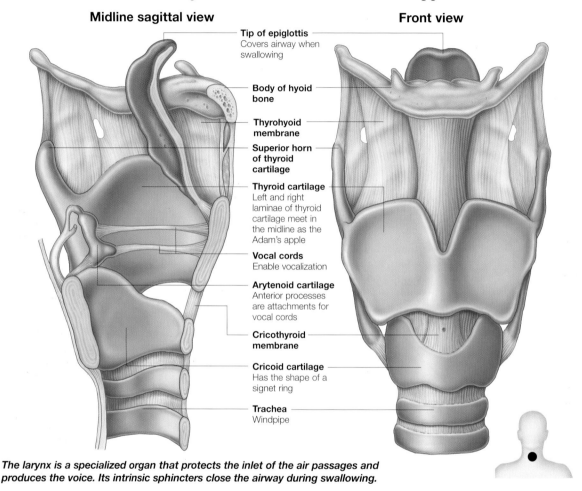

Tip of epiglottis
Covers airway when swallowing

Body of hyoid bone

Thyrohyoid membrane

Superior horn of thyroid cartilage

Thyroid cartilage
Left and right laminae of thyroid cartilage meet in the midline as the Adam's apple

Vocal cords
Enable vocalization

Arytenoid cartilage
Anterior processes are attachments for vocal cords

Cricothyroid membrane

Cricoid cartilage
Has the shape of a signet ring

Trachea
Windpipe

The larynx is a specialized organ that protects the inlet of the air passages and produces the voice. Its intrinsic sphincters close the airway during swallowing.

The larynx is composed of five cartilages (three single and one paired), connected by membranes, ligaments and muscles.

In adult men, the larynx lies opposite the third to sixth cervical vertebrae (slightly higher in women and children), between the base of the tongue and the trachea (windpipe).

The larynx serves as an inlet to the airways, taking air from the nose and mouth to the trachea. Because air and food share a common pathway, the primary function of the larynx is to prevent food and liquid from entering the airway. This is achieved by three 'sphincters' and by elevation. The larynx has

also evolved as an organ of phonation – the act of producing sounds – allowing vocalization.

LARYNGEAL CARTILAGES
The laryngeal prominence (thyroid cartilage protrusion, or Adam's apple) is readily visible in most men. Its greater protrusion in men compared to women is due

to the influence of the hormone testosterone. The thyroid cartilage has two rear extensions, a superior and an inferior 'horn'. The cricoid cartilage, the only complete ring of cartilage in the airway, is partly overlapped above by the thyroid cartilage. Above it sit a pair of mobile, pyramid-shaped arytenoid cartilages.

Muscles of the larynx

The muscles of the larynx act to close the laryngeal inlet while swallowing and move the vocal cords to enable vocalization.

During swallowing, the epiglottis, along with the rest of the larynx, is raised. As the front surface hits the rear part of the tongue, it flips backwards over the laryngeal inlet.

ARYEPIGLOTTIC FOLDS

The aryepiglottic folds of tissue are the free upper margins of the membranes that run between the epiglottis and the arytenoid cartilages. They contain a pair of transverse aryepiglottic and oblique aryepiglottic muscles. These arise from the muscular process of the opposite arytenoid cartilage, and attach to the sides of the epiglottis. They act like a 'purse-string', closing the laryngeal inlet. The lower ends of each of the quadrangular membranes form the vestibular folds, or 'false' vocal cords.

MUCOUS GLANDS

The quadrangular membranes are covered by a mucosa, and a submucosa rich in mucous glands. These are connected to the inner walls of the thyroid and cricoid cartilages. They keep the vocal cords moist, as the vocal folds have no submucosa themselves, and therefore rely upon these secretions from above. A groove, the piriform fossae, slopes backwards and serves to channel liquids toward the oesophagus and away from the larynx.

Rear view

Epiglottis
Leaf-shaped elastic cartilage connected to the hyoid bone and thyroid cartilage; closes over larynx during swallowing

Hyoid bone
Gives attachment to ligaments of the larynx, the epiglottis and muscles of the tongue and pharynx

Arytenoid cartilage
Gives attachment to vocal cords and intrinsic muscles of the larynx that control the laryngeal inlet and vocal cord movement

Thyrohyoid membrane
Connects hyoid bone to thyroid cartilage

Aryepiglottic muscle
Approximates the arytenoid cartilages

Oblique arytenoid muscles
Act with aryepiglottic muscles to close laryngeal inlet

Thyroid cartilage
Largest laryngeal cartilage providing structural framework and attachment for muscles and ligaments

Transverse arytenoid muscle
Closes posterior part of glottis

Posterior cricoarytenoid muscle
Opens glottis by abducting (opening) the vocal cords

Cricoid cartilage
Cartilage in the shape of a signet ring; the only complete cartilage ring in the airway

Tracheal cartilage
Deficient posteriorly, the tracheal cartilages allow passage of food boluses within the oesophagus, which lies directly behind the trachea

A rear view of the larynx reveals the muscles and cartilages related to vocalization and movement of the epiglottis.

Thyroid and parathyroid glands

The thyroid and parathyroid glands are situated in the neck. Together, they produce important hormones responsible for regulating growth, metabolism and calcium levels in the blood.

ANTERIOR (FRONT) VIEW

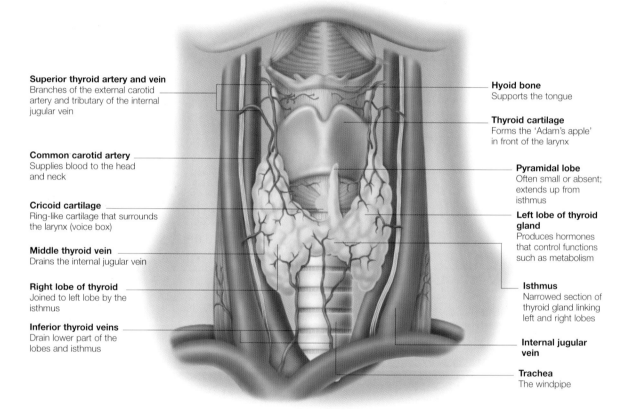

Superior thyroid artery and vein
Branches of the external carotid artery and tributary of the internal jugular vein

Common carotid artery
Supplies blood to the head and neck

Cricoid cartilage
Ring-like cartilage that surrounds the larynx (voice box)

Middle thyroid vein
Drains the internal jugular vein

Right lobe of thyroid
Joined to left lobe by the isthmus

Inferior thyroid veins
Drain lower part of the lobes and isthmus

Hyoid bone
Supports the tongue

Thyroid cartilage
Forms the 'Adam's apple' in front of the larynx

Pyramidal lobe
Often small or absent; extends up from isthmus

Left lobe of thyroid gland
Produces hormones that control functions such as metabolism

Isthmus
Narrowed section of thyroid gland linking left and right lobes

Internal jugular vein

Trachea
The windpipe

The thyroid gland is an endocrine gland situated in the neck, lying to the front and side of the larynx and trachea. It is similar to a bow-tie in shape, and it produces two iodine-dependent hormones: tri-iodothyronine and thyroxine. These are responsible for controlling metabolism through promotion of metabolic enzyme production.

In addition, the gland secretes calcitonin which is involved in the regulation of calcium levels in the blood. In children, growth is dependent upon this gland, through its stimulation of the metabolism of fats, carbohydrates and proteins.

PYRAMIDAL LOBE
The gland has two conical-shaped lobes connected by an 'isthmus' (a band of tissue), which usually lies in front of the second and third tracheal cartilage rings. The entire gland is surrounded by a thin connective tissue capsule and a layer of deep cervical fascia. There is often a small third lobe – the pyramidal lobe – which extends upwards from near the isthmus and lies over the cricothyroid membrane and the median cricothyroid ligament.

Posterior view of thyroid

The posterior view of the thyroid reveals the small parathyroid glands, embedded within the lobes. A rich network of vessels supply the glands.

POSTERIOR (BACK) VIEW

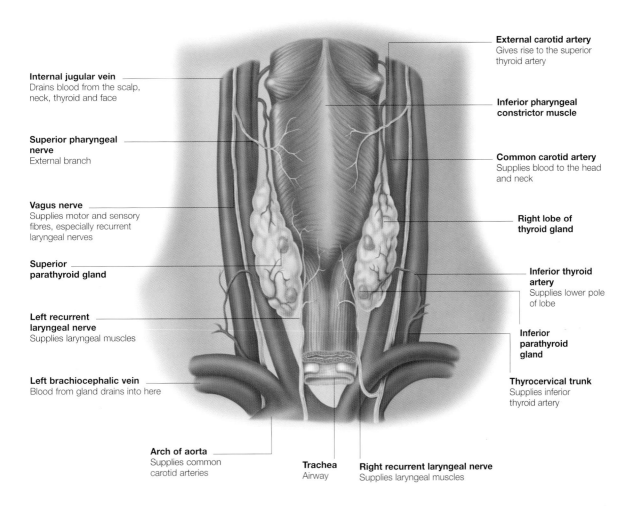

External carotid artery
Gives rise to the superior thyroid artery

Internal jugular vein
Drains blood from the scalp, neck, thyroid and face

Inferior pharyngeal constrictor muscle

Superior pharyngeal nerve
External branch

Common carotid artery
Supplies blood to the head and neck

Vagus nerve
Supplies motor and sensory fibres, especially recurrent laryngeal nerves

Right lobe of thyroid gland

Superior parathyroid gland

Inferior thyroid artery
Supplies lower pole of lobe

Left recurrent laryngeal nerve
Supplies laryngeal muscles

Inferior parathyroid gland

Left brachiocephalic vein
Blood from gland drains into here

Thyrocervical trunk
Supplies inferior thyroid artery

Arch of aorta
Supplies common carotid arteries

Trachea
Airway

Right recurrent laryngeal nerve
Supplies laryngeal muscles

The thyroid gland is well supplied by blood vessels. The upper pole receives arterial blood from the superior thyroid artery, a branch of the external carotid artery. The lower pole is supplied by the inferior thyroid artery, a branch of the thyrocervical trunk. The lobes of the gland are directly related to the common carotid arteries. The hormones are distributed to the bloodstream via a network (plexus) of veins in and around the gland, which ultimately drain into the internal jugular and brachiocephalic veins.

RELATION TO NERVES
In addition to the blood vessels, the gland is closely related to nerves. Posteriorly, the most important relation is the pair of recurrent laryngeal nerves from the vagus nerve. These ascend in the groove between the oesophagus and trachea, heading towards the larynx where they supply motor nerves to all laryngeal muscles (except the cricothyroid), and sensory nerves to the sub-glottic larynx. Hence, an enlarged thyroid may compress these nerves, causing a hoarseness in the voice.

83

Thoracic vertebrae

The 12 thoracic vertebrae are the bones of the spinal column to which the ribs are attached. The thoracic vertebrae sit between the cervical vertebrae of the neck and the lumbar vertebrae of the lower back.

Each thoracic vertebrae has two components, a cylindrical body in front and a vertebral arch behind. The body and vertebral arch enclose a hole, called the vertebral foramen, which is rounded. When all the vertebrae are articulated together, the space formed by the linked vertebral foramina forms the vertebral canal. This houses the spinal cord surrounded by three protective layers, called the meninges.

BONY PROCESSES
The part of the vertebral arch that attaches to the body on each side is called the pedicle and the arch is completed behind by two laminae that meet in the midline to form the spinous process. These processes project downwards, the eighth is the longest and most vertical. At the junction of the pedicles and laminae are the projecting transverse processes. These decrease in size from the top down.

MUSCLE ATTACHMENTS
Muscles and ligaments are attached to the spines and transverse processes. The thoracic vertebrae articulate with each other at the intervertebral joints. Between the vertebral bodies are intervertebral discs acting as shock absorbers.

Each vertebra has four surfaces (facets), which form moveable synovial joints with the adjacent vertebrae – one pair of facets articulates with the vertebra above, the other pair with the vertebra below. All of these joints are strengthened by ligaments.

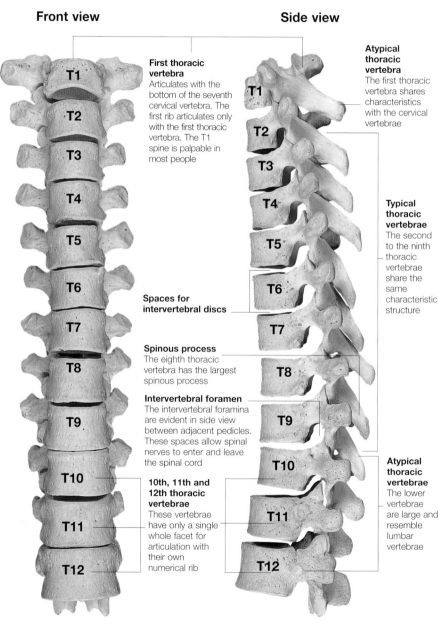

Front view

T1
T2
T3
T4
T5
T6
T7
T8
T9
T10
T11
T12

First thoracic vertebra
Articulates with the bottom of the seventh cervical vertebra. The first rib articulates only with the first thoracic vertebra. The T1 spine is palpable in most people

Spaces for intervertebral discs

Spinous process
The eighth thoracic vertebra has the largest spinous process

Intervertebral foramen
The intervertebral foramina are evident in side view between adjacent pedicles. These spaces allow spinal nerves to enter and leave the spinal cord

10th, 11th and 12th thoracic vertebrae
These vertebrae have only a single whole facet for articulation with their own numerical rib

Side view

T1
T2
T3
T4
T5
T6
T7
T8
T9
T10
T11
T12

Atypical thoracic vertebra
The first thoracic vertebra shares characteristics with the cervical vertebrae

Typical thoracic vertebrae
The second to the ninth thoracic vertebrae share the same characteristic structure

Atypical thoracic vertebrae
The lower vertebrae are large and resemble lumbar vertebrae

Examining the thoracic vertebrae

The thoracic vertebrae can be distinguished easily from the typical cervical vertebrae.

The thoracic vertebrae differ from the cervical vertebrae in several ways:
- An absence of the transverse process foramen (the foramen transversarium, through which nerves and blood vessels pass in the cervical vertebrae)
- A single, rather than bifid (two-part), spine
- The vertebral canal, through which the spinal cord runs, is smaller and more circular.
- The most distinguishing feature of the thoracic vertebrae, however, is the presence of facets enabling the ribs to articulate with the spine. Each typical thoracic vertebra has six facets for rib articulation – three on each side.

The head of the rib lies in the region of the intervertebral disc, at the back, and has two hemi- (half) facets that articulate with its own numbered vertebra (upper border) and the vertebra above (lower border).

ATYPICAL VERTEBRAE

The exceptions to the above rule are the first, 10th, 11th and 12th thoracic vertebrae. In the first thoracic vertebra, the facet on the upper border is a whole facet (rather than a half), as the first rib articulates only with its own vertebra.

Each of the 10th, 11th and 12th vertebrae has only one single whole facet to articulate with its own numerical rib. The 11th and 12th vertebrae have no articulation with the tubercle of the corresponding rib (so no articular facet). The last two ribs are called 'floating ribs' as they have no connections to the ribs above.

Fifth (typical) thoracic vertebra (front view)

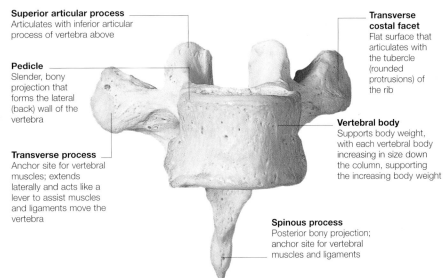

Superior articular process
Articulates with inferior articular process of vertebra above

Pedicle
Slender, bony projection that forms the lateral (back) wall of the vertebra

Transverse process
Anchor site for vertebral muscles; extends laterally and acts like a lever to assist muscles and ligaments move the vertebra

Transverse costal facet
Flat surface that articulates with the tubercle (rounded protrusions) of the rib

Vertebral body
Supports body weight, with each vertebral body increasing in size down the column, supporting the increasing body weight

Spinous process
Posterior bony projection; anchor site for vertebral muscles and ligaments

First (atypical) thoracic vertebra (side view)

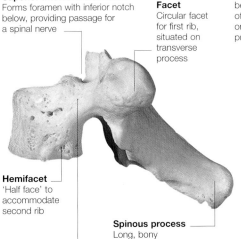

Superior intervertebral notch
Forms foramen with inferior notch below, providing passage for a spinal nerve

Facet
Circular facet for first rib, situated on transverse process

Hemifacet
'Half face' to accommodate second rib

Facet
Circular 'face' to accommodate head of first rib

Spinous process
Long, bony protrusion; points sharply inferiorly (downwards)

12th (atypical) thoracic vertebra (side view)

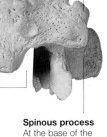

Body of vertebra
Structure of the lower thoracic vertebrae begins to resemble that of the lumbar vertebrae; only one round facet is present each side

Transverse process
11th and 12th thoracic vertebrae lack facet on transverse process

Inferior intervertebral notch
Forms the intervertebral foramen, though which a spinal nerve passes

Spinous process
At the base of the thoracic vertebrae, the spinous processes are small and rounded, resembling those of the lumbar vertebrae

Lumbar vertebrae

The five lumbar vertebrae of the lower back are
the strongest vertebrae of the spinal column.

Side view

Front view

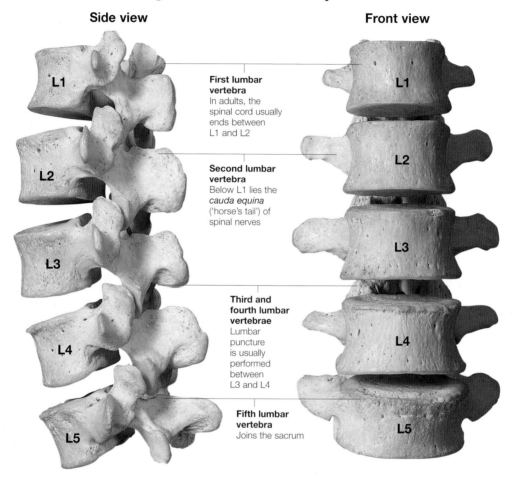

First lumbar vertebra
In adults, the spinal cord usually ends between L1 and L2

Second lumbar vertebra
Below L1 lies the *cauda equina* ('horse's tail') of spinal nerves

Third and fourth lumbar vertebrae
Lumbar puncture is usually performed between L3 and L4

Fifth lumbar vertebra
Joins the sacrum

The front of the lumbar vertebrae form a convex curve when viewed from the side, known as lumbar lordosis. This increases strength and helps to absorb shock.

The five lumbar vertebrae are subject to greater vertical compression forces than the rest of the spine. For this reason, these vertebrae are large and strong.

The individual lumbar vertebrae are the largest and strongest in the vertebral column. This is important, as the lower the position of the bones of the spinal column, the more body weight they must bear. The arrangement of the lumbar vertebral joints is designed to allow maximum flexion (allowing us to touch our toes), and some lateral flexion (allowing us to reach sideways), but little rotation (this occurs at the thoracic level).

BASIC STRUCTURE
As with the cervical and thoracic vertebrae, each lumbar vertebra has the same basic plan, consisting of a cylindrical body in the front and a vertebral arch behind which enclose a space, known as the vertebral foramen.

Each vertebral arch comprises a number of processes. There are two laterally projecting transverse processes, a centrally positioned spinous process and two pairs of articular facets, one pair above and one pair below. The transverse processes and spines are shorter and thicker than those of other vertebrae and are well adapted for the attachment of the large back muscles and strong ligaments.

Lumbar ligaments

The intervertebral discs and connecting ligaments support the bones of the spine. They act as shock absorbers, reducing wear on the vertebrae.

The intervertebral discs link the bones of adjacent vertebrae, prevent dislocation of the vertebral column and also act as shock absorbers between the vertebrae. Intervertebral discs contribute about one-fifth of the length of the vertebral column. They are thickest in the lumbar region where the forces of vertical compression are greatest.

STRENGTH AND STABILITY

To reinforce stability, the vertebral bodies are strengthened by tough, longitudinally running ligaments, consisting of fibrous tissue, in the front and rear. These ligaments are firmly attached to the intervertebral disc and adjacent edges of the vertebral body, but loosely attached to the rest of the body.

Movement between vertebrae is the result of the action of muscles attached to the processes of the vertebral arches. The joints associated with the articular processes are synovial joints, allowing the adjacent surfaces to glide smoothly over each other.

Each synovial joint is surrounded by a loose joint capsule. The joints of the vertebral arches are strengthened by various ligaments. The ligamenta flava join the laminae of the adjacent vertebra and contain elastic tissue.

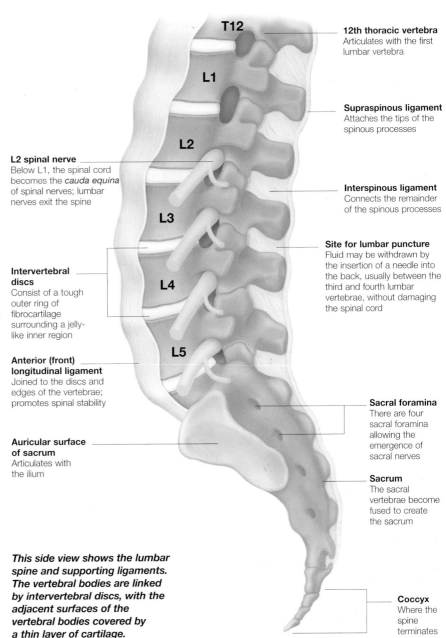

T12

L1

L2

L3

L4

L5

12th thoracic vertebra
Articulates with the first lumbar vertebra

Supraspinous ligament
Attaches the tips of the spinous processes

Interspinous ligament
Connects the remainder of the spinous processes

Site for lumbar puncture
Fluid may be withdrawn by the insertion of a needle into the back, usually between the third and fourth lumbar vertebrae, without damaging the spinal cord

Sacral foramina
There are four sacral foramina allowing the emergence of sacral nerves

Sacrum
The sacral vertebrae become fused to create the sacrum

Coccyx
Where the spine terminates

L2 spinal nerve
Below L1, the spinal cord becomes the *cauda equina* of spinal nerves; lumbar nerves exit the spine

Intervertebral discs
Consist of a tough outer ring of fibrocartilage surrounding a jelly-like inner region

Anterior (front) longitudinal ligament
Joined to the discs and edges of the vertebrae; promotes spinal stability

Auricular surface of sacrum
Articulates with the ilium

This side view shows the lumbar spine and supporting ligaments. The vertebral bodies are linked by intervertebral discs, with the adjacent surfaces of the vertebral bodies covered by a thin layer of cartilage.

Sacrum and coccyx

The sacrum and coccyx form the tail end of the spinal column.
Both are formed from fused vertebrae, allowing attachment for weight-bearing ligaments and muscles, and helping to protect pelvic organs.

Pelvic (inner) surface of sacrum (disarticulated)

Side view of sacrum

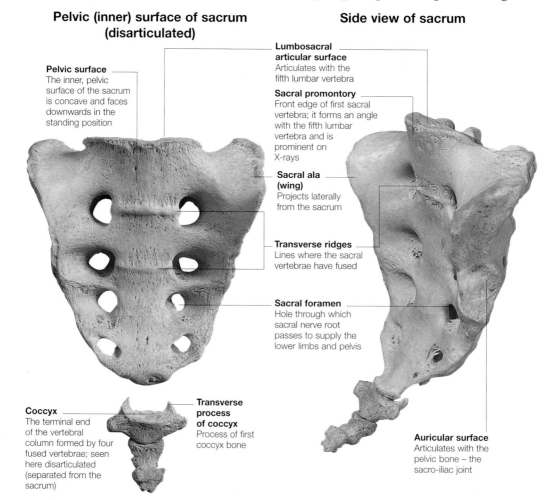

Pelvic surface
The inner, pelvic surface of the sacrum is concave and faces downwards in the standing position

Lumbosacral articular surface
Articulates with the fifth lumbar vertebra

Sacral promontory
Front edge of first sacral vertebra; it forms an angle with the fifth lumbar vertebra and is prominent on X-rays

Sacral ala (wing)
Projects laterally from the sacrum

Transverse ridges
Lines where the sacral vertebrae have fused

Sacral foramen
Hole through which sacral nerve root passes to supply the lower limbs and pelvis

Coccyx
The terminal end of the vertebral column formed by four fused vertebrae; seen here disarticulated (separated from the sacrum)

Transverse process of coccyx
Process of first coccyx bone

Auricular surface
Articulates with the pelvic bone – the sacro-iliac joint

The sacrum is a bony mass composed of five sacral vertebrae which fuse between puberty and the age of 30 years. It performs several functions: it attaches the vertebral column to the pelvic girdle, supporting the body's weight and transmitting it to the legs; it protects pelvic organs, such as the uterus and bladder; and it allows attachment of muscles that move the thigh.

The sacrum is shaped like an upside-down triangle, the five fused vertebral bodies diminishing in size from the wide base above (formed by the first sacral vertebra and the sacral alae, or 'wings') towards the apex below, where the coccyx is attached.

Centrally, horizontal bony ridges indicate the junctions between individual vertebrae; these are the remnants of intervertebral discs. On either side, sacral foramina (holes running through the bone) allow the passage of the ventral sacral motor nerve roots.

THE COCCYX
The coccyx, attached to the base of the sacrum, is the remains of the tail seen in our primate relatives. It consists of a small, pyramid-shaped bone formed from four fused vertebrae, and allows the attachment of ligaments and muscles, forming the anal sphincter.

Spinal nerve roots

The genitals, buttocks and lower limbs are supplied by nerve roots that emerge from the lumbar and sacral spine.

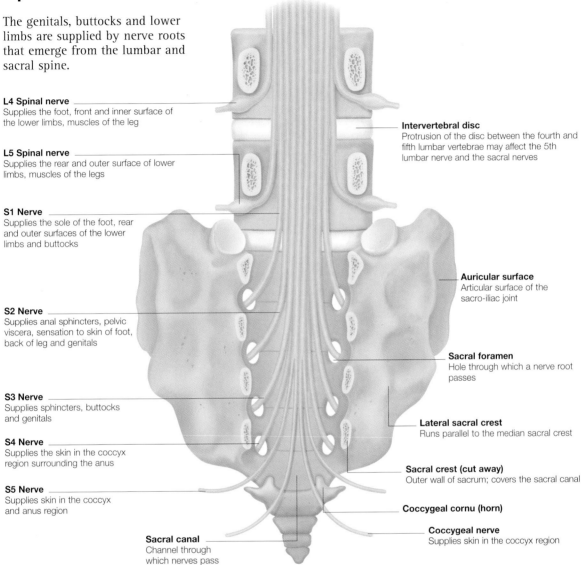

L4 Spinal nerve
Supplies the foot, front and inner surface of the lower limbs, muscles of the leg

L5 Spinal nerve
Supplies the rear and outer surface of lower limbs, muscles of the legs

S1 Nerve
Supplies the sole of the foot, rear and outer surfaces of the lower limbs and buttocks

S2 Nerve
Supplies anal sphincters, pelvic viscera, sensation to skin of foot, back of leg and genitals

S3 Nerve
Supplies sphincters, buttocks and genitals

S4 Nerve
Supplies the skin in the coccyx region surrounding the anus

S5 Nerve
Supplies skin in the coccyx and anus region

Sacral canal
Channel through which nerves pass

Intervertebral disc
Protrusion of the disc between the fourth and fifth lumbar vertebrae may affect the 5th lumbar nerve and the sacral nerves

Auricular surface
Articular surface of the sacro-iliac joint

Sacral foramen
Hole through which a nerve root passes

Lateral sacral crest
Runs parallel to the median sacral crest

Sacral crest (cut away)
Outer wall of sacrum; covers the sacral canal

Coccygeal cornu (horn)

Coccygeal nerve
Supplies skin in the coccyx region

SACRAL PLEXUS
The sensory and motor nerve supply to and from the pelvis and legs is derived from a network of nerve roots called the sacral plexus. This lies on the rear wall of the pelvic cavity in front of the piriformis muscle. Contributions to the sacral plexus come from the lumbosacral trunk, representing the 4th and 5th lumbar nerve roots and the sacral nerve roots.

At the sacral plexus these nerve roots exchange nerve fibres and re-form into major nerves. These include the superior and inferior gluteal nerves, supplying the buttocks, and the sciatic nerve, which supplies the muscles of the leg. The parasympathetic splanchnic nerves (S1, S2, S3) regulate urination and defecation by controlling the internal sphincters, and also erection by dilating penile arterioles.

SACRAL FORAMINA
The convex outer sacral surface has a ridge called the median crest in the midline, where the spinous processes fuse. The four posterior sacral foramina transmit the dorsal nerve roots. Nerves pass down the sacrum through the sacral canal.

A normal defect in the fusion of the fifth sacral vertebra posteriorly causes the canal to open out at the sacral hiatus. This is useful to doctors, who can anaesthetize the lower spinal nerves by passing a needle through the open space.

Spinal cord

The spinal cord is the communication pathway between the brain and the body. It allows signals to pass down to control body function and up to inform the brain of what is happening in the body.

The spinal cord is a slightly flattened cylindrical structure of 42–45 cm length in adults, with an average diameter of about 2.5 cm. It begins as a continuation of the medulla oblongata, the lowest part of the brainstem, at the level of the foramen magnum, the largest opening in the base of the skull. It then runs down the length of the neck and back in the vertebral canal, protected by the bony vertebrae which make up the vertebral column.

DEVELOPMENT

Up to the third month of development in the womb, the spinal cord runs the entire length of the vertebral column. Later on, however, the vertebral column outgrows the cord, which by birth ends at the level of the third lumbar vertebra. This more rapid growth of the vertebral column continues so that in the adult, the spinal cord ends at about the level of the disc between the first and second lumbar vertebrae.

ANATOMY OF THE CORD

The cord is enlarged in the region of the neck and lower back. The lower end of the cord tapers off into a cone-shaped region – the conus medullaris. From this, the filum terminale – a thin strand of modified pia mater (one of the membranes that surround the brain and spinal cord) – continues downwards to be attached to the back of the coccyx, anchoring the spinal cord.

Posterior view of spinal cord

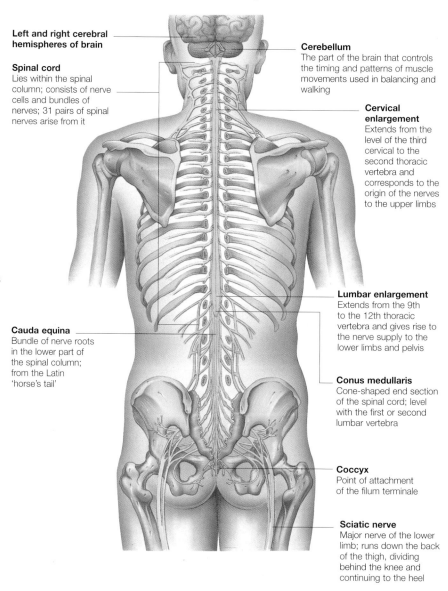

Left and right cerebral hemispheres of brain

Spinal cord
Lies within the spinal column; consists of nerve cells and bundles of nerves; 31 pairs of spinal nerves arise from it

Cerebellum
The part of the brain that controls the timing and patterns of muscle movements used in balancing and walking

Cervical enlargement
Extends from the level of the third cervical to the second thoracic vertebra and corresponds to the origin of the nerves to the upper limbs

Cauda equina
Bundle of nerve roots in the lower part of the spinal column; from the Latin 'horse's tail'

Lumbar enlargement
Extends from the 9th to the 12th thoracic vertebra and gives rise to the nerve supply to the lower limbs and pelvis

Conus medullaris
Cone-shaped end section of the spinal cord; level with the first or second lumbar vertebra

Coccyx
Point of attachment of the filum terminale

Sciatic nerve
Major nerve of the lower limb; runs down the back of the thigh, dividing behind the knee and continuing to the heel

Cross-sections through the spinal cord

The appearance of the spinal cord varies at different levels, according to the amount of muscle supplied by the nerves that emanate from it.

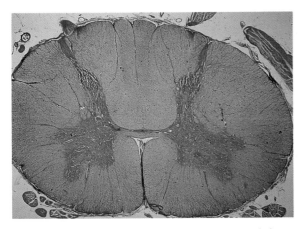

Cervical: the cord is relatively large and has an oval shape. Grey matter (dark red) is prominent, corresponding to the cervical enlargement supplying the upper limbs.

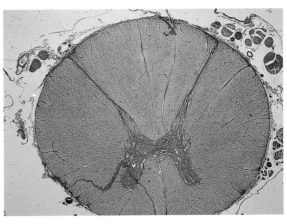

Thoracic: the cord is almost circular and has a smaller diameter. There is an intermediate amount of white matter. The grey matter is not as prominent here.

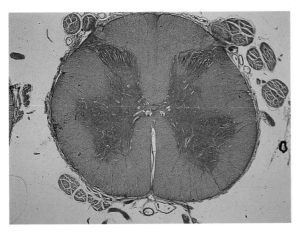

Lumbar: the cord has a larger diameter, corresponding to the increased amount of grey matter in the lumbar enlargement supplying the lower limbs. The white matter is less prominent.

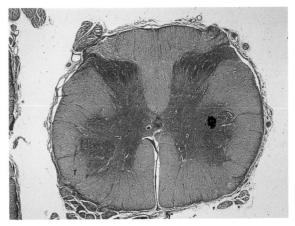

Sacral: in the region of the conus medullaris, the grey matter takes the form of two oval-shaped masses which occupy most of the cord with very little white matter.

The spinal cord is made up of an inner core of grey matter, which consists mainly of nerve cells and their supporting cells (neuroglia), surrounded by white matter, which is primarily myelinated nerve fibres – nerves with an insulating sheath of the fatty substance myelin.

In cross-section, the grey matter typically has the shape of a letter H or a butterfly, with two anterior columns or horns, two posterior columns and a thin grey commissure connecting the grey matter in the two halves. There is a small central canal containing cerebrospinal fluid which at its uppermost limit runs into the fourth ventricle in the region of the lower brainstem and cerebellum.

There is some variation in the appearance of a cross-section of the spinal cord at different levels. The amount of grey matter corresponds to the bulk of muscle, the nerve supply of which comes off at that level.

91

Spinal nerves

There are 31 pairs of spinal nerves, arranged on each side of the spinal cord along its length. The pairs are grouped by region: eight cervical, twelve thoracic, five lumbar, five sacral and one coccygeal.

Anterior view

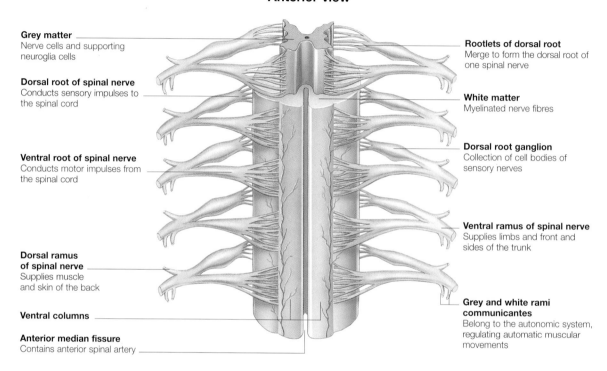

Grey matter
Nerve cells and supporting neuroglia cells

Dorsal root of spinal nerve
Conducts sensory impulses to the spinal cord

Ventral root of spinal nerve
Conducts motor impulses from the spinal cord

Dorsal ramus of spinal nerve
Supplies muscle and skin of the back

Ventral columns

Anterior median fissure
Contains anterior spinal artery

Rootlets of dorsal root
Merge to form the dorsal root of one spinal nerve

White matter
Myelinated nerve fibres

Dorsal root ganglion
Collection of cell bodies of sensory nerves

Ventral ramus of spinal nerve
Supplies limbs and front and sides of the trunk

Grey and white rami communicantes
Belong to the autonomic system, regulating automatic muscular movements

Each spinal nerve has two roots. The anterior, or ventral root, contains the axons of motor nerves which send impulses to control muscle movement. The posterior or dorsal root contains the axons of sensory nerves which send sensory information from the body into the spinal cord on its way to the brain.

SEGMENTS
Each root is formed by a series of small rootlets which attach it to the cord. The portion of the spinal cord which provides the rootlets for one dorsal root is referred to as a segment. In the lumbar and cervical regions, the rootlets are bunched closely, with the cord segments being only about 1 cm long. However, in the thoracic region they are more spread out, with segments more than 2 cm long.

NERVE FORMATION
The ventral and dorsal roots come together to form a single spinal nerve within the intervertebral foramina – small openings between the vertebrae through which the spinal nerves pass.

Just before the point of fusion with the ventral root, there is an enlargement of each dorsal root. This is known as the dorsal root ganglion – it is a collection of cell bodies of sensory nerves.

RAMI
Shortly after passing through its intervertebral foramen, each spinal nerve divides into several branches, or rami, including:
• Ventral ramus – supplying the limbs and front and sides of the trunk
• Dorsal ramus – supplying the deep muscles as well as the skin of the back
• Rami communicantes – part of the autonomic nervous system.

CAUDA EQUINA
Because the spinal cord is shorter than the vertebral column, the lower spinal nerve roots exit and travel downwards at quite an oblique angle. The lumbosacral nerve roots are bunched together and pass downwards almost vertically. This gives rise to the name cauda equina – Latin for horse's tail – which these lower nerve roots resemble.

Blood supply of the spinal cord

The spinal cord is supplied by a complex arrangement of arteries. This blood supply is vital for the normal functioning of the nervous system.

Membranes that protect the spinal cord

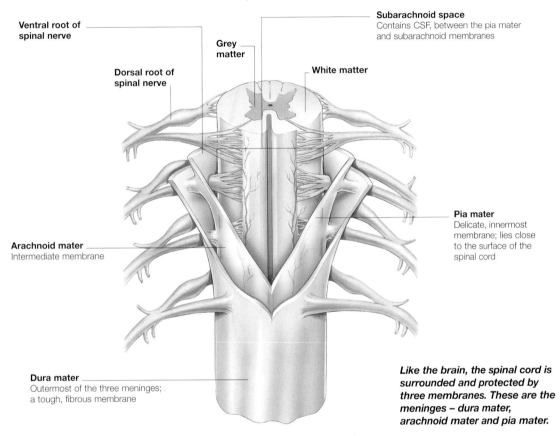

Ventral root of spinal nerve

Grey matter

Dorsal root of spinal nerve

White matter

Subarachnoid space
Contains CSF, between the pia mater and subarachnoid membranes

Pia mater
Delicate, innermost membrane; lies close to the surface of the spinal cord

Arachnoid mater
Intermediate membrane

Dura mater
Outermost of the three meninges; a tough, fibrous membrane

Like the brain, the spinal cord is surrounded and protected by three membranes. These are the meninges – dura mater, arachnoid mater and pia mater.

The bones of the vertebral column provide the major protection for the spinal cord, just as the skull does for the brain. However, like the brain, the cord has additional protection from three membranes, which continue down through the foramen magnum from inside the skull.

The dura mater is the tough, fibrous, outer membrane. The extradural or epidural space separates the dura from the bone of the vertebral bodies and contains fatty tissue and a plexus of veins.

The middle membrane is the arachnoid mater, which is much thinner and more delicate, with an arrangement of connective tissue fibres resembling a spider's web. There is a potential subdural space between the dura and the arachnoid, normally containing only a very thin film of fluid.

PIA MATER
The innermost membrane is the fine pia mater, which is closely applied to the surface of the spinal cord. It is transparent and richly supplied with fine blood vessels, which carry oxygen and nutrients to the cord. Between the arachnoid and the pia is the subarachnoid space, which contains cerebrospinal fluid (CSF), which cushions the spinal cord, as well as helping to remove chemical waste products produced by nerve activity and metabolism. CSF is formed by the choroid plexuses inside the cerebral ventricles and circulates around the brain and spinal cord.

About 21 triangular extensions of the pia – the denticulate ligaments – pass outwards between the anterior and posterior nerve roots to join with the arachnoid and inner surface of the dura. The spinal cord is suspended by these in its dural sheath.

93

Muscles of the back

The muscles of the back give us our upright posture and allow flexibility and mobility of the spine. The superficial back muscles also act with other muscles to move the shoulders and upper arms.

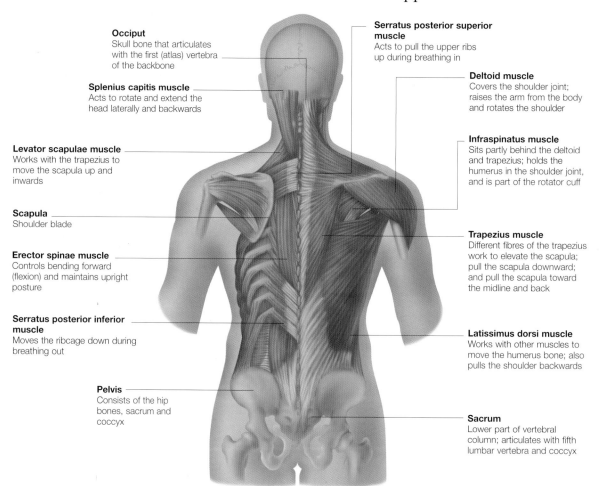

Occiput
Skull bone that articulates with the first (atlas) vertebra of the backbone

Splenius capitis muscle
Acts to rotate and extend the head laterally and backwards

Levator scapulae muscle
Works with the trapezius to move the scapula up and inwards

Scapula
Shoulder blade

Erector spinae muscle
Controls bending forward (flexion) and maintains upright posture

Serratus posterior inferior muscle
Moves the ribcage down during breathing out

Pelvis
Consists of the hip bones, sacrum and coccyx

Serratus posterior superior muscle
Acts to pull the upper ribs up during breathing in

Deltoid muscle
Covers the shoulder joint; raises the arm from the body and rotates the shoulder

Infraspinatus muscle
Sits partly behind the deltoid and trapezius; holds the humerus in the shoulder joint, and is part of the rotator cuff

Trapezius muscle
Different fibres of the trapezius work to elevate the scapula; pull the scapula downward; and pull the scapula toward the midline and back

Latissimus dorsi muscle
Works with other muscles to move the humerus bone; also pulls the shoulder backwards

Sacrum
Lower part of vertebral column; articulates with fifth lumbar vertebra and coccyx

The deep muscles of the back are concerned with support and movement of the spine, while the superficial muscles act to move the arm and shoulder.

SUPERFICIAL MUSCLES
The trapezius is a large, fan-shaped muscle whose top edge forms the visible slope from neck to shoulder.

It attaches to the skull and helps to hold up and rotate the head and enables us to brace the shoulders back. The latissimus dorsi, the largest and most powerful back muscle, is attached to the spine from above the lower edge of the trapezius and runs down to the back of the pelvis. The latissimus dorsi allows a lifted arm to

be pulled back into line with the trunk, even against great force.

Smaller muscles also contribute to this superficial muscle layer. Levator scapulae, rhomboid major and rhomboid minor run between the spine and the scapula and act to move the scapula up and inwards. The 'rotator cuff' is a group

of muscles that run between the scapula and head of the humerus (bone of the upper arm) at the shoulder joint. Together, they hold the head of the humerus tightly into the shoulder joint. Serratus posterior runs from the vertebrae to the ribs and moves the ribcage up during breathing.

Deep muscles of the back

The deep muscles of the back attach to underlying bones of the spine, pelvis and ribs. They act together to allow smooth movements of the spine.

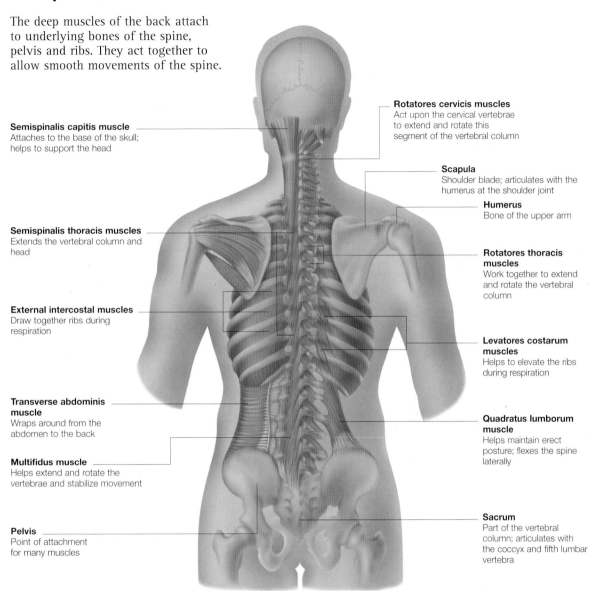

Rotatores cervicis muscles
Act upon the cervical vertebrae to extend and rotate this segment of the vertebral column

Semispinalis capitis muscle
Attaches to the base of the skull; helps to support the head

Scapula
Shoulder blade; articulates with the humerus at the shoulder joint

Humerus
Bone of the upper arm

Semispinalis thoracis muscles
Extends the vertebral column and head

Rotatores thoracis muscles
Work together to extend and rotate the vertebral column

External intercostal muscles
Draw together ribs during respiration

Levatores costarum muscles
Helps to elevate the ribs during respiration

Transverse abdominis muscle
Wraps around from the abdomen to the back

Quadratus lumborum muscle
Helps maintain erect posture; flexes the spine laterally

Multifidus muscle
Helps extend and rotate the vertebrae and stabilize movement

Pelvis
Point of attachment for many muscles

Sacrum
Part of the vertebral column; articulates with the coccyx and fifth lumbar vertebra

Muscles need to be attached to bone in order to give them the leverage they need to perform their functions. The bony attachments of the deep muscles of the back include the vertebrae, the ribs, the base of the skull and the pelvis.

DEEP MUSCLE LAYERS
The deep muscles of the backare built up in layers; the most deeply located muscles are very short, running from each vertebra obliquely to the one above. Over these lie muscles which are longer and run vertically between several vertebrae and the ribs. More superficially, the muscles become longer and some are attached to the pelvic bones and the occiput (back of the base of the skull) as well as to the vertebrae.

There are numerous muscles in these layers. Although each muscle is individually named according to its position, in practice they act in varying combinations rather than individually. Together, they form the large group of deep muscles which lie on either side of the spine and act in conjunction to maintain the spine in an S-shaped curve, enabling the fluid movements of the spine.

Pectoral girdle

The pectoral, or shoulder, girdle is the bony structure that articulates with and supports the upper limbs. It consists of the clavicles at the front of the chest and the scapulae that lie flat against the back.

Pectoral girdle from above

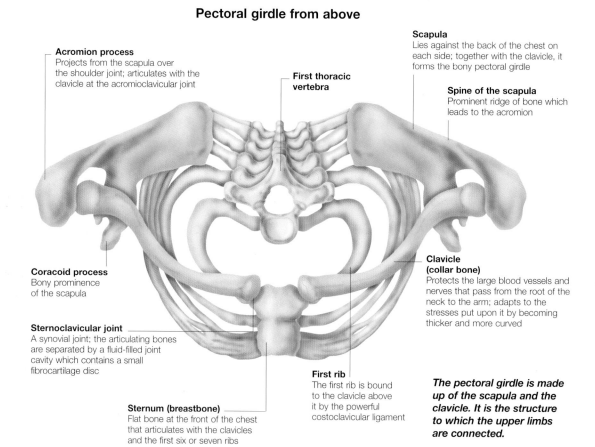

Acromion process
Projects from the scapula over the shoulder joint; articulates with the clavicle at the acromioclavicular joint

First thoracic vertebra

Scapula
Lies against the back of the chest on each side; together with the clavicle, it forms the bony pectoral girdle

Spine of the scapula
Prominent ridge of bone which leads to the acromion

Coracoid process
Bony prominence of the scapula

Sternoclavicular joint
A synovial joint; the articulating bones are separated by a fluid-filled joint cavity which contains a small fibrocartilage disc

Clavicle (collar bone)
Protects the large blood vessels and nerves that pass from the root of the neck to the arm; adapts to the stresses put upon it by becoming thicker and more curved

First rib
The first rib is bound to the clavicle above it by the powerful costoclavicular ligament

Sternum (breastbone)
Flat bone at the front of the chest that articulates with the clavicles and the first six or seven ribs

The pectoral girdle is made up of the scapula and the clavicle. It is the structure to which the upper limbs are connected.

The upper limb is connected to the skeleton by the pectoral or shoulder girdle, made up of the clavicle (collar bone) and the scapula (shoulder blade). The pectoral girdle has only one joint with the central skeleton, at the inner end of the clavicle where it articulates with the sternum (breastbone). The stability of the pectoral girdle is provided by muscles and ligaments attached to the skull, ribs, sternum and vertebrae.

THE CLAVICLE
The clavicle is an S-shaped bone that lies horizontally at the upper border of the chest. The front and upper surfaces of the clavicle are mostly smooth, while the under-surfaces are roughened and grooved by the attachments of muscles and ligaments.

The medial (inner) end of the clavicle has a large oval facet for connecting with the sternum at the sternoclavicular joint. A smaller facet lies at the other end where the clavicle articulates with the acromion (a bony prominence of the scapula) at the acromio-clavicular joint.

The clavicle acts as a strut to brace the upper limb away from the body, thereby allowing a wide range of free movement. Along with the scapula and its muscular connections, it also transmits the force of impacts on the upper limb to the skeleton.

Scapula

The scapula is a flat, triangular-shaped bone which lies against the back of the chest. With the clavicle, it forms the bony pectoral girdle.

The scapula, or shoulder blade, lies against the back of the chest on each side overlying the second to seventh ribs. As a rough triangle, the scapula has three borders: medial (inner), lateral (outer) and superior, with three angles between them.

SURFACES

The scapula has two surfaces: anterior (front) and posterior (back). The anterior or costal (rib) surface lies against the ribs at the back of the chest and is concave, having a large hollow called the subscapular fossa that provides a large surface area for the attachment of muscles.

The posterior surface is divided by a prominent spine. The supraspinous fossa is the small area above the spine, while the infraspinous fossa lies below. These hollows also provide sites of attachment for muscles of the same name.

BONY PROCESSES

The spine of the scapula is a thick projecting ridge, continuous with the bony outcrop called the acromion. This is a flattened prominence that forms the tip of the shoulder. The lateral angle, the thickest part of the scapula, contains the glenoid cavity, the depression into which the head of the humerus fits at the shoulder joint. The coracoid process – an important site of attachment of muscles and ligaments – is also palpable in this area.

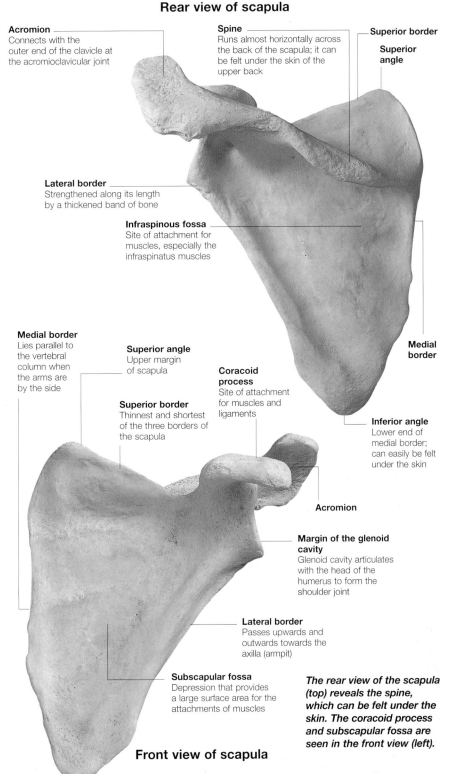

Rear view of scapula

Acromion
Connects with the outer end of the clavicle at the acromioclavicular joint

Spine
Runs almost horizontally across the back of the scapula; it can be felt under the skin of the upper back

Superior border

Superior angle

Lateral border
Strengthened along its length by a thickened band of bone

Infraspinous fossa
Site of attachment for muscles, especially the infraspinatus muscles

Medial border

Medial border
Lies parallel to the vertebral column when the arms are by the side

Superior angle
Upper margin of scapula

Coracoid process
Site of attachment for muscles and ligaments

Inferior angle
Lower end of medial border; can easily be felt under the skin

Superior border
Thinnest and shortest of the three borders of the scapula

Acromion

Margin of the glenoid cavity
Glenoid cavity articulates with the head of the humerus to form the shoulder joint

Lateral border
Passes upwards and outwards towards the axilla (armpit)

Subscapular fossa
Depression that provides a large surface area for the attachments of muscles

Front view of scapula

The rear view of the scapula (top) reveals the spine, which can be felt under the skin. The coracoid process and subscapular fossa are seen in the front view (left).

Muscles of the pectoral girdle

The pectoral girdle consists of the scapulae and clavicles, and is responsible for attaching the upper limbs to the central skeleton. The pectoral girdle muscles hold the scapulae and clavicles in place.

Muscles of the pectoral girdle from the front

Superficial   **Deep**

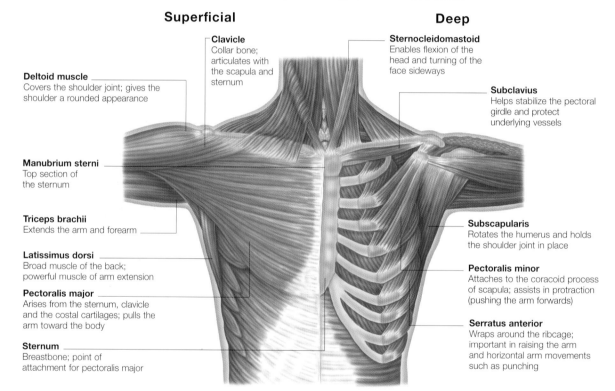

Clavicle
Collar bone; articulates with the scapula and sternum

Sternocleidomastoid
Enables flexion of the head and turning of the face sideways

Deltoid muscle
Covers the shoulder joint; gives the shoulder a rounded appearance

Subclavius
Helps stabilize the pectoral girdle and protect underlying vessels

Manubrium sterni
Top section of the sternum

Triceps brachii
Extends the arm and forearm

Subscapularis
Rotates the humerus and holds the shoulder joint in place

Latissimus dorsi
Broad muscle of the back; powerful muscle of arm extension

Pectoralis minor
Attaches to the coracoid process of scapula; assists in protraction (pushing the arm forwards)

Pectoralis major
Arises from the sternum, clavicle and the costal cartilages; pulls the arm toward the body

Serratus anterior
Wraps around the ribcage; important in raising the arm and horizontal arm movements such as punching

Sternum
Breastbone; point of attachment for pectoralis major

The pectoral girdle is defined as the structure which attaches the upper limbs to the axial skeleton, namely the clavicles and scapulae, to which the muscles of the pectoral girdle attach; however, there are a few muscles which connect the upper limb directly to the central skeleton, and cause indirect movements of the pectoral girdle. This group of muscles lie superficially on the trunk; pectoralis major at the front and latissimus dorsi at the back.

PECTORALIS MAJOR
Pectoralis major arises via two heads, one from the sternum (breastbone) and adjacent rib (costal) cartilages, and another from the middle third of the clavicle. Its tendon twists anti-clockwise as it courses towards the outer lip of the bicipital groove, on the upper end of the humerus. This twisting gives the clavicular head a greater mechanical advantage during flexion (bending) of the arm.

Pectoralis major derives its nerve supply and blood supply from a wide source. The sternocostal head of pectoralis major is a powerful adductor of the arm (pulling the limb towards the body), and is hence well-developed in climbers and weight-lifters. If the arm is kept fixed, this muscle can elevate the ribs as an accessory muscle of inspiration – breathing in.

Pectoral girdle from the back

The large trapezius and latissimus dorsi are superficial muscles of the back which attach to and influence the movement of the pectoral girdle.

Muscles of the pectoral girdle from the back

Superficial

Deep

Trapezius
Stabilizes, raises and rotates the scapula

Spine of scapula

Rhomboid minor
Pulls scapula backward

Supraspinatus
Stabilizes shoulder joint and raises arm away from body

Humerus
Long bone of upper arm

Infraspinatus
Holds humerus in glenoid cavity; rotates humerus laterally, as in a backhand stroke in tennis

Teres major
Acts with latissimus dorsi to extend the arm

Spines of vertebrae

Teres minor
Holds humerus in glenoid cavity

Infraspinatus
Holds humerus in glenoid cavity

Rhomboid major
Acts with rhomboid minor to pull scapula backward

Latissimus dorsi
Broad muscle of the back and powerful muscle of arm extension; assists during forceful expiration, for example, during coughing

Thoracolumbar fascia
Connective tissue to which the latissimus dorsi muscle is joined

Latissimus dorsi arises from the lower thoracic and the lumbar and sacral vertebrae. It also arises from the thoracolumbar fascia and posterior part of the iliac crest, with a few fibres attaching to the lower four ribs. From this broad base (latissimus means 'broadest' in Latin), it converges onto the floor of the bicipital groove at the upper end of the humerus.

This muscle assists pectoralis major in pulling the arm towards the body (adduction). Since the muscle wraps itself around the lower ribs, it assists during forceful expiration (breathing out), for example during coughing.

TRAPEZIUS
Partly overlapping the latissimus dorsi is the lower part of the trapezius muscle. The trapezius also has a broad origin from the base of the skull (occipital protuberance) to the spines of the twelve thoracic vertebrae. The lower fibres attach to the spine of the scapula; intermediate fibres to the acromion process; and upper fibres to the outer third of the clavicle. The upper part serves to shrug the shoulders; the middle and lower parts serve to laterally rotate the scapula.

Ribcage

The ribcage protects the vital organs of the thorax, as well as providing sites for the attachment of muscles of the back, chest and shoulders. It is also light enough to move during breathing.

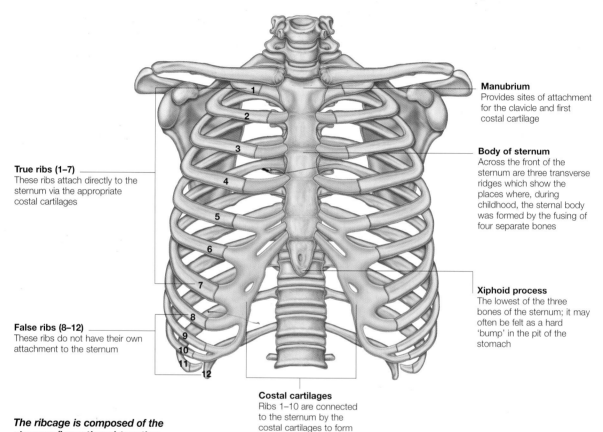

Manubrium
Provides sites of attachment for the clavicle and first costal cartilage

Body of sternum
Across the front of the sternum are three transverse ridges which show the places where, during childhood, the sternal body was formed by the fusing of four separate bones

True ribs (1–7)
These ribs attach directly to the sternum via the appropriate costal cartilages

Xiphoid process
The lowest of the three bones of the sternum; it may often be felt as a hard 'bump' in the pit of the stomach

False ribs (8–12)
These ribs do not have their own attachment to the sternum

Costal cartilages
Ribs 1–10 are connected to the sternum by the costal cartilages to form the costal margin

The ribcage is composed of the sternum (breastbone) together with 12 pairs of ribs and their associated costal cartilages.

The ribcage is supported at the back by the 12 thoracic vertebrae of the spinal column and is formed by the 12 paired ribs, the costal cartilages and the bony sternum, or breastbone, at the front.

Each of the 12 pairs of ribs is attached posteriorly (at the back) to the corresponding numbered thoracic vertebra. The ribs then curve down and around the chest towards the anterior (front) surface of the body.

The 12 ribs can be divided into two groups according to their anterior (front) site of attachment:

• True (vertebrosternal) ribs
The first seven pairs of ribs attach anteriorly directly to the sternum via individual costal cartilages.

• False ribs
These do not attach directly to the sternum. Rib pairs eight to 10 (vertebrochondral ribs) attach indirectly to the sternum via fused costal cartilages. Rib pairs 11 and 12 do not have attachments to bone or cartilage and so are known as 'vertebral' or 'floating' ribs. Their anterior ends lie buried within the musculature of the lateral abdominal wall.

The sternum

The sternum (breastbone) is a long, flat bone which lies vertically at the centre of the anterior (front) surface of the ribcage.

The sternum has three parts:
- The manubrium. This bone forms the upper part of the sternum and is in the shape of a rough triangle with a prominent notch in the centre of its superior surface, the 'suprasternal notch'
- The body. The manubrium and the body of the sternum lie in slightly different planes, angled so that their junction, the manubriosternal joint, projects forwards forming the 'sternal angle of Louis'. The body of the sternum is longer than the manubrium, forming the greater length of the breastbone
- The xiphoid process. This is a small pointed bone which projects downwards and slightly backwards from the lower end of the body of the sternum. In young people it may be cartilaginous, but it usually becomes completely ossified (changed to bone) by 40–50 years of age.

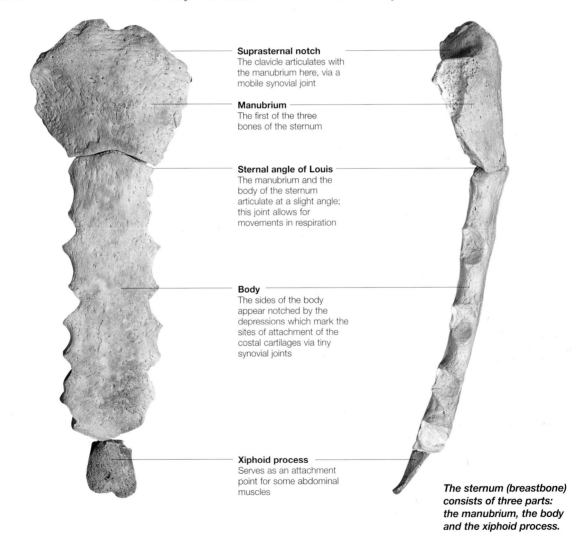

Suprasternal notch
The clavicle articulates with the manubrium here, via a mobile synovial joint

Manubrium
The first of the three bones of the sternum

Sternal angle of Louis
The manubrium and the body of the sternum articulate at a slight angle; this joint allows for movements in respiration

Body
The sides of the body appear notched by the depressions which mark the sites of attachment of the costal cartilages via tiny synovial joints

Xiphoid process
Serves as an attachment point for some abdominal muscles

The sternum (breastbone) consists of three parts: the manubrium, the body and the xiphoid process.

Muscles and movements of the ribcage

The bony skeleton of the ribcage is sheathed in several layers of muscle which include many of the powerful muscles of the upper limb and back, as well as those which act upon the ribcage alone.

Internal view of the chest wall

External intercostal muscles
Their fibres run downwards and forwards and act to lift the ribs during inspiration

Internal intercostal muscles
Their fibres run downwards and backwards from the upper to each lower rib below

Innermost intercostal muscles
These muscles lie deep to the internal intercostal muscles

Anterior scalene muscle
Elevates the first rib when breathing in; not an integral muscle but an accessory muscle of respiration

Sternum
Breastbone; forms a point of attachment for various muscles

Transversus thoracis muscles
Originates from the junction of rib shaft and costal cartilage

Transversus abdominis muscle
Fibres run horizontally

The internal view of the chest wall reveals the transversus thoracis muscles which attach to the sternum and the layers of intercostal muscles.

The integral muscles of the ribcage are concerned with respiration (breathing). They attach only to the ribcage and the thoracic spine. They form the structure of the thoracic wall, enclosing and protecting the vital internal organs of the thorax.

INTERCOSTAL MUSCLES
The intercostal muscles fill the 11 intercostal spaces between the ribs. They lie in three layers, the external intercostals lying superficially, then the internal intercostals, with the innermost intercostals at the deepest level.
- External intercostal muscles
  The fibres of each external intercostal muscle run downwards and forwards to the rib below and their

contraction acts to lift the ribs during inspiration (breathing in).
- Internal intercostal muscles
  The internal intercostal muscles lie just deep to the external intercostals and at right angles to them; that is, their fibres run downwards and backwards from the upper to the lower rib. Like the

external intercostal muscles, they act to assist in inspiration.
- Innermost intercostal muscles These lie deep to the internal intercostal muscles, their fibres running in the same direction. They are separated from the internal intercostals by connective tissue containing the nerves and blood vessels.

Movements of ribcage

During the action of breathing, the chest cavity expands and contracts, causing air to enter and leave. Expansion of the chest cavity is achieved by contraction of the diaphragm and movements of the ribcage.

Ribcage when breathing in

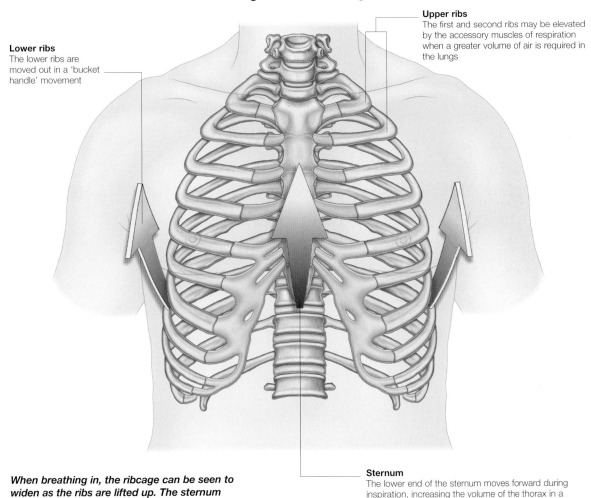

Upper ribs
The first and second ribs may be elevated by the accessory muscles of respiration when a greater volume of air is required in the lungs

Lower ribs
The lower ribs are moved out in a 'bucket handle' movement

When breathing in, the ribcage can be seen to widen as the ribs are lifted up. The sternum also moves out, further expanding the chest.

Sternum
The lower end of the sternum moves forward during inspiration, increasing the volume of the thorax in a 'pump handle' movement

Movements of the ribcage during quiet respiration are due to the action of the respiratory muscles, the most important being the intercostals. Contraction of the intercostal muscles causes the ribcage to expand both sideways and from front to back.

EXPANDING THE CHEST
The lower ribs, especially, are lifted up and out to the sides in what had been described as a 'bucket handle' movement that increases the width of the ribcage. When the upper ribs are elevated, the sternum (breastbone) is also pulled up and rotated slightly so that its lower end moves forward. This action increases the depth of the ribcage and is described as a 'pump handle' movement.

Together, these movements act to increase the volume of the ribcage which, in turn, expands the underlying lungs and draws air in.

When the intercostal muscles relax at the end of inspiration (breathing in), the ribcage descends again due to gravity and the natural elasticity of the lungs.

Female breast

The breast undergoes structural changes throughout the life of a woman. The most obvious changes occur during pregnancy as the breast prepares for its function as the source of milk for the baby.

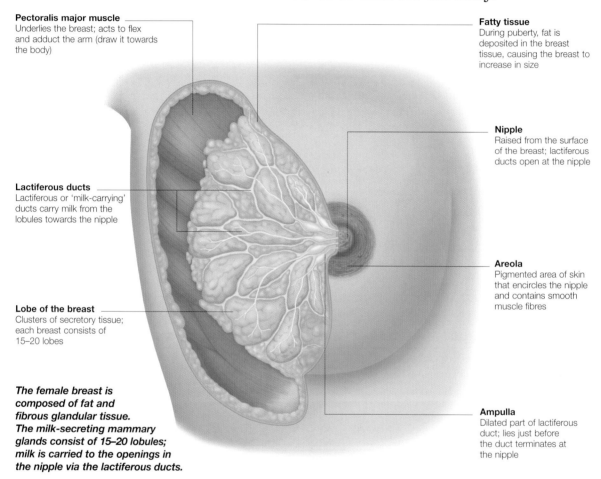

Pectoralis major muscle
Underlies the breast; acts to flex and adduct the arm (draw it towards the body)

Fatty tissue
During puberty, fat is deposited in the breast tissue, causing the breast to increase in size

Nipple
Raised from the surface of the breast; lactiferous ducts open at the nipple

Lactiferous ducts
Lactiferous or 'milk-carrying' ducts carry milk from the lobules towards the nipple

Lobe of the breast
Clusters of secretory tissue; each breast consists of 15–20 lobes

Areola
Pigmented area of skin that encircles the nipple and contains smooth muscle fibres

Ampulla
Dilated part of lactiferous duct; lies just before the duct terminates at the nipple

The female breast is composed of fat and fibrous glandular tissue. The milk-secreting mammary glands consist of 15–20 lobules; milk is carried to the openings in the nipple via the lactiferous ducts.

Men and women both have breast tissue, but the breast is normally a well-developed structure only in women. The two female breasts are roughly hemispherical and are composed of fat and glandular tissue which overlie the muscle layer of the front of the chest wall on either side of the sternum (breastbone).

BREAST STRUCTURE
The base of the breast is roughly circular in shape and extends from the level of the second rib above to the sixth rib below. In addition, there may be an extension of breast tissue towards the axilla (armpit), known as the 'axillary tail'.

Breast size varies greatly between women; this is mainly due to the amount of fatty tissue present, as there is generally the same amount of glandular tissue in every breast.

The mammary glands consist of 15 to 20 lobules – clusters of secretory tissue from which milk is produced. Milk is carried to the surface of the breast from each lobule by a tube known as a 'lactiferous duct', which has its opening at the nipple.

The nipple is a protruding structure surrounded by a circular, pigmented area, called the areola. The skin of the nipple is very thin and delicate and has no hair follicles or sweat glands.

Lymphatic drainage of the breast

Lymph, the fluid which leaks out of blood vessels into the spaces between cells, is returned to the blood circulation by the lymphatic system. Lymph passes through a series of lymph nodes, which act as filters to raemove bacteria, cells and other particles.

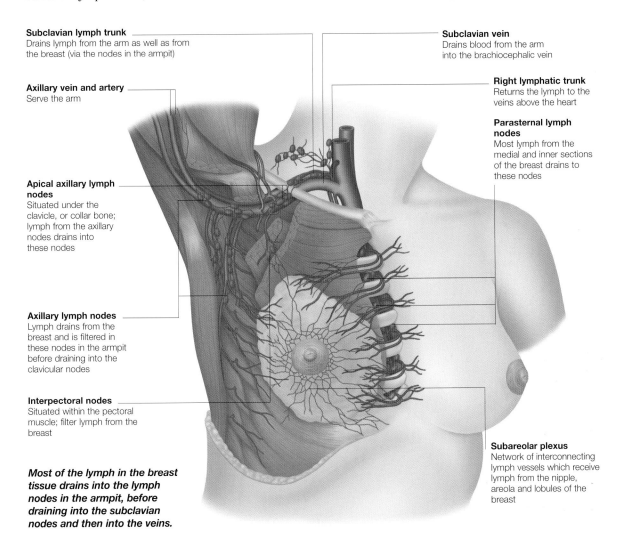

Subclavian lymph trunk
Drains lymph from the arm as well as from the breast (via the nodes in the armpit)

Axillary vein and artery
Serve the arm

Apical axillary lymph nodes
Situated under the clavicle, or collar bone; lymph from the axillary nodes drains into these nodes

Axillary lymph nodes
Lymph drains from the breast and is filtered in these nodes in the armpit before draining into the clavicular nodes

Interpectoral nodes
Situated within the pectoral muscle; filter lymph from the breast

Subclavian vein
Drains blood from the arm into the brachiocephalic vein

Right lymphatic trunk
Returns the lymph to the veins above the heart

Parasternal lymph nodes
Most lymph from the medial and inner sections of the breast drains to these nodes

Subareolar plexus
Network of interconnecting lymph vessels which receive lymph from the nipple, areola and lobules of the breast

Most of the lymph in the breast tissue drains into the lymph nodes in the armpit, before draining into the subclavian nodes and then into the veins.

Tiny lymphatic vessels arise from the tissue spaces and converge to form larger vessels which carry the (usually) clear lymph away from the tissues and into the venous system.

Lymph drains from the nipple, areola and mammary gland lobules into a network of small lymphatic vessels, the 'subareolar lymphatic plexus'. From this plexus the lymph may be carried in several different directions.

PATTERN OF DRAINAGE
About 75 per cent of the lymph from the subareolar plexus drains to the lymph nodes of the armpit, mostly from the outer quadrants of the breast. The lymph passes through a series of nodes in the region of the armpit draining into the subclavian lymph trunk, and ultimately into the right lymphatic trunk, which returns the lymph to the veins above the heart.

Most of the remaining lymph, mainly from the inner quadrants of the breast, is carried to the 'parasternal' lymph nodes, which lie towards the midline of the front of the chest. A small percentage of lymphatic vessels from the breast take another route and travel to the posterior intercostal nodes.

Diaphragm

The diaphragm is a sheet of muscle that separates the thorax from the abdominal cavity. It is essential for breathing as its contraction expands the chest cavity, allowing air to enter.

Abdominal surface of the diaphragm

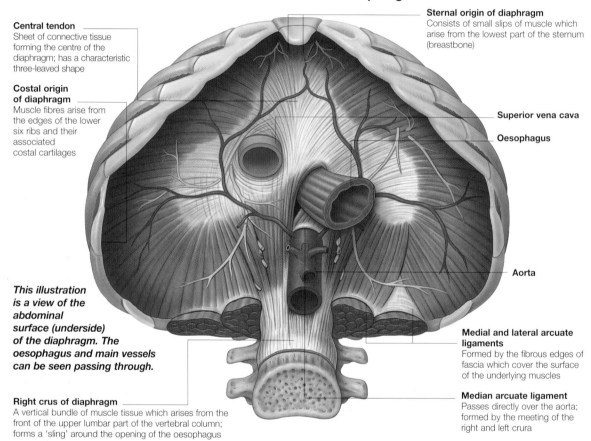

Central tendon
Sheet of connective tissue forming the centre of the diaphragm; has a characteristic three-leaved shape

Costal origin of diaphragm
Muscle fibres arise from the edges of the lower six ribs and their associated costal cartilages

Sternal origin of diaphragm
Consists of small slips of muscle which arise from the lowest part of the sternum (breastbone)

Superior vena cava

Oesophagus

Aorta

This illustration is a view of the abdominal surface (underside) of the diaphragm. The oesophagus and main vessels can be seen passing through.

Medial and lateral arcuate ligaments
Formed by the fibrous edges of fascia which cover the surface of the underlying muscles

Median arcuate ligament
Passes directly over the aorta; formed by the meeting of the right and left crura

Right crus of diaphragm
A vertical bundle of muscle tissue which arises from the front of the upper lumbar part of the vertebral column; forms a 'sling' around the opening of the oesophagus

The diaphragm is the main muscle involved in respiration and has several apertures for the passage of important structures which must pass between thorax and abdomen. It is made up of peripheral muscle fibres inserting into a central sheet of tendon which, unlike most tendons, does not have any attachment to bone.

MUSCLE OF THE DIAPHRAGM '
The muscle tissue of the diaphragm arises from three areas of the chest wall, merges to form a continuous sheet and converges on the central tendon, which acts as a site of muscular attachment.

The three areas of origin of the diaphragm give rise to three separately named parts: the sternal part, the costal part and the lumbar or vertebral part, which arises from the crus and arcuate ligaments.

CENTRAL TENDON
Muscle fibres of the diaphragm insert into the central tendon, which has a three-leaved shape. The central part lies just beneath, and is depressed by, the heart. It is attached by ligaments to the pericardium, the membrane surrounding the heart. The two lateral leaves lie towards the back and help form the right and left domes (cupolae) of the diaphragm.

Thoracic surface of the diaphragm

The upper aspect of the diaphragm is convex and forms the floor of the thoracic (chest) cavity. It is perforated by major vessels and structures which must pass through the muscle sheet in order to reach the abdomen.

Diaphragm from above

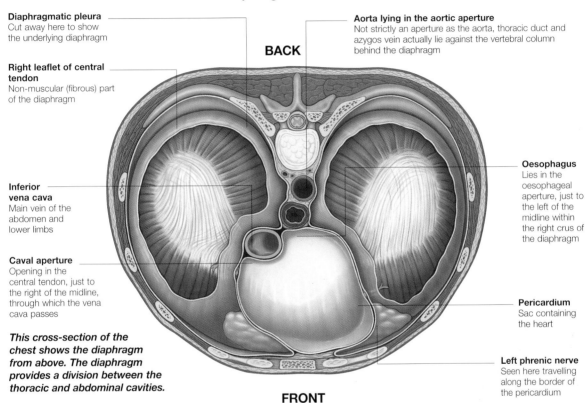

Diaphragmatic pleura
Cut away here to show the underlying diaphragm

Aorta lying in the aortic aperture
Not strictly an aperture as the aorta, thoracic duct and azygos vein actually lie against the vertebral column behind the diaphragm

BACK

Right leaflet of central tendon
Non-muscular (fibrous) part of the diaphragm

Oesophagus
Lies in the oesophageal aperture, just to the left of the midline within the right crus of the diaphragm

Inferior vena cava
Main vein of the abdomen and lower limbs

Caval aperture
Opening in the central tendon, just to the right of the midline, through which the vena cava passes

Pericardium
Sac containing the heart

This cross-section of the chest shows the diaphragm from above. The diaphragm provides a division between the thoracic and abdominal cavities.

Left phrenic nerve
Seen here travelling along the border of the pericardium

FRONT

The central part of the surface of the diaphragm is covered by the pericardium, the membrane which surrounds the heart.
To either side, the upper surface of the diaphragm is lined with the diaphragmatic part of the parietal pleura (the thin membrane which lines the chest cavity). This is continuous around the edges of the diaphragm with the costal pleura, which covers the inside of the chest wall.

DIAPHRAGMATIC APERTURES
Although the diaphragm acts to separate the chest and abdominal cavities, certain structures do pass through 'diaphragmatic apertures'. The three largest of these are:

• The caval aperture is an opening in the central tendon of the diaphragm which allows passage of the inferior vena cava, the main vein of the abdomen and lower limbs. As the opening is in the central tendon rather than the muscle of the diaphragm it will not close when the diaphragm contracts during inspiration; in fact the opening widens and blood flow increases. The opening also contains branches of the right phrenic nerve and lymphatic vessels.
• The oesophageal aperture allows the passage of the oesophagus (gullet) through the diaphragm to reach the stomach. The muscle fibres of the right crus act as a sphincter, closing off the oesophageal opening when the diaphragm contracts during inspiration. As well as the oesophagus, the aperture also gives passage to nerves (vagus), arteries and lymphatic vessels.
• The aortic aperture
This opening lies behind the diaphragm. As the aorta does not actually pierce the diaphragm, the flow of blood within it is not affected by diaphragmatic contractions while breathing. The aorta emerges under the median arcuate ligament, in front of the vertebral column. The aortic aperture also transmits the thoracic duct (major lymphatic channel) and the azygos vein.

Lungs

The paired lungs are cone-shaped organs of respiration which occupy the thoracic cavity, lying to either side of the heart, great blood vessels and other structures of the central mediastinum.

Anterior view of the lungs

Right lung   Left lung

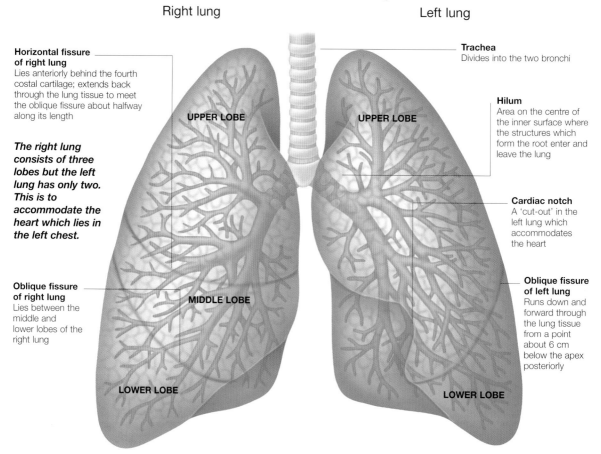

Horizontal fissure of right lung
Lies anteriorly behind the fourth costal cartilage; extends back through the lung tissue to meet the oblique fissure about halfway along its length

The right lung consists of three lobes but the left lung has only two. This is to accommodate the heart which lies in the left chest.

Oblique fissure of right lung
Lies between the middle and lower lobes of the right lung

UPPER LOBE

MIDDLE LOBE

LOWER LOBE

UPPER LOBE

LOWER LOBE

Trachea
Divides into the two bronchi

Hilum
Area on the centre of the inner surface where the structures which form the root enter and leave the lung

Cardiac notch
A 'cut-out' in the left lung which accommodates the heart

Oblique fissure of left lung
Runs down and forward through the lung tissue from a point about 6 cm below the apex posteriorly

The right and left lungs are separate entities, each enclosed within a bag of membranes, the right and left pleural sacs. Each lung lies free within the thoracic cavity, attached to the mediastinum only by a root made up of the main bronchus and large blood vessels.

The lung tissue is soft and spongy and has great elasticity. In children the lungs are pink in colour, but they usually become darkener and mottled later in life as they are exposed to dust which is taken in by the defence cells of the lining of the airways.

Each lung has:
• An apex, which projects up into the base of the neck behind the clavicle (collarbone)
• A base, with a concave surface, which rests on the superior surface of the diaphragm
• A concave mediastinal surface, which lies against various structures of the mediastinum.

LOBES AND FISSURES
The lungs are divided into sections known as lobes by deep fissures. The right lung has three lobes while the left lung, which is slightly smaller (due to the position of the heart), has two. Each lobe is independent of the others, receiving air via its own lobar bronchus and blood from lobar arteries.

The fissures are deep, extending right through the structure of the lung, and are lined by the pleural membrane.

The pleura

The lungs are covered by a thin membrane known as the pleura.
The pleura lines both the outer surface of the lung and the inner
surface of the thoracic cage.

Position of the lungs and pleura

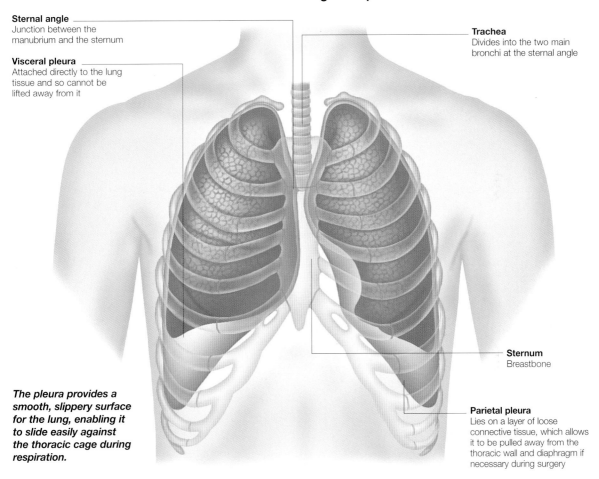

Sternal angle
Junction between the
manubrium and the sternum

Visceral pleura
Attached directly to the lung
tissue and so cannot be
lifted away from it

Trachea
Divides into the two main
bronchi at the sternal angle

Sternum
Breastbone

Parietal pleura
Lies on a layer of loose
connective tissue, which allows
it to be pulled away from the
thoracic wall and diaphragm if
necessary during surgery

*The pleura provides a
smooth, slippery surface
for the lung, enabling it
to slide easily against
the thoracic cage during
respiration.*

The layer of pleura covering
the lung is called the
visceral pleura, while that
lining the thoracic cage is
the parietal pleura.

VISCERAL PLEURA
This thin membrane covers
the lung surface, dipping
down into the fissures
between the lobes
of the lung.

PARIETAL PLEURA
This is continuous with the
visceral pleura at the hilum
of the lung. Here, the
membrane reflects back and
lines all the inner surfaces of
the thoracic cavity. The
parietal pleura is one
continuous membrane
divided into areas that are
named after the surfaces they
cover:

- Costal pleura – lines the
  inside of the ribcage, the
  back of the sternum and
  the sides of the vertebral
  bodies of the spine
- Mediastinal pleura – covers
  the mediastinum, the
  central area of the thoracic
  cavity
- Diaphragmatic pleura –
  lines the upper surface of
  the diaphragm, except

where it is covered by the
pericardium
- Cervical pleura – covers
  the tip of the lung as it
  projects up into the base
  of the neck.

Respiratory airways

The airways form a network along which air travels to,
from and within the lungs. The airways branch repeatedly, each branch
narrowing until the end terminals – the alveoli – are reached.

As a breath is taken, air enters through the nose and mouth, then passes down through the larynx to enter the trachea (windpipe). The air is carried down into the chest by the trachea, which then divides into smaller tubes – the bronchi – which take the air into the lungs.

The bronchi divide to form progressively smaller tubes that reach all areas of the lung. These tubes terminate in the alveolar sacs, which form the substance of the lung. It is in these thin-walled sacs that gas exchange with the blood occurs.

TRACHEA

The trachea extends down from the cricoid cartilage just below the larynx in the neck to enter the chest. At the level of the sternal angle it ends by dividing into two branches, the right and left main bronchi.

The trachea is composed of strong fibroelastic tissue, within which are embedded a series of incomplete rings of hyaline cartilage, the tracheal cartilages. In adults the trachea is quite wide (approximately 2.5 cm), but it is much narrower in infants – about the width of a pencil.

The posterior (back) surface of the trachea has no cartilaginous support and instead consists of fibrous tissue and trachealis muscle fibres. This posterior wall lies in contact with the oesophagus, which is directly behind the trachea.

The major airways

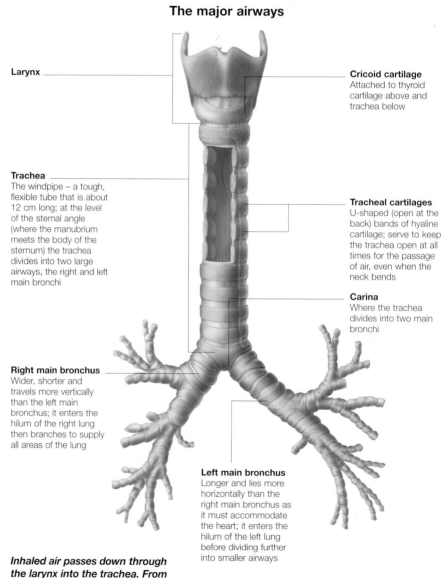

Larynx

Cricoid cartilage
Attached to thyroid cartilage above and trachea below

Trachea
The windpipe – a tough, flexible tube that is about 12 cm long; at the level of the sternal angle (where the manubrium meets the body of the sternum) the trachea divides into two large airways, the right and left main bronchi

Tracheal cartilages
U-shaped (open at the back) bands of hyaline cartilage; serve to keep the trachea open at all times for the passage of air, even when the neck bends

Carina
Where the trachea divides into two main bronchi

Right main bronchus
Wider, shorter and travels more vertically than the left main bronchus; it enters the hilum of the right lung then branches to supply all areas of the lung

Left main bronchus
Longer and lies more horizontally than the right main bronchus as it must accommodate the heart; it enters the hilum of the left lung before dividing further into smaller airways

Inhaled air passes down through the larynx into the trachea. From here it is channelled into two streams, each of which supplies a lung via a main bronchus.

Smaller airways and alveoli

On entering the lung the main bronchus divides again and again, forming the 'bronchial tree', which takes air to all parts of the lung.

The branching of the bronchial tree ends in numerous terminal bronchioles from which arise the respiratory bronchioles, alveolar ducts and alveoli. The clusters of alveoli provide a large surface area for gaseous exchange.

Bronchioles and alveoli

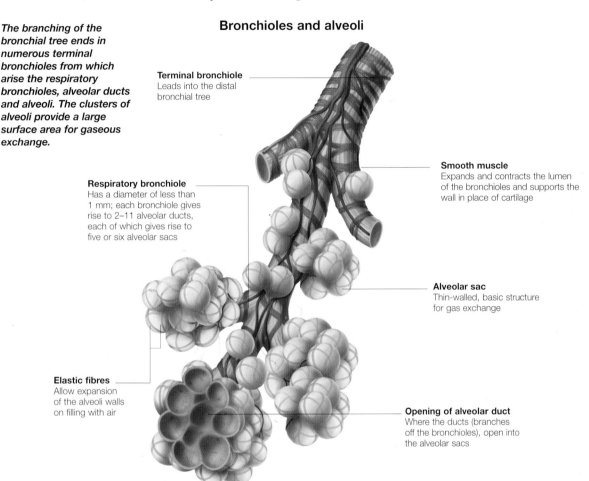

Terminal bronchiole
Leads into the distal bronchial tree

Respiratory bronchiole
Has a diameter of less than 1 mm; each bronchiole gives rise to 2–11 alveolar ducts, each of which gives rise to five or six alveolar sacs

Smooth muscle
Expands and contracts the lumen of the bronchioles and supports the wall in place of cartilage

Alveolar sac
Thin-walled, basic structure for gas exchange

Elastic fibres
Allow expansion of the alveoli walls on filling with air

Opening of alveolar duct
Where the ducts (branches off the bronchioles), open into the alveolar sacs

The first division of the main bronchus gives rise to the lobar bronchi, three on the right and two on the left, which each supply one lobe of the lung. Each of these lobar bronchi divides to form the smaller bronchi, which supply each of the independent bronchopulmonary segments.

BRONCHI STRUCTURE
The bronchi have a similar structure to the trachea, being very elastic and flexible, having cartilage in their walls and being lined with respiratory epithelium. There are also numerous muscle fibres, which allow for changes in diameter of these tubes.

BRONCHIOLES
Within the bronchopulmonary segments the bronchi continue to divide, perhaps as many as 25 times, before they terminate in the blind-ended alveolar sacs.

At each division the tubes become smaller, although the total cross-sectional area increases. When the air tubes have an internal diameter of less than 1 mm they become known as bronchioles.

Bronchioles differ from bronchi in that they have no cartilage in their walls nor any mucus-secreting cells in their lining. They do, however, still have muscle fibres in their walls.

Further divisions lead to the formation of terminal bronchioles, which in turn divide to form a series of respiratory bronchioles, the smallest and finest air passages. Respiratory bronchioles are so named because they have a few alveoli (air sacs) opening directly into them. Most of the alveoli, however, arise in clusters from alveolar ducts, which are formed from division of the respiratory bronchioles.

Vessels of the lung

The primary function of the lungs is to reoxygenate the blood used by the tissues of the body and to remove accumulated waste carbon dioxide. This is effected via the pulmonary blood circulation.

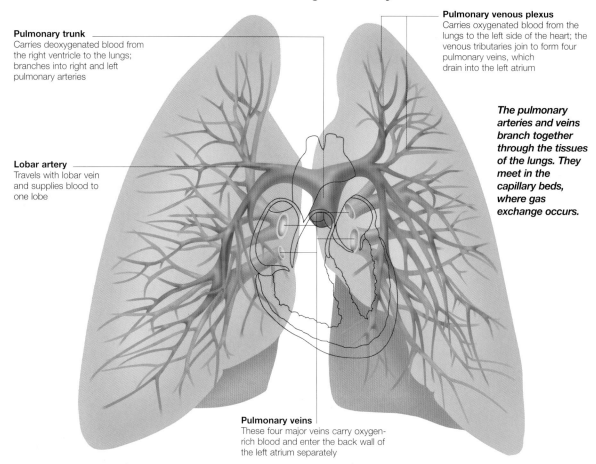

Pulmonary trunk
Carries deoxygenated blood from the right ventricle to the lungs; branches into right and left pulmonary arteries

Lobar artery
Travels with lobar vein and supplies blood to one lobe

Pulmonary venous plexus
Carries oxygenated blood from the lungs to the left side of the heart; the venous tributaries join to form four pulmonary veins, which drain into the left atrium

The pulmonary arteries and veins branch together through the tissues of the lungs. They meet in the capillary beds, where gas exchange occurs.

Pulmonary veins
These four major veins carry oxygen-rich blood and enter the back wall of the left atrium separately

Blood from the body returns to the right side of the heart and from there passes directly to the lungs via the pulmonary arteries.

Having been oxygenated by its passage through the lungs, the blood returns to the left side of the heart in the pulmonary veins. The oxygen-rich blood is then pumped around the body. Collectively, the pulmonary

arteries and veins and their branches are referred to as the pulmonary circulation.

PULMONARY VESSELS
A large artery known as the pulmonary trunk arises from the heart's right ventricle carrying dark-red deoxygenated blood from the body into the lungs.

The pulmonary trunk divides into two smaller

branches, the right and left pulmonary arteries, which run horizontally and enter the lungs at the hilum, alongside the bronchi (main airways).

Within the lungs the arteries divide to supply each lobe of their respective lung; two on the left and three on the right. The lobar arteries divide further to give the segmental arteries, which

supply the broncho-pulmonary segments (structural units of the lung). Each segmental artery ends in a network of capillaries.

Oxygenated blood returns to the left side of the heart through a system of pulmonary veins running alongside the arteries.

Lymphatics of the lung

Lymphatic drainage of the lung originates in two main networks, or plexuses: the superficial (subpleural) plexus and the deep lymphatic plexus. These communicate freely with each other.

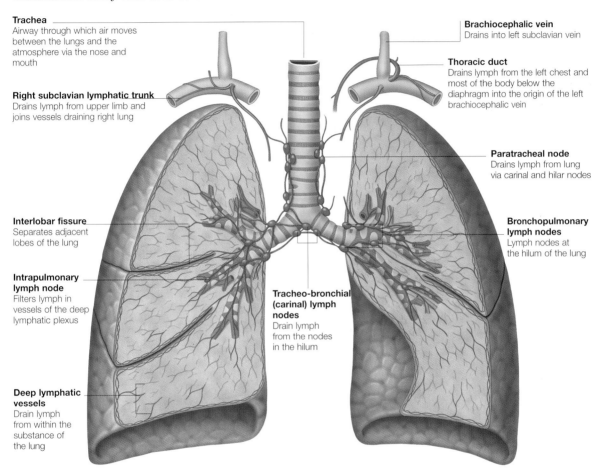

Trachea
Airway through which air moves between the lungs and the atmosphere via the nose and mouth

Right subclavian lymphatic trunk
Drains lymph from upper limb and joins vessels draining right lung

Interlobar fissure
Separates adjacent lobes of the lung

Intrapulmonary lymph node
Filters lymph in vessels of the deep lymphatic plexus

Deep lymphatic vessels
Drain lymph from within the substance of the lung

Brachiocephalic vein
Drains into left subclavian vein

Thoracic duct
Drains lymph from the left chest and most of the body below the diaphragm into the origin of the left brachiocephalic vein

Paratracheal node
Drains lymph from lung via carinal and hilar nodes

Bronchopulmonary lymph nodes
Lymph nodes at the hilum of the lung

Tracheo-bronchial (carinal) lymph nodes
Drain lymph from the nodes in the hilum

Lymph is a fluid which is collected from the spaces between cells and carried in lymphatic vessels back to the venous circulation. On its way, the lymph must pass through a series of lymph nodes, which act as filters to remove particulate matter and any invading micro-organisms.

SUPERFICIAL PLEXUS
This network of fine lymphatic vessels extends over the surface of the lung, just beneath the visceral pleura (covering of the lung). The superficial plexus drains lymph from the lung towards the bronchi and trachea, where the main groups of lymph nodes are found.

Lymph from the superficial plexus arrives first at the bronchopulmonary group of lymph nodes, which lie at the hilum of the lung.

DEEP PLEXUS
The lymphatic vessels of the deep plexus originate in the connective tissue surrounding the small airways, bronchioles and bronchi (the alveoli have no lymphatic vessels). There are also small lymphatic vessels within the lining of the larger airways.

These lymphatic vessels join and run back along the route of the bronchi and pulmonary blood vessels, passing through intrapulmonary nodes which lie within the lung. From these nodes lymph passes through vessels which drain towards the hilum into the broncho-pulmonary lymph nodes.

The bronchopulmonary nodes at the hilum of the lung therefore receive lymph from both superficial and deep lymphatic plexuses.

BRONCHI ANDTRACHEA
From the bronchopulmonary nodes, lymph drains to the tracheobronchial (carinal) lymph nodes. From here, lymph passes up through the paratracheal nodes, into the paired bronchomediastinal lymph trunks, which return the fluid to the venous system in the neck.

113

Heart

The adult heart is about the size of a clenched fist and lies within the mediastinum in the thoracic cavity. It rests on the central tendon of the diaphragm and is flanked on either side by the lungs.

Surrounding the heart is a protective sac of connective tissue called the pericardium.

The heart is hollow and is composed almost entirely of muscle. The typical weight of the normal heart is only about 250 to 350 grams yet it has incredible power and stamina, beating over 70 times every minute to pump blood around the body.

SURFACES OF THE HEART

Roughly the shape of a pyramid on its side, the heart is said to have a base, three surfaces and an apex:

- The base of the heart lies posteriorly (at the back) and is formed mainly by the left atrium, the chamber of the heart that receives oxygenated blood from the lungs
- The inferior or diaphragmatic surface lies on the underside and is formed by the left and right ventricles separated by the posterior interventricular groove. The right and left ventricles are the large chambers which pump blood around the lungs and the body respectively
- The anterior, or sternocostal, surface lies at the front of the heart just behind the sternum and the ribs and is formed mainly by the right ventricle
- The left, or pulmonary, surface is formed mainly by the large left ventricle, which lies in a concavity of the left lung.

Position of the heart

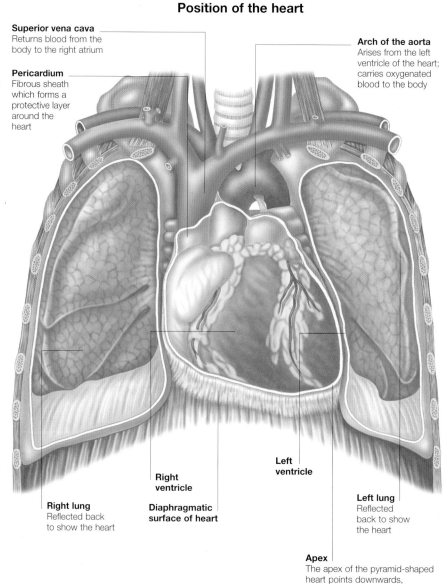

Superior vena cava
Returns blood from the body to the right atrium

Pericardium
Fibrous sheath which forms a protective layer around the heart

Arch of the aorta
Arises from the left ventricle of the heart; carries oxygenated blood to the body

Right lung
Reflected back to show the heart

Right ventricle

Diaphragmatic surface of heart

Left ventricle

Left lung
Reflected back to show the heart

Apex
The apex of the pyramid-shaped heart points downwards, forwards and to the left; it is formed by the left ventricle

The pericardium

The heart is enclosed within a protective triple- walled bag of connective tissue called the pericardium. The pericardium is composed of two parts, the fibrous pericardium and the serous pericardium.

FIBROUS PERICARDIUM

The fibrous pericardium forms the outer part of the bag and is composed of tough fibrous connective tissue. It has three main functions:

- Protection. The fibrous pericardium is strong enough to provide some protection from trauma for such a vital structure as the heart
- Attachment. There are fibrous attachments between this part of the pericardium and both the sternum and the diaphragm.
  In addition, the fibrous pericardium fuses with the strong walls of the arteries which pass through it from the heart. These attachments help to anchor the heart to its surrounding structures
- Prevention of overfilling of the heart. Because the fibrous pericardium is non-elastic, it does not allow the heart to expand with blood beyond a certain safe limit.

SEROUS PERICARDIUM

The serous pericardium covers and surrounds the heart in the same way as the pleura does the lungs. This part of the pericardium is a thin membrane which has two parts that are continuous with each other, the parietal and the visceral layers.

The parietal pericardium lines the inner surface of the fibrous pericardium and reflects back onto the surface

The pericardial sac with the heart removed

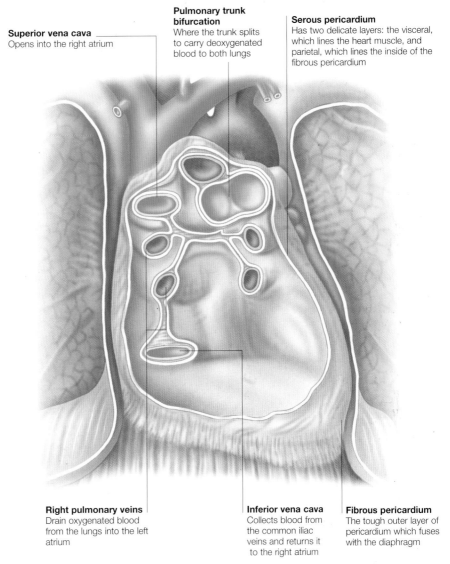

Superior vena cava
Opens into the right atrium

Pulmonary trunk bifurcation
Where the trunk splits to carry deoxygenated blood to both lungs

Serous pericardium
Has two delicate layers: the visceral, which lines the heart muscle, and parietal, which lines the inside of the fibrous pericardium

Right pulmonary veins
Drain oxygenated blood from the lungs into the left atrium

Inferior vena cava
Collects blood from the common iliac veins and returns it to the right atrium

Fibrous pericardium
The tough outer layer of pericardium which fuses with the diaphragm

of the heart at the roots of the large blood vessels to form the visceral pericardium.

Between the two layers of serous pericardium lies a slit-like cavity, the pericardial cavity, which is filled with a very small amount of fluid. The presence of this fine fluid layer, together with the slipperiness of the layers of serous pericardium, allows the chambers of the heart to move freely within the pericardium as the heart beats.

If the pericardial cavity becomes filled with an abnormally large amount of fluid, as may happen in infection or inflammation, the heart becomes compressed within the confines of the fibrous pericardium and is unable to function properly. In extreme cases, when it is known as 'cardiac tamponade', this is life-threatening.

Chambers of the heart

The heart is divided into four chambers: two thin-walled atria, which receive venous blood, and two larger, thick-walled ventricles, which pump blood into the arterial system.

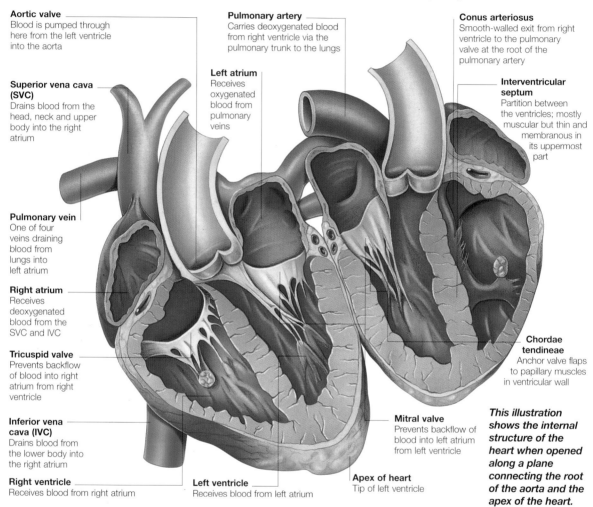

Aortic valve
Blood is pumped through here from the left ventricle into the aorta

Pulmonary artery
Carries deoxygenated blood from right ventricle via the pulmonary trunk to the lungs

Conus arteriosus
Smooth-walled exit from right ventricle to the pulmonary valve at the root of the pulmonary artery

Superior vena cava (SVC)
Drains blood from the head, neck and upper body into the right atrium

Left atrium
Receives oxygenated blood from pulmonary veins

Interventricular septum
Partition between the ventricles; mostly muscular but thin and membranous in its uppermost part

Pulmonary vein
One of four veins draining blood from lungs into left atrium

Right atrium
Receives deoxygenated blood from the SVC and IVC

Tricuspid valve
Prevents backflow of blood into right atrium from right ventricle

Chordae tendineae
Anchor valve flaps to papillary muscles in ventricular wall

Inferior vena cava (IVC)
Drains blood from the lower body into the right atrium

Mitral valve
Prevents backflow of blood into left atrium from left ventricle

This illustration shows the internal structure of the heart when opened along a plane connecting the root of the aorta and the apex of the heart.

Right ventricle
Receives blood from right atrium

Left ventricle
Receives blood from left atrium

Apex of heart
Tip of left ventricle

The heart is divided into left and right sides, each having an atrium and ventricle.

THE VENTRICLES
The two ventricles make up the bulk of the muscle of the heart, the left being larger and more powerful than the right. The right ventricle lies in front, forming much of the anterior surface of the heart, while the left lies behind and below, comprising the greater part of the inferior surface. The apex of the heart is formed by the tip of the left ventricle.

The right ventricle receives blood from the right atrium, back flow being prevented by the tricuspid valve. Blood is then pumped by contraction of the ventricular muscle up through the pulmonary valve into the pulmonary trunk and from there into the lungs.

The left ventricle receives blood from the left atrium through the left atrioventricular orifice, which bears the mitral valve. Powerful contractions of the left ventricle then pump the blood up through the aortic valve into the aorta, the main artery of the body.

The atria

The atria are the two smaller, thin-walled chambers of the heart. They sit above the ventricles separated by the atriovbentricular valves.

Right atrium of the heart

Superior vena cava
Carries blood from the head, neck and upper body back to the heart

Crista terminalis
Shallow, vertical groove that separates the rough and smooth atrial walls

Right auricle
Conical, muscular pouch projecting from the right atrium

Pulmonary vessels
Artery and veins which travel to and from the lungs

Opening of coronary sinus
Short venous trunk which returns blood from the cardiac veins to the heart

Fossa ovalis
Remnant of the fetal foramen ovale: an opening between the right and left atria which closes at birth

Inferior vena cava
Carries blood from the lower body and legs back to the heart

The right atrium of the heart receives venous blood from the superior and inferior venae cavae. Blood is pumped from here into the right ventricle.

All the venous blood from the body is delivered to the right atrium by the two great veins, the superior and inferior vena cavae (SVC and IVC respectively). The coronary sinus, the vessel which collects venous blood from the heart tissues, also drains into the right atrium.

The interior has a smooth-walled posterior part and a rough-walled anterior section. These two areas are separated by a ridge of tissue known as the crista terminalis.

The roughened anterior wall is thicker than the posterior part, being composed of the pectinate muscles, which give a comb-like appearance to the inner surface. The fossa ovalis is a depression on the wall adjoining the left atrium.

The pectinate muscles extend into a small, ear-like outpouching of the right atrium called the auricle. This conical chamber wraps around the outside of the main artery from the heart – the aorta – and acts to increase the capacity of the right atrium.

OPENINGS INTO THE RIGHT ATRIUM
The SVC, which receives blood from the upper half of the body, opens into the upper part of the smooth area of the right atrium.

The IVC, which receives blood from the lower half of the body, enters the lower part of the right atrium. The SVC has no valve to prevent backflow of blood; the IVC only bears a rudimentary non-functional valve.

The opening of the coronary sinus lies between the IVC opening and the opening that allows blood through into the right ventricle (the right atrioventricular orifice).

THE LEFT ATRIUM
The left atrium is smaller than the right, and forms the main part of the base of the heart. It is roughly cuboid in shape and has smooth walls, except for the lining of the left auricle, which is roughened by muscle ridges. The four pulmonary veins, which bring oxygenated blood back from the lungs, open into the posterior part of the left atrium. There are no valves in these orifices.

In the wall adjoining the right atrium lies the oval fossa, which corresponds to the oval fossa on the right side.

117

Valves of the heart

The heart is a powerful muscular pump through which blood flows in a forward direction only. Backflow is prevented by the four heart valves, which have a vital role in maintaining the circulation.

Each of the two sides of the heart has two valves. On the right side of the heart, the tricuspid valve lies between the atrium and the ventricle, and the pulmonary valve lies at the junction of the ventricle and the pulmonary trunk. On the left side, the mitral valve separates the atrium and ventricle while the aortic valve lies between the ventricle and the aorta.

THE TRICUSPID AND MITRAL VALVES

The tricuspid and mitral valves are also known as the atrioventricular valves as they lie between the atria and the ventricles on each side. They are composed of tough connective tissue covered with endocardium, the thin layer of cells which lines the entire heart. The upper surface of the valves is smooth whereas the lower surface carries the attachments of the chordae tendineae.

The tricuspid valve has three cusps, or flaps. In contrast, the mitral valve has only two and is consequently also known as the bicuspid valve; the name 'mitral' comes from its supposed likeness to a bishop's mitre.

THE HEARTBEAT

During its contraction, the normal heart makes a two-component sound (a 'lub-dup') which can be heard using a stethoscope. The first of these sounds comes from the closure of theatrio ventricular valves, the second is due to the closure of the pulmonary and aortic valves.

Diastolic heart with the atria removed

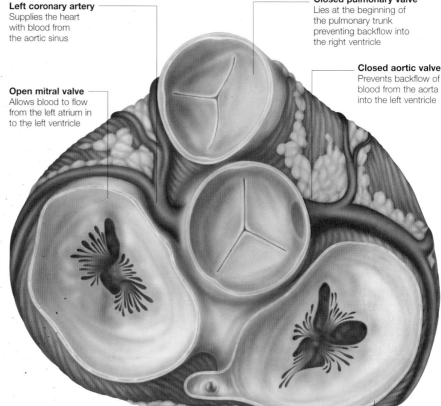

Left coronary artery
Supplies the heart with blood from the aortic sinus

Open mitral valve
Allows blood to flow from the left atrium in to the left ventricle

Closed pulmonary valve
Lies at the beginning of the pulmonary trunk preventing backflow into the right ventricle

Closed aortic valve
Prevents backflow of blood from the aorta into the left ventricle

Open tricuspid valve
Allows blood to flow from the right atrium into the right ventricle

When the heart is in systole the ventricles are contracting and the aortic and pulmonary valves open, allowing blood to be pumped out of the heart.

During diastole the heart muscle of the ventricles is relaxing. The tricuspid and mitral valves are open allowing blood to flow from the atria to fill the ventricles.

Aortic and pulmonary valves

The pulmonary and aortic valves are also known as the semilunar valves. They guard the route of exit of blood from the heart, preventing backflow of blood into the ventricles as they relax after a contraction.

Each of these two valves is composed of three semilunar pocket-like cusps, which have a core of connective tissue covered by a lining of endothelium. This lining ensures a smooth surface for the passage of blood.

AORTIC VALVE

The aortic valve lies between the left ventricle and the aorta, the main artery that carries oxygenated blood to the body. It is stronger and more robust than the pulmonary valve as it has to cope with the higher pressures of the systemic circulation (to the body).

Above each cusp of the valve, formed by bulges of the aortic wall, lie the aortic sinuses. From two of these sinuses arise the right and left coronary arteries, which carry blood to the muscle and coverings of the heart itself.

PULMONARY VALVE

The pulmonary valve separates the ventricle from the pulmonary trunk, the large artery that carries blood from the heart towards the lungs. Just above each cusp of the valve the pulmonary trunk bulges slightly to form the pulmonary sinuses, blood-filled spaces that prevent the cusps from sticking to the arterial wall behind them when they open.

View of the left ventricle opened up

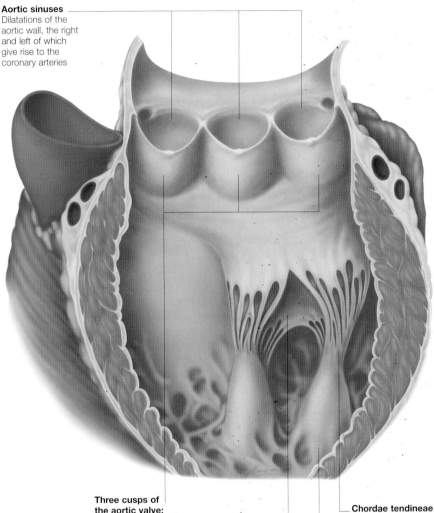

Aortic sinuses
Dilatations of the aortic wall, the right and left of which give rise to the coronary arteries

Three cusps of the aortic valve:
Right cusp
Posterior cusp
Left cusp

Mitral valve
Lies between the left atrium and the left ventricle

Chordae tendineae
Fibrous strands that attach the cusps to the anchoring papillary muscles

Papillary muscle
Larger in the left ventricle than in the right due to having a greater force to overcome

The aortic valve lies between the left ventricle and the aorta. The three cusps of the valve prevent blood expelled from the ventricle from re-entering.

119

Vessels of the heart

Blood is delivered to the heart by two large veins – the superior and inferior venae cavae – and pumped out into the aorta. The venae cavae and aorta are collectively known as the great vessels.

THE VENAE CAVAE

The superior vena cava is the large vein that drains blood from the upper body to the right atrium of the heart. It is formed by the union of the right and left brachiocephalic veins which, in turn, have been formed by smaller veins that receive blood from the head, neck and upper limbs.

The inferior vena cava is the widest vein in the body, but only its last part lies within the thorax as it passes up through the diaphragm to deliver blood to the right atrium.

THE AORTA

The aorta is the largest artery in the body, having an internal diameter of about 2.5 cm in adults. Its relatively thick walls contain elastic connective tissue which allows the vessel to expand slightly, as blood is pumped into it under pressure, and then recoil, thus maintaining blood pressure between heart beats.

The aorta passes upwards initially, then curves around to the left and travels down into the abdomen. It consists of the ascending aorta, the arch of the aorta and the descending (thoracic) aorta. The various sections of the aorta are named for their shape or the positions in which they lie, and each has branches which carry blood to the tissues of the body.

The heart and great vessels

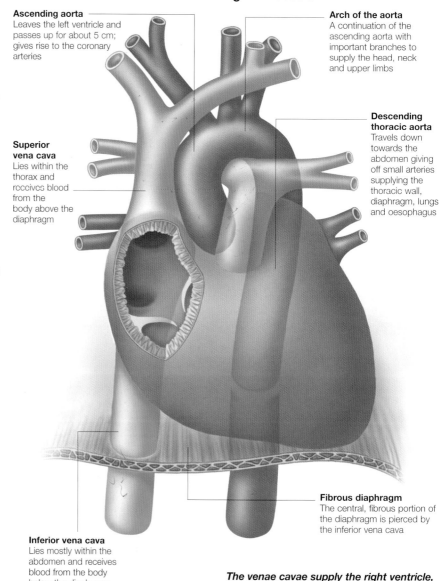

Ascending aorta
Leaves the left ventricle and passes up for about 5 cm; gives rise to the coronary arteries

Arch of the aorta
A continuation of the ascending aorta with important branches to supply the head, neck and upper limbs

Superior vena cava
Lies within the thorax and receives blood from the body above the diaphragm

Descending thoracic aorta
Travels down towards the abdomen giving off small arteries supplying the thoracic wall, diaphragm, lungs and oesophagus

Inferior vena cava
Lies mostly within the abdomen and receives blood from the body below the diaphragm

Fibrous diaphragm
The central, fibrous portion of the diaphragm is pierced by the inferior vena cava

The venae cavae supply the right ventricle, which pumps deoxygenated blood to the lungs. The left ventricle pumps oxygenated blood into the aorta.

Supplying blood to the heart

The heart muscle itself and the coverings of the heart need their own blood supply, which is provided by the coronary arteries.

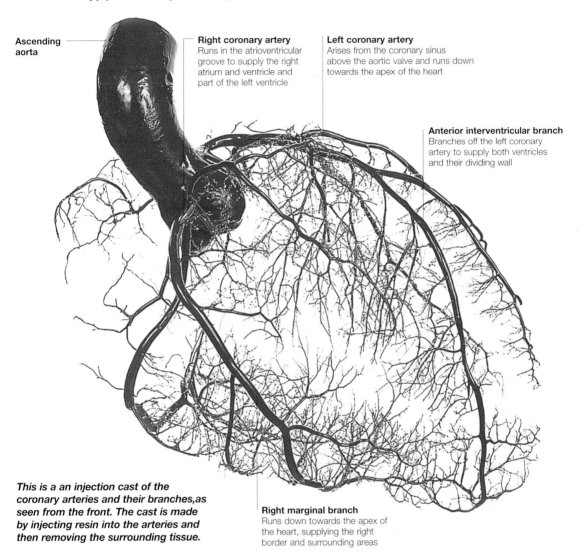

Ascending aorta

Right coronary artery
Runs in the atrioventricular groove to supply the right atrium and ventricle and part of the left ventricle

Left coronary artery
Arises from the coronary sinus above the aortic valve and runs down towards the apex of the heart

Anterior interventricular branch
Branches off the left coronary artery to supply both ventricles and their dividing wall

This is a an injection cast of the coronary arteries and their branches,as seen from the front. The cast is made by injecting resin into the arteries and then removing the surrounding tissue.

Right marginal branch
Runs down towards the apex of the heart, supplying the right border and surrounding areas

There are two coronary arteries: right and left. These arise from the ascending aorta just above the aortic valve and run around the heart just beneath the epicardium, embedded in fat.
• The right coronary artery This arises within the right aortic sinus, a small outpouching of the arterial wall just behind the aortic valve. It runs down and to the right, along the groove between the right atrium and the right ventricle until it lies along the inferior surface of the heart. Here it terminates in an anastomosis (connecting network) with branches of the left coronary artery. The right coronary artery gives off several branches.
• The left coronary artery This arises from the coronary sinus above the aortic valve and runs down towards the apex of the heart. The left coronary artery divides early on into two branches.

VENOUS DRAINAGE
The main vein of the heart is the coronary sinus. It receives blood from the cardiac veins and empties into the right atrium. In general, cardiac veins follow the routes of the coronary arteries.

121

Conducting system of the heart

When the body is at rest, the heart beats at a rate of about 70 to 80 beats per minute. Within its muscular walls, a conducting system sets the pace and ensures that the muscle contracts in a co-ordinated way.

SINO-ATRIAL NODE

The sino-atrial (SA) node is a collection of cells within the wall of the right atrium.

Each contraction of the cells of the SA node generates an electrical impulse, which is passed to the other muscle cells of the right and left atria and then to the atrio-ventricular (AV) node.

ATRIOVENTRICULAR NODE

The cells of the AV node will initiate contractions of their own, and pass on impulses at a slower rate, if not stimulated by the SA node. Impulses from the AV node are passed to the ventricles through the next stage of conducting tissue.

ATRIOVENTRICULAR BUNDLE

The AV bundle passes from the atria to the ventricles through an insulating layer of fibrous tissue. It then divides into two parts, the right and left bundle branches, which supply the right and left ventricles respectively.

The intrinsic conduction system of the heart carries a wave of nerve impulses, which create synchronized contraction of the heart muscle.

The intrinsic conduction system of the heart

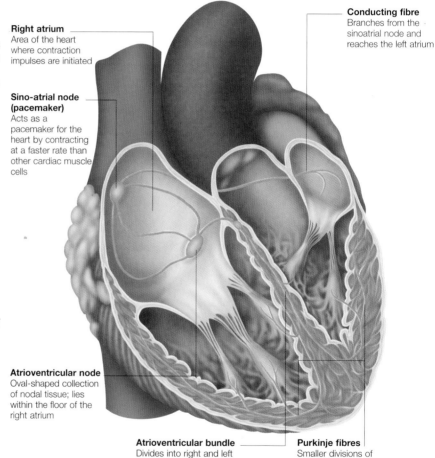

Conducting fibre
Branches from the sinoatrial node and reaches the left atrium

Right atrium
Area of the heart where contraction impulses are initiated

Sino-atrial node (pacemaker)
Acts as a pacemaker for the heart by contracting at a faster rate than other cardiac muscle cells

Atrioventricular node
Oval-shaped collection of nodal tissue; lies within the floor of the right atrium

Atrioventricular bundle
Divides into right and left bundle branches, which supply first the papillary muscles, and then the rest of the ventricular myocardium

Purkinje fibres
Smaller divisions of the bundle branches

The cardiac cycle

The cardiac cycle is the series of changes within the heart which causes blood to be pumped around the body. It is divided into a period when heart muscle contracts, known as systole and a period when it is relaxed, known as diastole.

VENTRICULAR FILLING
During diastole the tricuspid and mitral valves are open. Blood from the great veins fills the atria and then passes through these open valves to fill the relaxing ventricles.

ATRIAL CONTRACTION
As diastole ends and systole begins, the SA node sparks off a contraction of the atrial muscle which forces more blood into the ventricles.

VENTRICULAR CONTRACTION
The wave of contraction reaches the ventricles via the AV bundles and the Purkinje fibres. The tricuspid and mitral valves snap shut as pressure increases. The blood pushes against the closed pulmonary and aortic valves and causes them to open.

As the wave of contraction dies away, the ventricles relax. The cycle begins again with the next SA node impulse about a second later.

The movements of the heart cause the circulation of blood. The sequence of contraction is repeated, in normal adults, about 70 to 90 times a minute.

Events of the cardiac cycle

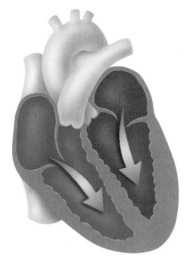

Ventricular filling **1**
The heart muscle is relaxed allowing blood to fill the chambers

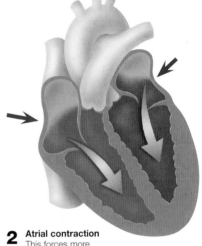

2 Atrial contraction
This forces more blood into the ventricles, filling them to capacity

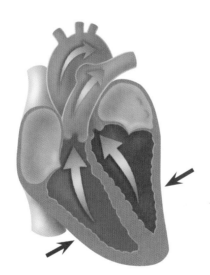

3 Ventricular contraction
Pulmonary and aortic valves open letting blood pass up and out into the pulmonary trunk and the aorta

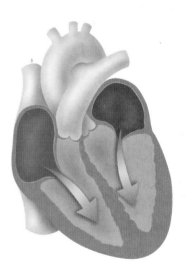

4 Ventricular filling
As the wave of contraction dies away the ventricles relax and allow blood to enter again

Shoulder joint

The glenohumeral, or shoulder joint, is a ball-and-socket joint at the point of articulation of the humerus and the scapula. The construction of this joint allows the arm a wide range of movement.

The glenohumeral, or shoulder joint, is the point of articulation between the glenoid cavity of the scapula (shoulder blade) and the head of the humerus (bone of the upper arm). It is a ball-and-socket synovial (fluid-filled) joint constructed to allow the upper limb a very wide range of movement.

ARTICULAR SURFACE

To permit a wide range of movement, the head of the humerus provides a large articular surface. The glenoid cavity of the scapula, deepened by a ring of tough fibrocartilage (the glenoid labrum), offers only a shallow socket. The resulting ball-and-socket is so shallow that the joint needs to be held firmly together by the surrounding muscles and ligaments.

A thin layer of smooth articular (or hyaline) cartilage allows the bones to slip over each other with minimum friction.

JOINT CAPSULE

The shoulder joint is surrounded by a loose capsule of fibrous tissue. This capsule is lined by the synovial membrane which covers all the inner surfaces of the joint except those covered with articular cartilage.

The cells of this synovial membrane secrete synovial fluid, a viscous liquid which lubricates and also nourishes the joint.

Shoulder joint viewed from the front

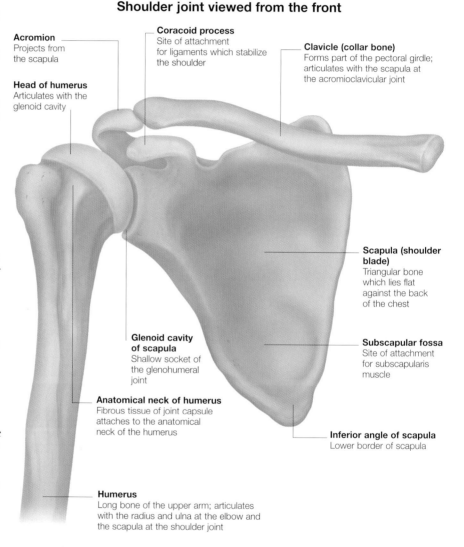

Acromion
Projects from the scapula

Head of humerus
Articulates with the glenoid cavity

Coracoid process
Site of attachment for ligaments which stabilize the shoulder

Clavicle (collar bone)
Forms part of the pectoral girdle; articulates with the scapula at the acromioclavicular joint

Scapula (shoulder blade)
Triangular bone which lies flat against the back of the chest

Subscapular fossa
Site of attachment for subscapularis muscle

Inferior angle of scapula
Lower border of scapula

Glenoid cavity of scapula
Shallow socket of the glenohumeral joint

Anatomical neck of humerus
Fibrous tissue of joint capsule attaches to the anatomical neck of the humerus

Humerus
Long bone of the upper arm; articulates with the radius and ulna at the elbow and the scapula at the shoulder joint

Ligaments of the shoulder joint

The ligaments of the shoulder joint, along with the surrounding muscles, are crucial for the stability of this shallow ball-and-socket joint.

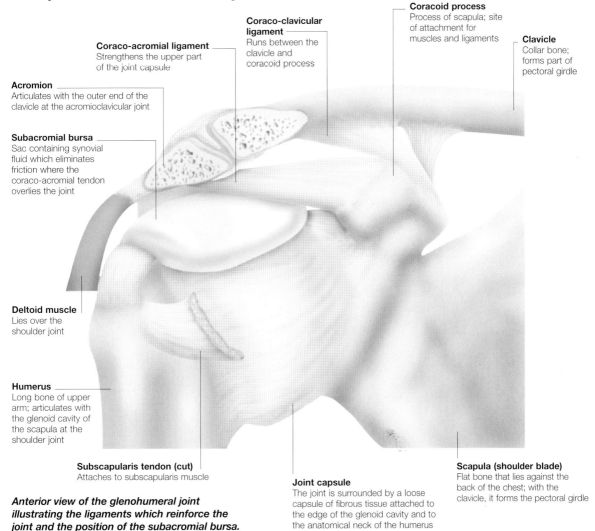

Coracoid process
Process of scapula; site of attachment for muscles and ligaments

Coraco-clavicular ligament
Runs between the clavicle and coracoid process

Coraco-acromial ligament
Strengthens the upper part of the joint capsule

Clavicle
Collar bone; forms part of pectoral girdle

Acromion
Articulates with the outer end of the clavicle at the acromioclavicular joint

Subacromial bursa
Sac containing synovial fluid which eliminates friction where the coraco-acromial tendon overlies the joint

Deltoid muscle
Lies over the shoulder joint

Humerus
Long bone of upper arm; articulates with the glenoid cavity of the scapula at the shoulder joint

Subscapularis tendon (cut)
Attaches to subscapularis muscle

Joint capsule
The joint is surrounded by a loose capsule of fibrous tissue attached to the edge of the glenoid cavity and to the anatomical neck of the humerus

Scapula (shoulder blade)
Flat bone that lies against the back of the chest; with the clavicle, it forms the pectoral girdle

Anterior view of the glenohumeral joint illustrating the ligaments which reinforce the joint and the position of the subacromial bursa.

The ligaments around any joint contribute to its stability by holding the bones firmly together. In the shoulder joint, the main stabilizers are the surrounding muscles, but ligaments also play a role.

STABILIZING LIGAMENTS
The fibrous joint capsule has ligaments within it which help to strengthen the joint:
- The glenohumeral ligaments are three weak, fibrous bands which reinforce the front of the capsule
- The coracohumeral ligament is a strong, broad band which strengthens the upper aspect of the capsule. Although not actually part of the glenohumeral joint itself, the coraco-acromial ligament is important as it spans the gap between the acromion and the coracoid process of the scapula. The arch of bone and ligament is so strong that even if the humerus is forcibly pushed up, it will not break; the clavicle or the humerus will give way first
- The transverse humeral ligament runs from the greater to the lesser tuberosity of the humerus, creating a tunnel for the passage of the biceps brachii tendon in its synovial sheath.

125

Movements of the shoulder joint

The shoulder joint is a ball-and-socket joint which allows 360° of movement to give maximum flexibility. In addition to enabling these movements, the muscles of the pectoral girdle add stability.

Front view of muscles of the shoulder

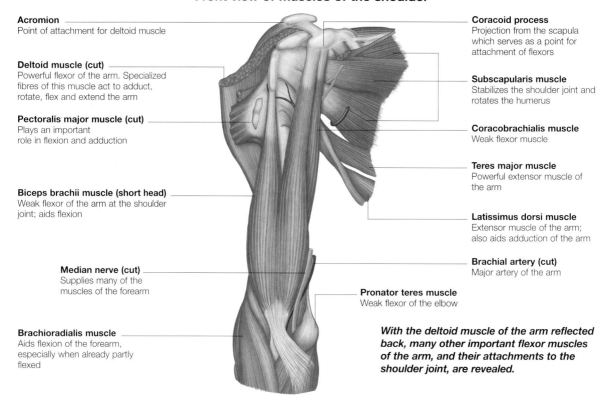

Acromion
Point of attachment for deltoid muscle

Deltoid muscle (cut)
Powerful flexor of the arm. Specialized fibres of this muscle act to adduct, rotate, flex and extend the arm

Pectoralis major muscle (cut)
Plays an important role in flexion and adduction

Biceps brachii muscle (short head)
Weak flexor of the arm at the shoulder joint; aids flexion

Median nerve (cut)
Supplies many of the muscles of the forearm

Brachioradialis muscle
Aids flexion of the forearm, especially when already partly flexed

Coracoid process
Projection from the scapula which serves as a point for attachment of flexors

Subscapularis muscle
Stabilizes the shoulder joint and rotates the humerus

Coracobrachialis muscle
Weak flexor muscle

Teres major muscle
Powerful extensor muscle of the arm

Latissimus dorsi muscle
Extensor muscle of the arm; also aids adduction of the arm

Brachial artery (cut)
Major artery of the arm

Pronator teres muscle
Weak flexor of the elbow

With the deltoid muscle of the arm reflected back, many other important flexor muscles of the arm, and their attachments to the shoulder joint, are revealed.

The movements of the shoulder joint take place around three axes: a horizontal axis through the centre of the glenoid fossa; axis perpendicular to this (front-back) through the humeral head; and a third axis running vertically through the shaft of the humerus. These give the axes of flexion and extension, adduction (movement towards the body) and abduction (movement away from the body), and medial (internal) and lateral (external) rotation respectively. A combination of these movements can allow a circular motion of the limb called circumduction.

MUSCLES OF SHOULDER MOVEMENT
Many of the muscles involved in these movements are attached to the pectoral girdle (the clavicles and scapulae). The scapula has muscles attached to its rear and front surfaces and the coracoid process, a bony projection. Some muscles arise directly from the trunk (pectoralis major and latissimus dorsi). Other muscles influence the movement of the humerus even though they are not attached to it directly (such as trapezius). They do this by moving the scapula, and hence the shoulder joint.

Rotation of the arm and 'rotator cuff'

The rotator cuff muscles include subscapularis, supraspinatus, infraspinatus and teres minor. These muscles act to strengthen and increase the stability of the shoulder joint. They also act individually to move the humerus and upper arm.

Muscles of shoulder movement (front)   Muscles of shoulder movement (back)

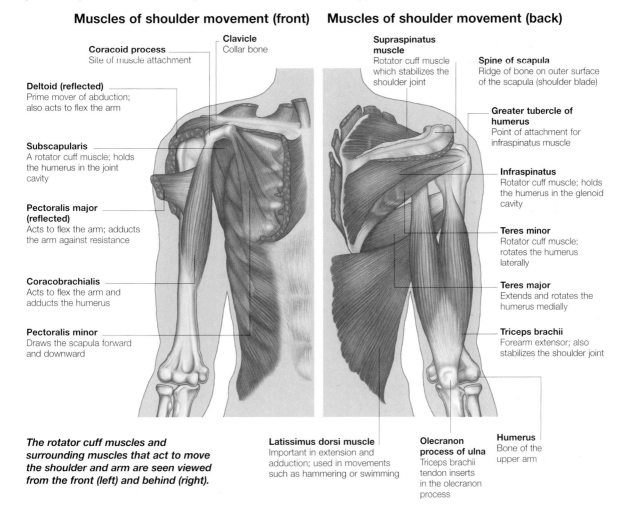

Coracoid process
Site of muscle attachment

Clavicle
Collar bone

Deltoid (reflected)
Prime mover of abduction; also acts to flex the arm

Subscapularis
A rotator cuff muscle; holds the humerus in the joint cavity

Pectoralis major (reflected)
Acts to flex the arm; adducts the arm against resistance

Coracobrachialis
Acts to flex the arm and adducts the humerus

Pectoralis minor
Draws the scapula forward and downward

Supraspinatus muscle
Rotator cuff muscle which stabilizes the shoulder joint

Spine of scapula
Ridge of bone on outer surface of the scapula (shoulder blade)

Greater tubercle of humerus
Point of attachment for infraspinatus muscle

Infraspinatus
Rotator cuff muscle; holds the humerus in the glenoid cavity

Teres minor
Rotator cuff muscle; rotates the humerus laterally

Teres major
Extends and rotates the humerus medially

Triceps brachii
Forearm extensor; also stabilizes the shoulder joint

The rotator cuff muscles and surrounding muscles that act to move the shoulder and arm are seen viewed from the front (left) and behind (right).

Latissimus dorsi muscle
Important in extension and adduction; used in movements such as hammering or swimming

Olecranon process of ulna
Triceps brachii tendon inserts in the olecranon process

Humerus
Bone of the upper arm

The pectoralis major, anterior fibres of deltoid, teres major and latissimus dorsi muscles also cause medial rotation of the humerus. The most powerful medial rotator, however, is subscapularis. This muscle occupies the entire front surface of the scapula, and attaches to the joint capsule around the lesser tuberosity of the humerus.

ROTATOR CUFF
Subscapularis is one of a set of four short muscles, collectively called the 'rotator cuff', which attach to and strengthen the joint capsule. In addition, they pull the humerus into the socket of the joint (glenoid fossa), increasing contact of the bony elements. This is the most important factor contributing to the

stability of the joint.
The other muscles of the group are supraspinatus, infraspinatus and teres minor. These latter three muscles attach to the three facets on the greater tuberosity of the humerus. Infraspinatus and teres minor are lateral rotators of the shoulder joint, together with the posterior fibres of the deltoid.

Injury to the rotator cuff muscles is disabling, because the stability of the humerus in the joint is lost. The other muscles of the arm lose the ability to move the humerus correctly, resulting in dislocation of the joint.

Axilla

The axilla, or armpit, is a roughly pyramidal space where the upper arm joins the thorax. It contains a number of important structures, such as blood vessels and nerves passing to and from the upper limb.

Front view of the shoulder showing the structures of the axilla

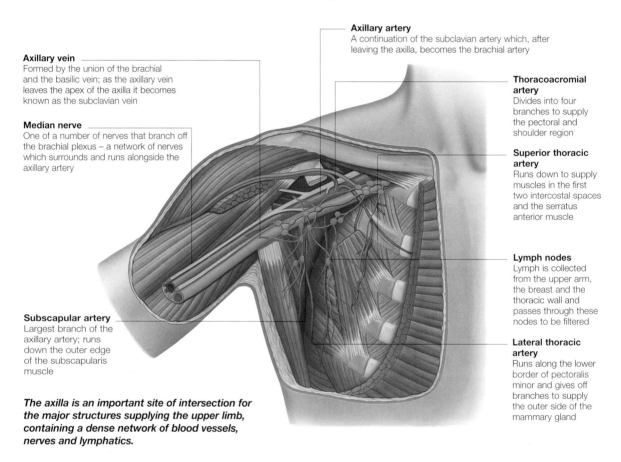

Axillary artery
A continuation of the subclavian artery which, after leaving the axilla, becomes the brachial artery

Axillary vein
Formed by the union of the brachial and the basilic vein; as the axillary vein leaves the apex of the axilla it becomes known as the subclavian vein

Median nerve
One of a number of nerves that branch off the brachial plexus – a network of nerves which surrounds and runs alongside the axillary artery

Thoracoacromial artery
Divides into four branches to supply the pectoral and shoulder region

Superior thoracic artery
Runs down to supply muscles in the first two intercostal spaces and the serratus anterior muscle

Lymph nodes
Lymph is collected from the upper arm, the breast and the thoracic wall and passes through these nodes to be filtered

Subscapular artery
Largest branch of the axillary artery; runs down the outer edge of the subscapularis muscle

Lateral thoracic artery
Runs along the lower border of pectoralis minor and gives off branches to supply the outer side of the mammary gland

The axilla is an important site of intersection for the major structures supplying the upper limb, containing a dense network of blood vessels, nerves and lymphatics.

Vessels, nerves and lymphatics serving the upper limb all pass through the axilla. The structures lie embedded in fatty connective tissue, which occupies the axillary space.

THE AXILLARY ARTERY
The axillary artery and its branches supply oxygenated blood to the upper limb.

As it passes through the axilla this artery gives off several branches which supply the surrounding structures of the shoulder and pectoral regions.

THE AXILLARY VEIN
The axillary vein runs through the axilla on the medial side of the axillary artery. The pattern of veins

and venous drainage varies but the axillary vein, in general, receives blood from tributary veins which correspond to the branches of the axillary artery.

NERVES IN THE AXILLA
The nerves which lie in the axilla are part of a complex network the 'brachial plexus'.

LYMPHATICS
Within the fatty connective tissue of the axilla lie a series of groups of lymph nodes which are connected by lymphatic vessels. Lymph nodes are scattered throughout the fat of the axilla.

The clavipectoral fascia

This is a sheet of strong connective tissue, which is attached at its upper border to the coracoid process of the scapula and the clavicle.

Front view of the axilla showing the clavipectoral fascia

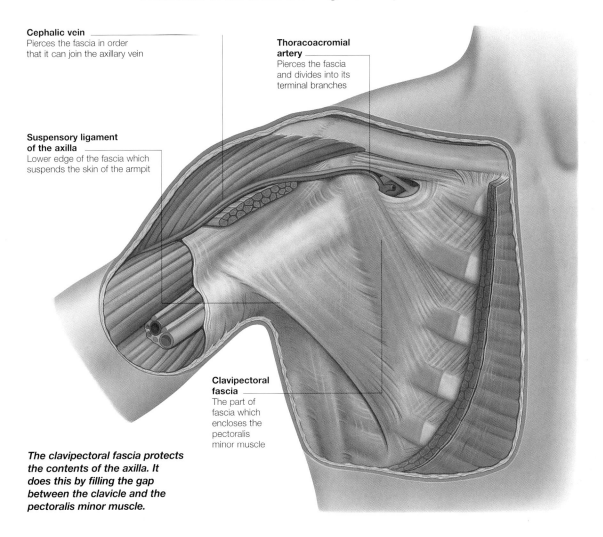

Cephalic vein
Pierces the fascia in order that it can join the axillary vein

Thoracoacromial artery
Pierces the fascia and divides into its terminal branches

Suspensory ligament of the axilla
Lower edge of the fascia which suspends the skin of the armpit

Clavipectoral fascia
The part of fascia which encloses the pectoralis minor muscle

The clavipectoral fascia protects the contents of the axilla. It does this by filling the gap between the clavicle and the pectoralis minor muscle.

The clavipectoral fascia descends to enclose the subclavius muscle and the pectoralis minor muscle and then joins with the overlying axillary fascia in the base of the axilla.

The part of the clavipectoral fascia that lies above the pectoralis minor muscle is known as the 'costocoracoid membrane' and is pierced by the nerve which supplies the overlying pectoralis minor muscle.

Below the pectoralis minor muscle, the fascia becomes the 'suspensory ligament of the axilla', which attaches to the skin of the armpit and is responsible for pulling that skin up when the arm is raised.

The clavipectoral fascia is continuous with the brachial fascia, which envelops the arm like a sleeve.

The fascia is pierced by a number of veins, arteries and nerves. These are the cephalic vein, the thoracoacromial artery (a branch of the axillary artery) and the lateral pectoral nerve.

Structure of the humerus

The humerus, a typical 'long bone', is found in the upper arm. It has a long shaft with expanded ends that connect with the scapula at the shoulder joint and the radius and ulna at the elbow.

At the top of the humerus (the proximal end) lies the smooth, hemispherical head that fits into the glenoid cavity of the scapula at the shoulder joint. Behind the head is a shallow constriction known as the 'anatomical neck' of the humerus, which separates the head from two bony prominences, the greater and the lesser tuberosities. These are sites for muscle attachment and are separated by the intertubercular (or bicipital) groove.

THE SHAFT

At the upper end of the shaft is the slightly narrowed 'surgical neck' of the humerus – a common site for fractures. The relatively smooth shaft has two distinctive features. About half way down the shaft, on the lateral (outer) side, lies the deltoid tuberosity, a raised site of attachment of the deltoid muscle. The second feature is the radial (or spiral) groove which runs across the back of the middle part of the shaft. This depression marks the path of the radial nerve and the profunda brachii artery.

Ridges at each side of the lower shaft pass down to end in the prominent medial (inner) and lateral epicondyles. There are two main parts to the articular surface: the trochlea, which articulates with the ulna; and the capitulum, which articulates with the radius.

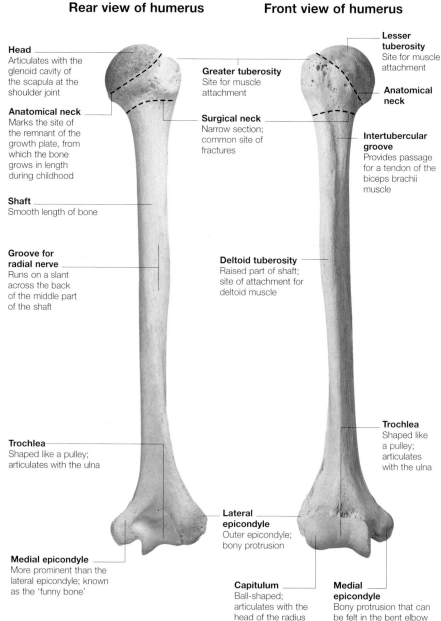

Rear view of humerus

Head
Articulates with the glenoid cavity of the scapula at the shoulder joint

Anatomical neck
Marks the site of the remnant of the growth plate, from which the bone grows in length during childhood

Shaft
Smooth length of bone

Groove for radial nerve
Runs on a slant across the back of the middle part of the shaft

Trochlea
Shaped like a pulley; articulates with the ulna

Medial epicondyle
More prominent than the lateral epicondyle; known as the 'funny bone'

Greater tuberosity
Site for muscle attachment

Surgical neck
Narrow section; common site of fractures

Lateral epicondyle
Outer epicondyle; bony protrusion

Capitulum
Ball-shaped; articulates with the head of the radius

Front view of humerus

Lesser tuberosity
Site for muscle attachment

Anatomical neck

Intertubercular groove
Provides passage for a tendon of the biceps brachii muscle

Deltoid tuberosity
Raised part of shaft; site of attachment for deltoid muscle

Trochlea
Shaped like a pulley; articulates with the ulna

Medial epicondyle
Bony protrusion that can be felt in the bent elbow

Inside the humerus

The structure of the humerus is typical of the long bones. The bone is divided into the diaphysis (shaft) and the epiphysis (head) at either end.

Long bones are elongated in shape and longer than they are wide. Most of the bones of the limbs are long bones, even the small bones of the fingers, and as such they have many features in common with the humerus.

The humerus consists of a diaphysis, or shaft, with an epiphysis (expanded head) at each end. The diaphysis is of tubular construction with an outer layer of dense, thick bone surrounding a central medulla (inner region) containing fat cells. The epiphyses of the humerus are, at the upper end, the head and at the lower end the condylar region. These are composed of a thin layer of compact bone covering cancellous (spongy) bone which makes up the greater volume.

BONE SURFACE

The surface of the humerus (and all long bones) is covered by a thick membrane, the periosteum. The articular surfaces at the joints are the only parts of the bone not covered by the periosteum. These surfaces are covered by tough articular (or hyaline) cartilage which is smooth, allowing the bones to glide over each other.

The outer compact bone receives its blood supply from the arteries of the periosteum, and will die if that periosteum is stripped off, the inner parts of the bone are supplied by nutrient arteries which pierce the compact bone.

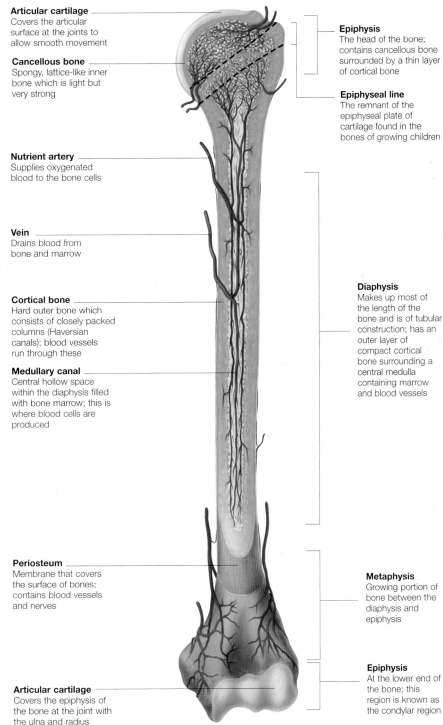

Articular cartilage
Covers the articular surface at the joints to allow smooth movement

Cancellous bone
Spongy, lattice-like inner bone which is light but very strong

Nutrient artery
Supplies oxygenated blood to the bone cells

Vein
Drains blood from bone and marrow

Cortical bone
Hard outer bone which consists of closely packed columns (Haversian canals); blood vessels run through these

Medullary canal
Central hollow space within the diaphysis filled with bone marrow; this is where blood cells are produced

Periosteum
Membrane that covers the surface of bones; contains blood vessels and nerves

Articular cartilage
Covers the epiphysis of the bone at the joint with the ulna and radius

Epiphysis
The head of the bone; contains cancellous bone surrounded by a thin layer of cortical bone

Epiphyseal line
The remnant of the epiphyseal plate of cartilage found in the bones of growing children

Diaphysis
Makes up most of the length of the bone and is of tubular construction; has an outer layer of compact cortical bone surrounding a central medulla containing marrow and blood vessels

Metaphysis
Growing portion of bone between the diaphysis and epiphysis

Epiphysis
At the lower end of the bone; this region is known as the condylar region

Ulna and radius

The ulna and the radius are the long bones of the forearm. They articulate with the humerus and the wrist bones and are uniquely adapted to enable rotation of the hand and forearm.

The ulna and radius are the two parallel long bones of the forearm and lie between the elbow and wrist joints. The ulna lies on the same side as the little finger (medially), while the radius lies on the same side as the thumb (laterally).

The radio-ulnar joints allow the ulna and radius to rotate around each other in the movements peculiar to the forearm known as 'pronation' (rotating the forearm so that the palm faces down), and 'supination' (rotating the forearm so that the palm faces up).

THE ULNA
The ulna is longer than the radius and is the main stabilizing bone of the forearm. It has a long shaft with two expanded ends. The upper end of the ulna has two prominent projections, the olecranon and the coronoid process, which are separated by the deep trochlear notch, which articulates with the trochlea of the humerus.

On the lateral (outer) side of the coronoid process, there is a small, rounded recess (the radial notch), which is the site of articulation of the upper end of the ulna with the neighbouring head of the radius. The head of the ulna is separated from the wrist joint by an articular disc and does not play much part in the wrist joint itself.

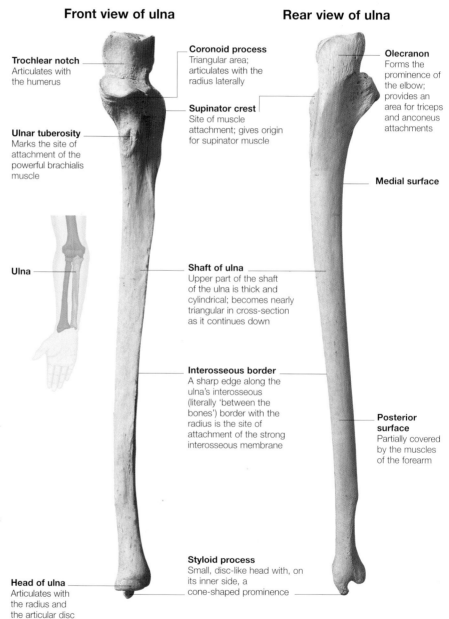

Front view of ulna

Rear view of ulna

Trochlear notch
Articulates with the humerus

Ulnar tuberosity
Marks the site of attachment of the powerful brachialis muscle

Coronoid process
Triangular area; articulates with the radius laterally

Supinator crest
Site of muscle attachment; gives origin for supinator muscle

Olecranon
Forms the prominence of the elbow; provides an area for triceps and anconeus attachments

Medial surface

Ulna

Shaft of ulna
Upper part of the shaft of the ulna is thick and cylindrical; becomes nearly triangular in cross-section as it continues down

Interosseous border
A sharp edge along the ulna's interosseous (literally 'between the bones') border with the radius is the site of attachment of the strong interosseous membrane

Posterior surface
Partially covered by the muscles of the forearm

Head of ulna
Articulates with the radius and the articular disc

Styloid process
Small, disc-like head with, on its inner side, a cone-shaped prominence

The radius

The radius is the shorter of the two bones of the forearm and articulates with the wrist. It is joined firmly to the ulna by a tough layer of connective tissue.

Like the ulna, the radius has a long shaft with upper and lower expanded ends. While the ulna is the forearm bone which contributes most to the elbow, the radius plays a major part in the wrist joint.

HEAD OF THE RADIUS
The disc-like head of the radius is concave above, where it articulates with the capitulum of the humerus in the elbow joint. The cartilage that covers this concavity continues down over the head, especially on the side nearest the ulna, to allow the smooth articulation of the head of the radius with the radial notch at the upper end of the ulna.

THE SHAFT
The shaft of the radius becomes progressively thicker as it continues down to the wrist. It also has a sharp edge for attachment of the interosseous membrane. On the inner side, next to the ulna, there is a concavity (the ulnar notch), which is the site for articulation with the head of the ulna.

Extending from the opposite side is the radial styloid process, a blunt cone which projects a little further down than the ulnar styloid process. At the back of the end of the radius, and easily felt at the back of the wrist, is the dorsal tubercle.

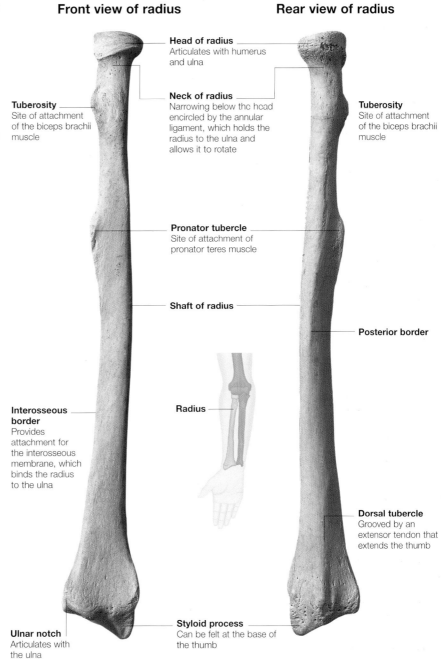

Front view of radius

Rear view of radius

Head of radius
Articulates with humerus and ulna

Neck of radius
Narrowing below the head encircled by the annular ligament, which holds the radius to the ulna and allows it to rotate

Tuberosity
Site of attachment of the biceps brachii muscle

Tuberosity
Site of attachment of the biceps brachii muscle

Pronator tubercle
Site of attachment of pronator teres muscle

Shaft of radius

Posterior border

Interosseous border
Provides attachment for the interosseous membrane, which binds the radius to the ulna

Radius

Dorsal tubercle
Grooved by an extensor tendon that extends the thumb

Ulnar notch
Articulates with the ulna

Styloid process
Can be felt at the base of the thumb

133

Elbow

The elbow is the fluid-filled joint where the humerus of the upper arm and the radius and ulna of the forearm articulate. The joint structure only allows hinge-like movement but is extremely stable.

View of right elbow from front **View of right elbow from behind**

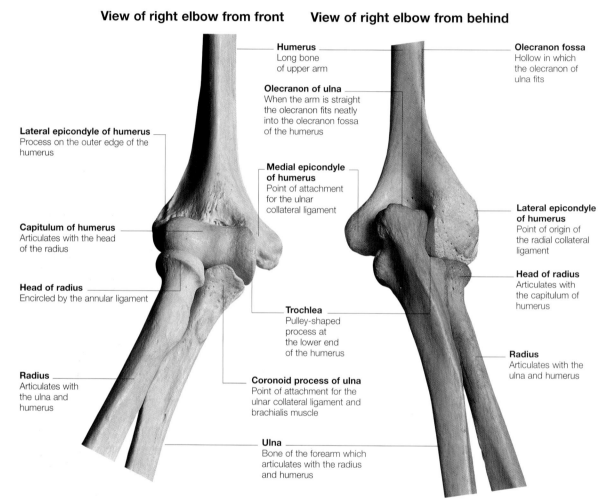

Humerus
Long bone
of upper arm

Olecranon fossa
Hollow in which
the olecranon of
ulna fits

Olecranon of ulna
When the arm is straight
the olecranon fits neatly
into the olecranon fossa
of the humerus

Lateral epicondyle of humerus
Process on the outer edge of the
humerus

**Medial epicondyle
of humerus**
Point of attachment
for the ulnar
collateral ligament

**Lateral epicondyle
of humerus**
Point of origin of
the radial collateral
ligament

Capitulum of humerus
Articulates with the head
of the radius

Head of radius
Articulates with
the capitulum of
humerus

Head of radius
Encircled by the annular ligament

Trochlea
Pulley-shaped
process at
the lower end
of the humerus

Radius
Articulates with the
ulna and humerus

Radius
Articulates with
the ulna and
humerus

Coronoid process of ulna
Point of attachment for the
ulnar collateral ligament and
brachialis muscle

Ulna
Bone of the forearm which
articulates with the radius
and humerus

The elbow is a synovial (fluid-filled) joint between the lower end of the humerus and the upper ends of the ulna and radius. It is the best example of a 'hinge' joint, where the only movements permitted are flexion (bending) and extension (straightening). Its structure gives the joint great stability.

STRUCTURE OF THE ELBOW
At the elbow, the pulley-shaped trochlea of the lower end of the humerus articulates with the deep trochlear notch of the ulna, while its hemispherical capitulum articulates with the head of the radius. All the opposing joint surfaces are covered by smooth

articular cartilage (hyaline) to reduce friction between the bony surfaces during movement.

The whole joint is surrounded by a fibrous capsule which extends down from the articular surfaces of the humerus to the upper end of the ulna. The capsule is loose at the

back of the elbow to allow flexion and extension. The capsule is lined with synovial membrane which secretes thick synovial fluid filling the joint cavity. This fluid nourishes the joint and acts as a lubricant. The joint cavity is continuous with that of the superior radioulnar joint below.

Ligaments of the elbow

The elbow is supported and strengthened at each side by the strong collateral ligaments. These are thickenings of the joint capsule.

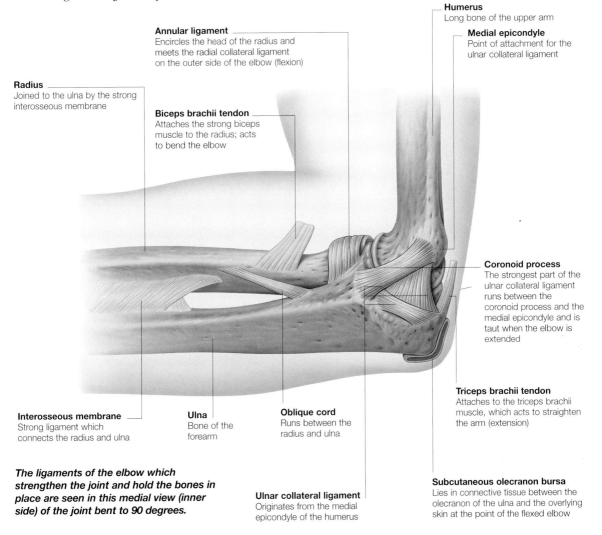

Humerus
Long bone of the upper arm

Medial epicondyle
Point of attachment for the ulnar collateral ligament

Annular ligament
Encircles the head of the radius and meets the radial collateral ligament on the outer side of the elbow (flexion)

Radius
Joined to the ulna by the strong interosseous membrane

Biceps brachii tendon
Attaches the strong biceps muscle to the radius; acts to bend the elbow

Coronoid process
The strongest part of the ulnar collateral ligament runs between the coronoid process and the medial epicondyle and is taut when the elbow is extended

Interosseous membrane
Strong ligament which connects the radius and ulna

Ulna
Bone of the forearm

Oblique cord
Runs between the radius and ulna

Triceps brachii tendon
Attaches to the triceps brachii muscle, which acts to straighten the arm (extension)

Ulnar collateral ligament
Originates from the medial epicondyle of the humerus

Subcutaneous olecranon bursa
Lies in connective tissue between the olecranon of the ulna and the overlying skin at the point of the flexed elbow

The ligaments of the elbow which strengthen the joint and hold the bones in place are seen in this medial view (inner side) of the joint bent to 90 degrees.

The radial collateral ligament is a fan-shaped ligament that originates from the lateral epicondyle – a bony prominence on the outer side of the lower end of the humerus – and runs down to blend with the annular ligament, which encircles the head of the radius. It is not attached to the radius itself so does not restrict movement of the radius during pronation (when the forearm is rotated so that the palm faces down) and supination (when the forearm is rotated so the palm faces up).

The ulnar collateral ligament runs between the medial (inner) epicondyle of the humerus and the upper end of the ulna and is in three parts, which form a rough triangle.

CARRYING ANGLE
When the arm is fully extended downwards with the palm facing forwards, the long axis of the forearm is not in line with the long axis of the upper arm, but deviates slightly outwards.

The angle so formed at the elbow is known as the 'carrying angle' and is greater in women than in men (by about 10 degrees), possibly to accommodate the wider hips of the female body. The carrying angle disappears when the forearm is pronated (turned so the palm faces in to the body).

Muscles of the upper arm

The musculature of the upper arm is divided into two distinct compartments. The muscles of the anterior compartment act to flex the arm and the muscles of the posterior compartment extend it.

The muscles of the anterior (front) compartment of the upper arm are all flexors:

- Biceps brachii. This muscle arises from two heads, which join together to form the body of the muscle. The bulging body then tapers as it runs down to form the strong tendon of insertion. When the elbow is straight, biceps acts to flex the forearm. However, when the elbow is already bent the biceps muscle is a powerful supinator of the forearm, rotating the forearm so that the hand is palm up.
- Brachialis. This arises from the lower half of the anterior surface of the humerus and passes down to cover the front of the elbow joint, its tendon inserting the coronoid process and tuberosity of the ulna.

Brachialis is the main flexor muscle of the elbow, whatever the position of the forearm.

- Coracobrachialis. This muscle arises from the tip of the coracoid process of the scapula and runs down and outwards to insert into the inner surface of the humerus. This muscle helps to flex the upper arm at the shoulder and to pull it back into line with the body (adduction).

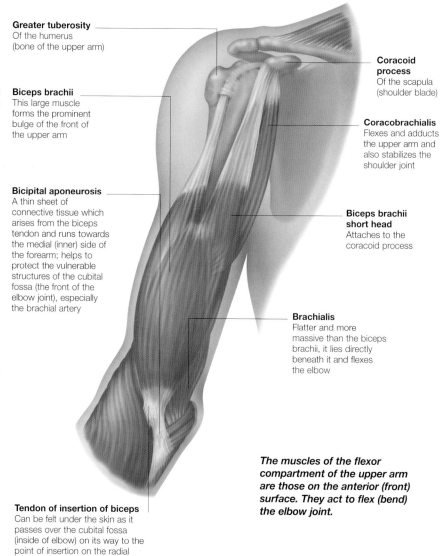

Greater tuberosity
Of the humerus
(bone of the upper arm)

Biceps brachii
This large muscle forms the prominent bulge of the front of the upper arm

Bicipital aponeurosis
A thin sheet of connective tissue which arises from the biceps tendon and runs towards the medial (inner) side of the forearm; helps to protect the vulnerable structures of the cubital fossa (the front of the elbow joint), especially the brachial artery

Coracoid process
Of the scapula
(shoulder blade)

Coracobrachialis
Flexes and adducts the upper arm and also stabilizes the shoulder joint

Biceps brachii short head
Attaches to the coracoid process

Brachialis
Flatter and more massive than the biceps brachii, it lies directly beneath it and flexes the elbow

The muscles of the flexor compartment of the upper arm are those on the anterior (front) surface. They act to flex (bend) the elbow joint.

Tendon of insertion of biceps
Can be felt under the skin as it passes over the cubital fossa (inside of elbow) on its way to the point of insertion on the radial tuberosity

Muscles of the posterior compartment

The muscles of the back of the upper arm act to extend the elbow, so straightening the forearm with the upper arm.

The posterior compartment has only one major muscle, the triceps brachii, which is a powerful extensor (straightens the arm). The only other muscle in this compartment is the small, relatively insignificant anconeus.

TRICEPS BRACHII

This is a large, bulky muscle which lies posterior to the humerus and, as its name implies, has three heads:
- The long head
- The lateral head
- The medial head.

The three heads converge in the middle of the upper arm on a wide, flattened tendon which passes down, over a small bursa, to attach to the olecranon process of the ulna.

The main action of triceps is to extend (straighten) the elbow joint. In addition, because of its position, the long head of the triceps muscle helps to stabilize the shoulder joint.

ANCONEUS

The small anconeus muscle lies behind and below the elbow joint and is triangular. As with the triceps, it extends the elbow and it also has a function in the stabilization of the elbow joint.

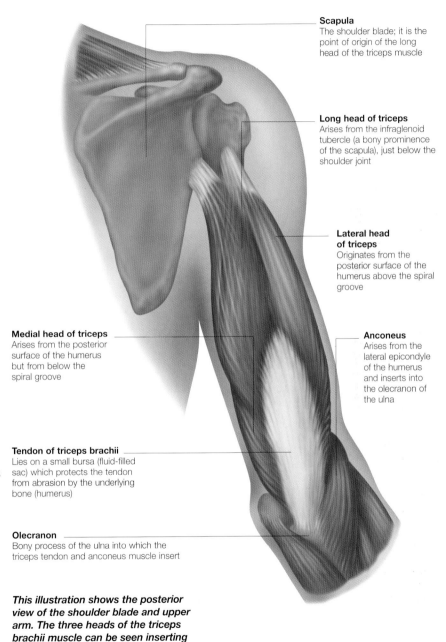

Scapula
The shoulder blade; it is the point of origin of the long head of the triceps muscle

Long head of triceps
Arises from the infraglenoid tubercle (a bony prominence of the scapula), just below the shoulder joint

Lateral head of triceps
Originates from the posterior surface of the humerus above the spiral groove

Medial head of triceps
Arises from the posterior surface of the humerus but from below the spiral groove

Anconeus
Arises from the lateral epicondyle of the humerus and inserts into the olecranon of the ulna

Tendon of triceps brachii
Lies on a small bursa (fluid-filled sac) which protects the tendon from abrasion by the underlying bone (humerus)

Olecranon
Bony process of the ulna into which the triceps tendon and anconeus muscle insert

This illustration shows the posterior view of the shoulder blade and upper arm. The three heads of the triceps brachii muscle can be seen inserting into a common tendon.

Muscles of the forearm

The flexor muscles of the front compartment of the forearm act to flex the hand, wrist and fingers. They are divided into superficial and deep muscles of the flexor and extensor compartments.

Superficial flexor muscles

Deep flexor muscles

This compartment, or section, of the forearm lies in the front of the forearm and contains muscles which flex the wrist and fingers as well as some which pronate the forearm (turn the hand palm down). They are sub-divided into superficial and deep layers according to position.

The superficial group contains five muscles which all originate at the medial epicondyle of the humerus, where their fibres merge to form the 'common flexor tendon':

- Pronator teres – pronates the forearm and flexes the elbow
- Flexor carpi radialis – acts to produce flexion and abduction (bending away from the midline of the body) of the wrist
- Palmaris longus – this small muscle is absent in 14 per cent of people; it acts to flex the wrist
- Flexor carpi ulnaris – this muscle flexes and adducts the wrist (bends away from the midline of the body); unlike the other muscles of the flexor compartment, this muscle is innervated by the ulnar nerve
- Flexor digitorum superficialis – this is the largest superficial muscle of the forearm and it acts, as its name suggests, to flex the fingers, or digits.

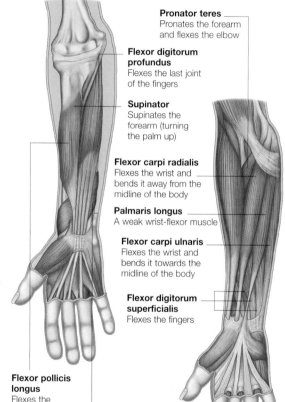

Deep flexor muscles

Superficial flexor muscles

Pronator teres
Pronates the forearm and flexes the elbow

Flexor digitorum profundus
Flexes the last joint of the fingers

Supinator
Supinates the forearm (turning the palm up)

Flexor carpi radialis
Flexes the wrist and bends it away from the midline of the body

Palmaris longus
A weak wrist-flexor muscle

Flexor carpi ulnaris
Flexes the wrist and bends it towards the midline of the body

Flexor digitorum superficialis
Flexes the fingers

Flexor pollicis longus
Flexes the thumb

Pronator quadratus
Pronates the forearm

The five main superficial flexor muscles of the forearm are shown in this illustration. These muscles originate from the humerus bone of the upper arm.

The deep flexor muscles lie close to the bones of the forearm (the ulna and radius). These act to flex the hand, wrist and fingers.

The deep layer of the flexor compartment consists of three muscles:

- Flexor digitorum profundus
This bulky muscle originates from a wide area of the ulna and neighbouring interosseous membrane (a strong sheet of tissue connecting the radius and ulna). It is the only muscle which flexes the last joint of the fingers and so acts with its more superficial counterpart to curl the fingers. Like the flexor digitorum superficialis muscle, this deeper muscle divides into four tendons, which pass through the carpal tunnel within the same synovial sheath. The tendons insert into the bases of the distal (far end) phalanges of the four fingers.
- Flexor pollicis longus
This muscle flexes the thumb. Its long, flat tendon passes through the carpal tunnel within its own synovial sheath and inserts into the base of the distal phalanx of the thumb (which, unlike the fingers, has only two phalanges).
- Pronator quadratus
The deepest muscle of the anterior compartment, the pronator quadratus acts to pronate the forearm and is the only muscle which attaches solely to the radius and ulna. It also assists the interosseous membrane in binding the radius and ulna tightly together.

Flexing the hand

The muscles of the forearm are divided into front and rear compartments. The front flexor muscles bend the wrist and fingers and the rear extensor muscles act to straighten them again.

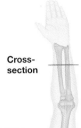

Cross-section

The muscles of the forearm are roughly divided into two groups, according to their function. These two groups are isolated from each other by the radius and ulna bones and by fascial layers (sheets of connective tissue) to form the 'anterior flexor compartment' and the 'posterior extensor compartment' of the forearm.

OPPOSING ACTIONS
The flexor muscles act to flex (bend) the wrist joint and the fingers, while the extensors act to extend (straighten) the same joints. Within these two groups are both deep and superficial muscles which act together to give the wide range of movements characteristic of the wrist and hand.

FOREARM TENDONS
So that the wrist and hand may move flexibly, the bulk of muscle around the lower end of the upper limb is kept to a minimum. This is achieved by using long tendons from muscles higher up in the forearm to work the wrist, and the fingers.

The muscles concerned are forearm muscles and need to be longer than the forearm will allow to work at maximum efficiency and thus many originate from the lower end of the humerus. The humerus has developed two projections called the medial (inner) and lateral (outer) epicondyles. The flexor muscles are attached to the medial epicondyle while the extensor muscles are attached to the lateral epicondyle.

Cross-section through the forearm

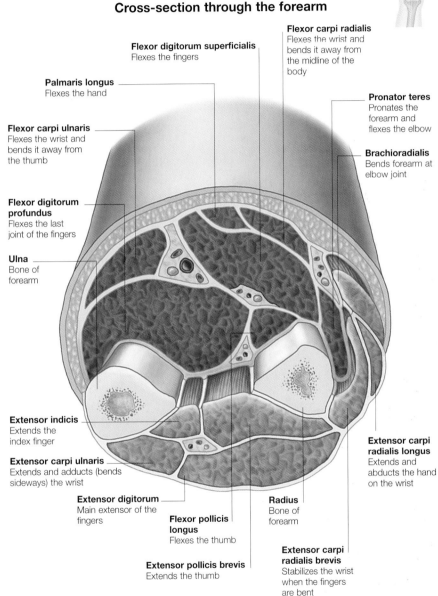

Flexor digitorum superficialis
Flexes the fingers

Flexor carpi radialis
Flexes the wrist and bends it away from the midline of the body

Palmaris longus
Flexes the hand

Pronator teres
Pronates the forearm and flexes the elbow

Flexor carpi ulnaris
Flexes the wrist and bends it away from the thumb

Brachioradialis
Bends forearm at elbow joint

Flexor digitorum profundus
Flexes the last joint of the fingers

Ulna
Bone of forearm

Extensor indicis
Extends the index finger

Extensor carpi ulnaris
Extends and adducts (bends sideways) the wrist

Extensor carpi radialis longus
Extends and abducts the hand on the wrist

Extensor digitorum
Main extensor of the fingers

Flexor pollicis longus
Flexes the thumb

Radius
Bone of forearm

Extensor pollicis brevis
Extends the thumb

Extensor carpi radialis brevis
Stabilizes the wrist when the fingers are bent

This illustration shows a cross-section through the forearm looking towards the hand with the palm upturned.

139

Blood vessels of the arm

The arteries of the arm supply blood to the soft tissues and bones. The main arteries divide to form many smaller vessels which communicate at networks – anastomoses – at the elbow and wrist.

The main blood supply to the arm is provided by the brachial artery, a continuation of the axillary artery, which runs down the inner side of the upper arm. It gives rise to many smaller branches that supply surrounding muscles and the humerus (upper bone of the arm). The largest of these is the profunda brachii artery, which supplies the muscles that straighten the elbow.

The profunda brachii artery and the other, smaller arteries given off by the lower part of the brachial artery run down around the elbow joint. They then form a network of connecting arteries before rejoining the main arteries of the forearm.

FOREARM AND HAND

The brachial artery divides below the elbow joint into the radial and the ulnar arteries. The radial artery runs from the cubital fossa along the length of the radius (bone of the forearm). At the lower end of the radius, it lies under the skin and connective tissue; pulsations can be felt here. The ulnar artery runs towards the base of the ulna (the other bone of the forearm).

The hand has a profuse blood supply from the end branches of the radial and ulnar arteries. Branches of the two arteries join together in the palm to form the deep and the superficial palmar arches, from which small arteries arise to supply the fingers.

Arteries of the arm

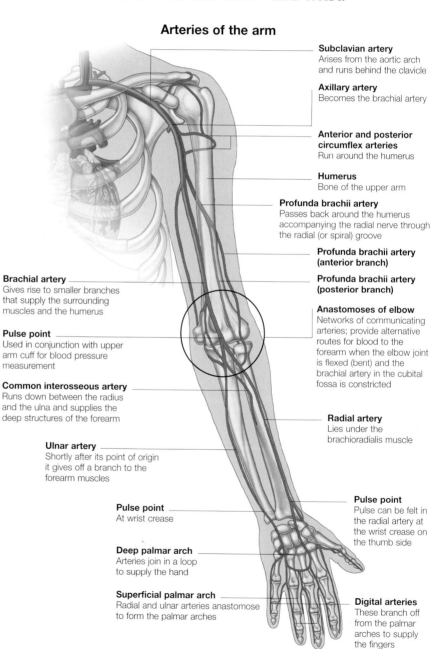

Subclavian artery
Arises from the aortic arch and runs behind the clavicle

Axillary artery
Becomes the brachial artery

Anterior and posterior circumflex arteries
Run around the humerus

Humerus
Bone of the upper arm

Profunda brachii artery
Passes back around the humerus accompanying the radial nerve through the radial (or spiral) groove

Profunda brachii artery (anterior branch)

Profunda brachii artery (posterior branch)

Anastomoses of elbow
Networks of communicating arteries; provide alternative routes for blood to the forearm when the elbow joint is flexed (bent) and the brachial artery in the cubital fossa is constricted

Radial artery
Lies under the brachioradialis muscle

Pulse point
Pulse can be felt in the radial artery at the wrist crease on the thumb side

Digital arteries
These branch off from the palmar arches to supply the fingers

Brachial artery
Gives rise to smaller branches that supply the surrounding muscles and the humerus

Pulse point
Used in conjunction with upper arm cuff for blood pressure measurement

Common interosseous artery
Runs down between the radius and the ulna and supplies the deep structures of the forearm

Ulnar artery
Shortly after its point of origin it gives off a branch to the forearm muscles

Pulse point
At wrist crease

Deep palmar arch
Arteries join in a loop to supply the hand

Superficial palmar arch
Radial and ulnar arteries anastomose to form the palmar arches

Veins of the arm

The veins of the upper limb are divided into deep and superficial veins. The superficial veins lie close to the skin's surface and are often easily visible.

Venous drainage of the upper limb is achieved by two interconnecting series of veins, the deep and the superficial systems. Deep veins run alongside the arteries, while superficial veins lie in the subcutaneous tissue. The layout of the veins is very variable but usually resembles the pattern detailed below.

DEEP VEINS

In most cases, the deep veins are paired or double veins (venae comitantes) that lie on either side of the artery they accompany, making frequent anastomoses and forming a network surrounding the artery. The pulsations of blood within the artery alternately compress and release the surrounding veins, helping blood return to the heart.

The radial and ulnar veins arise from the palmar venous arches of the hand and run up the forearm to merge at the elbow, forming the brachial vein. This, in turn, merges with the basilic vein to form the large axillary vein.

SUPERFICIAL VEINS

There are two main superficial veins of the arm, the cephalic and the basilic veins, which originate at the dorsal venous arch of the hand. The cephalic vein runs under the skin along the radial side of the forearm.

The basilic vein runs up the ulnar side of the forearm, crossing the elbow to lie along the border of the biceps muscle. About halfway up the upper arm it turns inwards to become a deep vein.

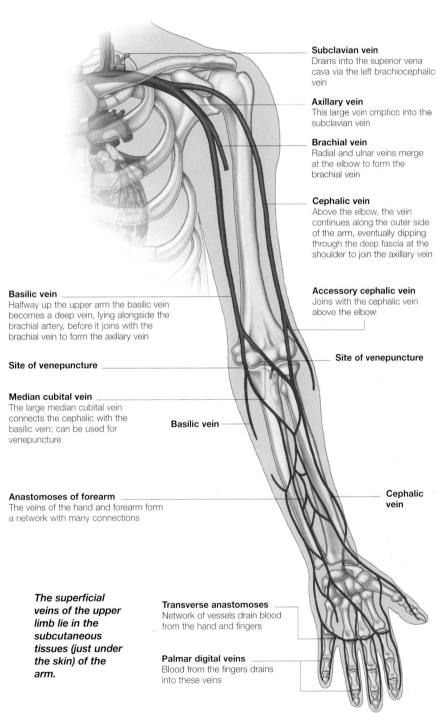

Subclavian vein
Drains into the superior vena cava via the left brachiocephalic vein

Axillary vein
This large vein empties into the subclavian vein

Brachial vein
Radial and ulnar veins merge at the elbow to form the brachial vein

Cephalic vein
Above the elbow, the vein continues along the outer side of the arm, eventually dipping through the deep fascia at the shoulder to join the axillary vein

Accessory cephalic vein
Joins with the cephalic vein above the elbow

Site of venepuncture

Basilic vein
Halfway up the upper arm the basilic vein becomes a deep vein, lying alongside the brachial artery, before it joins with the brachial vein to form the axillary vein

Site of venepuncture

Median cubital vein
The large median cubital vein connects the cephalic with the basilic vein; can be used for venepuncture

Basilic vein

Anastomoses of forearm
The veins of the hand and forearm form a network with many connections

Cephalic vein

The superficial veins of the upper limb lie in the subcutaneous tissues (just under the skin) of the arm.

Transverse anastomoses
Network of vessels drain blood from the hand and fingers

Palmar digital veins
Blood from the fingers drains into these veins

141

Nerves of the arm

The nerves of the arm supply the skin and muscles of the forearm and hand. There are four main nerves in the arm: the radial, musculocutaneous, median and ulnar nerves.

The nerve supply to the upper limb is provided by four main nerves and their branches. These receive sensory information from the hand and arm, and also innervate the numerous muscles of the upper limb. The radial and musculocutaneous nerves supply muscles and skin of all parts of the arm, while the median and ulnar nerves only supply structures below the elbow.

RADIAL NERVE

The radial nerve is of great importance as it is the main supplier of innervation to the extensor muscles which straighten the bent elbow, wrist and fingers. It arises as the largest branch of the 'brachial plexus', a network of nerves from the spinal cord in the neck.

Near the lateral epicondyle, the radial nerve divides into its two terminal branches:
• The superficial terminal branch – sensory nerve supply to the skin over the back of the hand, thumb and adjoining two and a half fingers
• The deep terminal branch – motor nerve supply to all of the extensor muscles of the forearm.

MUSCULOCUTANEOUS NERVE

The musculocutaneous nerve supplies both muscles and skin in the front of the upper arm. Below the elbow it becomes the lateral cutaneous nerve of the forearm, a sensory nerve which supplies a large area of forearm skin.

Rear view of the nerves of the arm

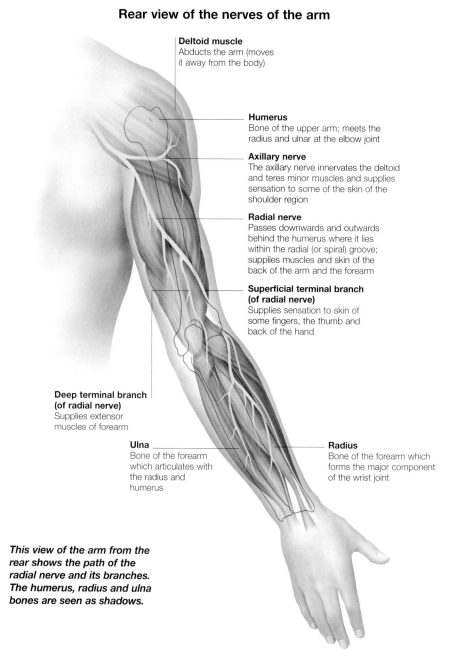

Deltoid muscle
Abducts the arm (moves it away from the body)

Humerus
Bone of the upper arm; meets the radius and ulnar at the elbow joint

Axillary nerve
The axillary nerve innervates the deltoid and teres minor muscles and supplies sensation to some of the skin of the shoulder region

Radial nerve
Passes downwards and outwards behind the humerus where it lies within the radial (or spiral) groove; supplies muscles and skin of the back of the arm and the forearm

Superficial terminal branch (of radial nerve)
Supplies sensation to skin of some fingers, the thumb and back of the hand

Deep terminal branch (of radial nerve)
Supplies extensor muscles of forearm

Ulna
Bone of the forearm which articulates with the radius and humerus

Radius
Bone of the forearm which forms the major component of the wrist joint

This view of the arm from the rear shows the path of the radial nerve and its branches. The humerus, radius and ulna bones are seen as shadows.

Median and ulnar nerves

The median nerve supplies the forearm muscles enabling the actions of flexion and pronation. The ulnar nerve passes behind the elbow – where it may be felt if the 'funny bone' is knocked – to supply some of the small muscles of the hand.

The median nerve of the upper limb arises from the brachial plexus and runs downwards centrally to the elbow. It is the main nerve of the front of the forearm, which contains the muscles of flexion and pronation.

At the wrist, the median nerve passes through the carpal tunnel. The median nerve ends in branches that are responsible for supplying some of the small muscles of the hand, as well as the skin over the thumb and some neighbouring fingers.

ULNAR NERVE

The ulnar nerve passes down along the humerus to the elbow where it loops behind the medial epicondyle, beneath the skin where it can easily be felt. It gives off branches to supply the elbow, two of the muscles of the forearm and several areas of overlying skin before entering the hand. In the hand, the ulnar nerve divides into deep and superficial branches.

MEDIAN NERVE DAMAGE

The median nerve can be damaged by fractures of the lower end of the humerus or compressed by swollen muscle tendons within the carpal tunnel (carpal tunnel syndrome). Median nerve injury can make it difficult to use the 'pincer grip' of the thumb and fingers, as the nerve supplies the small muscles of the thenar eminence (the fleshy prominence below the base of the thumb).

The ulnar nerve is most vulnerable to injury as it passes behind the medial epicondyle of the humerus.

Front view of the nerves of the arm

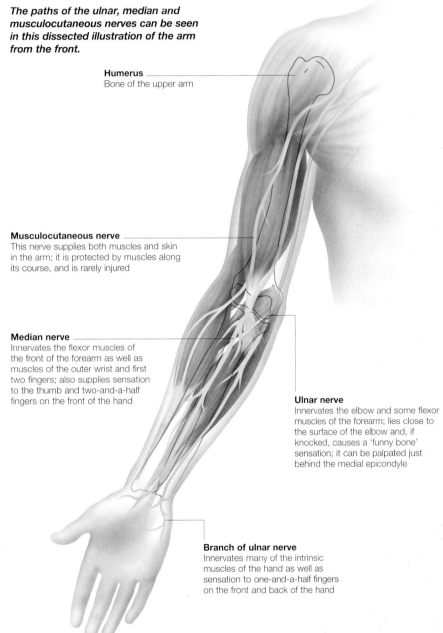

The paths of the ulnar, median and musculocutaneous nerves can be seen in this dissected illustration of the arm from the front.

Humerus
Bone of the upper arm

Musculocutaneous nerve
This nerve supplies both muscles and skin in the arm; it is protected by muscles along its course, and is rarely injured

Median nerve
Innervates the flexor muscles of the front of the forearm as well as muscles of the outer wrist and first two fingers; also supplies sensation to the thumb and two-and-a-half fingers on the front of the hand

Ulnar nerve
Innervates the elbow and some flexor muscles of the forearm; lies close to the surface of the elbow and, if knocked, causes a 'funny bone' sensation; it can be palpated just behind the medial epicondyle

Branch of ulnar nerve
Innervates many of the intrinsic muscles of the hand as well as sensation to one-and-a-half fingers on the front and back of the hand

Bones of the wrist

The wrist lies between the radius and ulna of the forearm and the bones of the fingers. It is made up of eight marble-sized bones which move together to allow flexibility of the wrist joint and the hand.

The area that we commonly think of as the wrist is the end area of the forearm overlying the lower ends of the radius and ulna bones of the forearm. The wrist, in fact, lies in the base of the hand, and comprises eight bones held together by ligaments. They move in relation to each other, thus allowing the wrist to be flexible.

The carpal bones form two rows of four bones each – the proximal row (nearer the forearm) and the distal row (nearer the fingers). The main joint of the wrist is between the first of these two rows and the lower end of the radius.

THE PROXIMAL ROW

The proximal row of the wrist consists of the following bones:

- Scaphoid – a 'boat-shaped' bone which has a large facet for articulation with the lower end of the radius; it articulates with three bones of the distal row
- Lunate – this is a moon-shaped bone that articulates with the lower end of the radius
- Triquetral – this pyramid-shaped bone articulates with the disc of the inferior radioulnar joint and the pisiform bone
- Pisiform – although usually considered part of the proximal row, this small bone plays no part in the wrist joint. It is about the size and shape of a pea and is a 'sesamoid' bone, a bone that lies within a muscle tendon.

Bones of the left wrist viewed from above

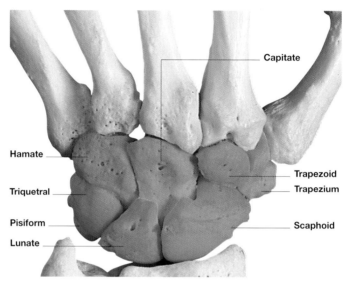

Capitate

Hamate

Triquetral

Pisiform

Lunate

Trapezoid

Trapezium

Scaphoid

The way that the bones of the wrist sit in relation to each other is seen in this image. The top (distal) bones – closest to the fingers – are tinted orange; the bottom (proximal) bones – closest to the forearm are purple.

Wrist bones

Bottom (proximal) row of wrist bones

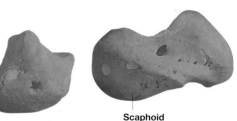

The proximal row of carpal bones includes two bones that can be easily felt: the pisiform and scaphoid bones.

Triquetral
Has a small facet which is the site of articulation with the pisiform bone

Pisiform
Lies within the tendon of the flexor carpi ulnaris muscle

Lunate
Articulates with the lower end of the radius

Scaphoid
Has a narrowed 'waist' which is of importance clinically as it may be the site of a fracture

The wrist joint

The bones of the wrist are covered in cartilage and enclosed by a synovial membrane. This secretes a viscous fluid which allows the bones to move in relation to one another with minimum friction.

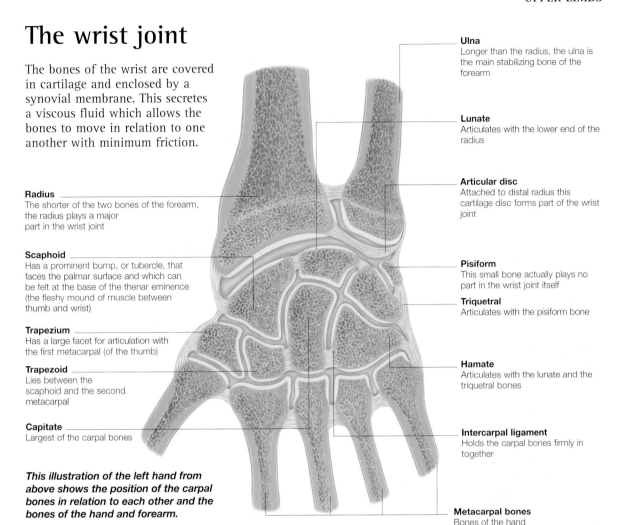

Ulna
Longer than the radius, the ulna is the main stabilizing bone of the forearm

Lunate
Articulates with the lower end of the radius

Articular disc
Attached to distal radius this cartilage disc forms part of the wrist joint

Pisiform
This small bone actually plays no part in the wrist joint itself

Triquetral
Articulates with the pisiform bone

Hamate
Articulates with the lunate and the triquetral bones

Intercarpal ligament
Holds the carpal bones firmly in together

Metacarpal bones
Bones of the hand

Radius
The shorter of the two bones of the forearm, the radius plays a major part in the wrist joint

Scaphoid
Has a prominent bump, or tubercle, that faces the palmar surface and which can be felt at the base of the thenar eminence (the fleshy mound of muscle between thumb and wrist)

Trapezium
Has a large facet for articulation with the first metacarpal (of the thumb)

Trapezoid
Lies between the scaphoid and the second metacarpal

Capitate
Largest of the carpal bones

This illustration of the left hand from above shows the position of the carpal bones in relation to each other and the bones of the hand and forearm.

The wrist, or radiocarpal, joint is a synovial (fluid-filled) joint. On one side lie the lower end of the radius and the articular disc of the inferior radioulnar joint; while on the other are three bones of the first row of carpal (wrist) bones: the scaphoid, lunate and triquetral bones. The fourth bone of this first row, the pisiform, plays no part in the wrist joint.

THE RADIOCARPAL JOINT

The radiocarpal joint is made up of three areas, which are separated by two low ridges:
• A lateral area (outer, on the same side as the thumb), which is formed by the lateral half of the end of the radius as it articulates with the scaphoid bone
• A middle area, where the medial (inner) half of the end of the radius articulates with the lunate bone
• A medial area (inner, on the side of the little finger), where the articular disc which separates the wrist joint from the inferior radioulnar joint articulates with the triquetral bone.
All the articular surfaces are covered with smooth, articular (hyaline) cartilage for reduction of friction

during movement. The joint is lined with synovial membrane which secretes thick, lubricating synovial fluid, and is surrounded by a fibrous capsule that is strengthened by ligaments.

The articular surfaces as a whole form an ellipsoid shape, with the long axis of the ellipse lying across the width of the wrist. The shape of the articular surfaces of a joint helps to determine the range of movements; an ellipsoid shape does not allow rotation of that joint. The joint is convex towards the hand.

THE INTERCARPAL JOINTS

As well as the joint between the lower end of the forearm and the first row of carpal bones, there is articulation between the carpal bones themselves. There is a large, irregular, 'midcarpal' joint which lies between the two rows of carpal bones. It is a synovial joint with a joint cavity which extends into the gaps between the eight bones and allows them to glide over one another, giving the flexibility which is needed in the wrist.

Carpal tunnel

The strong ligaments of the wrist bind together the carpal bones, allowing stability and flexibility. Within the wrist is a fibrous band through which important tendons and nerves run – the carpal tunnel

The eight carpal bones fit together in the wrist to form the shape of an arch. The back of the wrist, the dorsal surface, is gently convex upwards while the palmar surface is concave. The arch is deepened on the palmar aspect of the wrist by the presence of the prominent tubercles of the scaphoid and trapezium bones on one side and the hook of the hamate and the pisiform bone on the other.

STRUCTURE OF THE WRIST
This bony arch is converted into a tunnel by a tough band of fibrous tissue, the flexor retinaculum, which lies across the palmar surface and is attached on each side to the bony projections. This is called the carpal tunnel; through it run the long tendons of the muscles which flex (bend) the fingers. The presence of this band ensures that these tendons are held close to the wrist even when the wrist is bent, so allowing flexion of the fingers at every position of the wrist.

The flexor retinaculum holds the tendons tightly together, allowing flexion in all positions. This is called the carpal tunnel.

Compression of the carpal tunnel (blue) may affect the median nerve and therefore the functioning of the fingers.

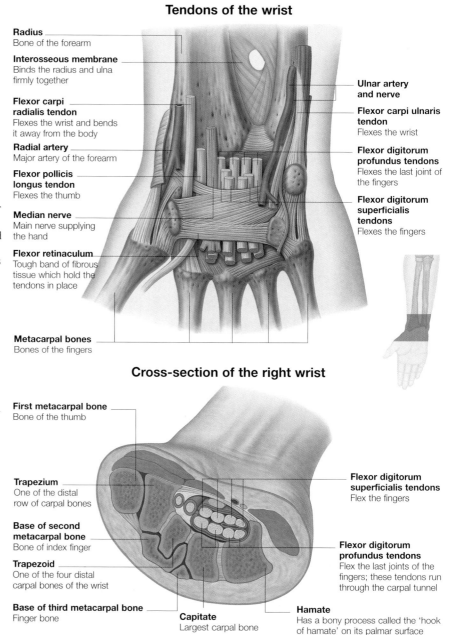

Tendons of the wrist

Radius
Bone of the forearm

Interosseous membrane
Binds the radius and ulna firmly together

Flexor carpi radialis tendon
Flexes the wrist and bends it away from the body

Radial artery
Major artery of the forearm

Flexor pollicis longus tendon
Flexes the thumb

Median nerve
Main nerve supplying the hand

Flexor retinaculum
Tough band of fibrous tissue which hold the tendons in place

Metacarpal bones
Bones of the fingers

Ulnar artery and nerve

Flexor carpi ulnaris tendon
Flexes the wrist

Flexor digitorum profundus tendons
Flexes the last joint of the fingers

Flexor digitorum superficialis tendons
Flexes the fingers

Cross-section of the right wrist

First metacarpal bone
Bone of the thumb

Trapezium
One of the distal row of carpal bones

Base of second metacarpal bone
Bone of index finger

Trapezoid
One of the four distal carpal bones of the wrist

Base of third metacarpal bone
Finger bone

Capitate
Largest carpal bone

Flexor digitorum superficialis tendons
Flex the fingers

Flexor digitorum profundus tendons
Flex the last joints of the fingers; these tendons run through the carpal tunnel

Hamate
Has a bony process called the 'hook of hamate' on its palmar surface

Ligaments of the wrist

The ligaments of the wrist joint are thickenings of the joint capsule which help tie the wrist strongly to the lower ends of the radius and ulna.

Dorsal view of the ligaments of the wrist

There are many ligamentous connections between the carpal bones which help them bind together in a stable formation while allowing flexibility of the wrist as a whole.

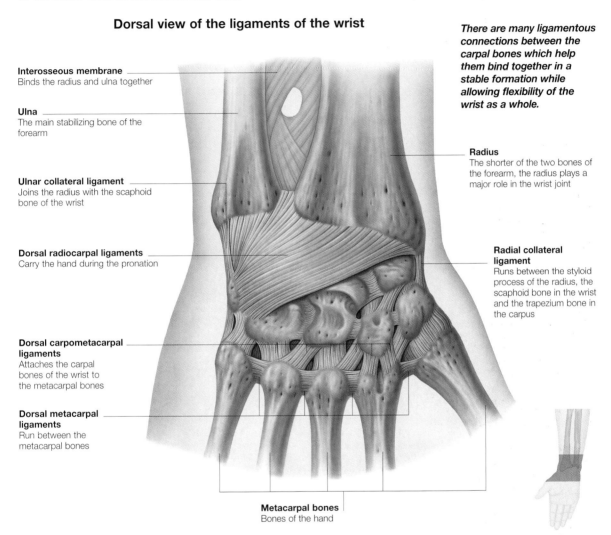

Interosseous membrane
Binds the radius and ulna together

Ulna
The main stabilizing bone of the forearm

Ulnar collateral ligament
Joins the radius with the scaphoid bone of the wrist

Dorsal radiocarpal ligaments
Carry the hand during the pronation

Dorsal carpometacarpal ligaments
Attaches the carpal bones of the wrist to the metacarpal bones

Dorsal metacarpal ligaments
Run between the metacarpal bones

Radius
The shorter of the two bones of the forearm, the radius plays a major role in the wrist joint

Radial collateral ligament
Runs between the styloid process of the radius, the scaphoid bone in the wrist and the trapezium bone in the carpus

Metacarpal bones
Bones of the hand

The wrist joint itself cannot rotate and so rotation of the hand is achieved by pronation and supination of the forearm. The strong ligaments between the carpal bones and the radius are important as they 'carry' the hand round with the forearm during these actions. These ligaments include:

- Palmar radiocarpal ligaments Run from the radius to the carpal bones on the palm side of the hand. The fibres are directed so that the hand will go with the forearm during supination.
- Dorsal radiocarpal ligaments Run at the back of the wrist from the radius to the carpal bones and carry the hand back during pronation.

COLLATERAL LIGAMENTS
Strong collateral ligaments run down each side of the wrist to strengthen the joint capsule and add to the stability of the wrist. These limit the movements of the wrist joint when it is bent.
- Radial collateral ligament Runs between the styloid process of the radius and the scaphoid bone in the wrist.
- Ulnar collateral ligament Runs between the styloid process of the ulna and the triquetral bone.

Bones of the hand

The bones of the hand are divided into the metacarpal bones which support the palm and the phalanges or finger bones. The joints of these bones allow the fingers and the thumb great mobility.

The skeleton of the hand is made up of eight carpal bones of the wrist, the five metacarpal bones, which support the palm, and the 14 phalanges, or finger bones.

THE METACARPALS

Five slender bones radiate out from the wrist bones towards the fingers to form the support of the palm of the hand. They are numbered from one to five, starting at the thumb.

Each of the metacarpals is made up of a body, or shaft, and two slightly bulbous ends. The proximal end (near to the wrist) or base articulates with one of the carpal bones. The distal end (away from the wrist), or head, articulates with the first phalanx of the corresponding finger. In a clenched fist, the heads of the metacarpals are the knuckles.

THE THUMB

The first metacarpal, at the base of the thumb, is the shortest and thickest of the five bones and is rotated slightly out of line. It is extremely mobile, allowing a wider range of movement to the thumb than to the fingers, including the action of opposition whereby the thumb can touch the tips of each of the fingers.

The Phalanges are the bones of the fingers, or digits. The digits are numbered one to five the thumb being one.

The metacarpal bones

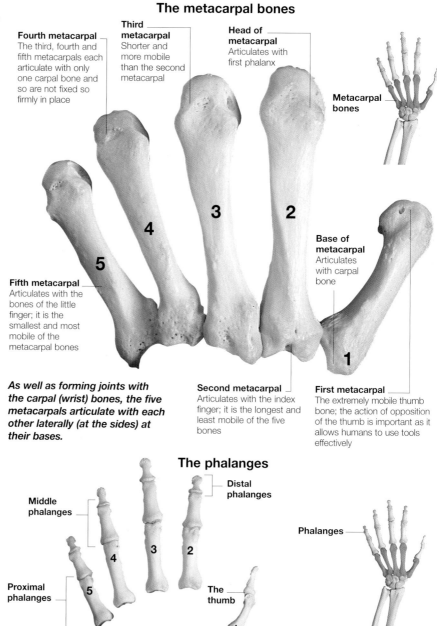

Fourth metacarpal
The third, fourth and fifth metacarpals each articulate with only one carpal bone and so are not fixed so firmly in place

Third metacarpal
Shorter and more mobile than the second metacarpal

Head of metacarpal
Articulates with first phalanx

Metacarpal bones

Base of metacarpal
Articulates with carpal bone

Fifth metacarpal
Articulates with the bones of the little finger; it is the smallest and most mobile of the metacarpal bones

As well as forming joints with the carpal (wrist) bones, the five metacarpals articulate with each other laterally (at the sides) at their bases.

Second metacarpal
Articulates with the index finger; it is the longest and least mobile of the five bones

First metacarpal
The extremely mobile thumb bone; the action of opposition of the thumb is important as it allows humans to use tools effectively

The phalanges

Middle phalanges

Distal phalanges

Proximal phalanges

Phalanges

The thumb

Finger joints

The joints between the phalanges are surrounded by fibrous capsules, lined with synovial membrane and supported by strong collateral ligaments.

The joints of the metacarpal bones with the carpal bones of the wrist – the carpometacarpal joints – are synovial (fluid-filled). The thumb has a saddle-shaped joint with the trapezium allowing a wide range of movement while the other metacarpals form 'plane' joints where the articulating surfaces are flat; they therefore have a limited range of movements.

The carpometacarpal joints are surrounded by fibrous joint capsules lined with synovial membrane. This secretes the lubricating synovial fluid which fills the joint cavity. In most people there is a single, continuous joint cavity for the second to fifth carpometacarpal joints. The joint of the first metacarpal with the trapezium has its own separate joint cavity.

METACARPOPHALANGEAL JOINTS

The joints between the metacarpals and the proximal phalanges are 'condyloid' synovial joints – this shape allows movement in two planes. The fingers may flex and extend (bend and straighten), or abduct and adduct (move apart and together sideways, spreading the fingers). This adds to the mobility and versatility of the hand, as the fingers can be placed in a wide variety of positions.

INTERPHALANGEAL JOINTS

The joints between each phalanx and the next are simple hinge-shaped joints which allow flexion and extension only.

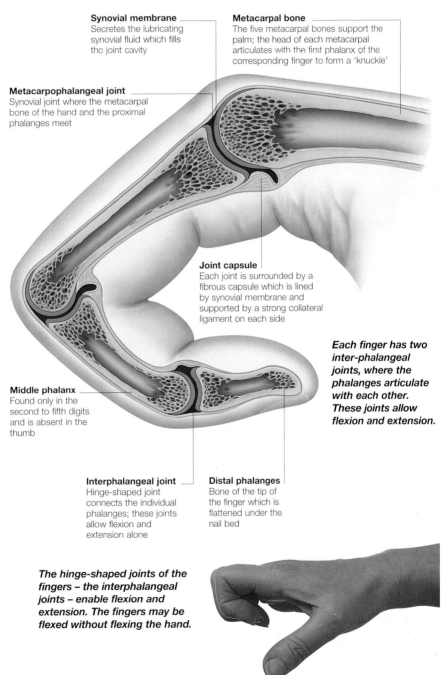

Synovial membrane
Secretes the lubricating synovial fluid which fills the joint cavity

Metacarpal bone
The five metacarpal bones support the palm; the head of each metacarpal articulates with the first phalanx of the corresponding finger to form a 'knuckle'

Metacarpophalangeal joint
Synovial joint where the metacarpal bone of the hand and the proximal phalanges meet

Joint capsule
Each joint is surrounded by a fibrous capsule which is lined by synovial membrane and supported by a strong collateral ligament on each side

Each finger has two inter-phalangeal joints, where the phalanges articulate with each other. These joints allow flexion and extension.

Middle phalanx
Found only in the second to fifth digits and is absent in the thumb

Interphalangeal joint
Hinge-shaped joint connects the individual phalanges; these joints allow flexion and extension alone

Distal phalanges
Bone of the tip of the finger which is flattened under the nail bed

The hinge-shaped joints of the fingers – the interphalangeal joints – enable flexion and extension. The fingers may be flexed without flexing the hand.

Muscles of the hand

The human hand is an exceptionally versatile structure, capable of powerful and delicate movements. These are produced by the actions and interactions of the numerous muscles which act upon it.

Many powerful movements of the hand, which need the contractile strength of a large bulk of muscle tissue, are controlled by the action of muscles in the forearm via tendons, rather than the hand.

Precise and delicate actions are produced by small, or 'intrinsic', muscles. These can be divided into three groups:
- The muscles of the thenar eminence (the bulge of muscle which lies between the base of the thumb and the wrist), which move the thumb
- The muscles of the hypothenar eminence (muscle between the little finger and the wrist), which move the little finger
- The short muscles that run deep in the palm of the hand.

There are two groups of muscles which run longitudinally deep within the hand – the lumbricals and the interossei.

THE LUMBRICALS

There are four lumbrical muscles arising in the palm from the tendons of the flexor digitorum profundus, a powerful muscle of the forearm. The four lumbrical muscles pass around the thumb side of the corresponding digit and insert into the area on the back of the finger which contains the extensor tendons (extensor expansion, or hood).

The lumbrical muscles

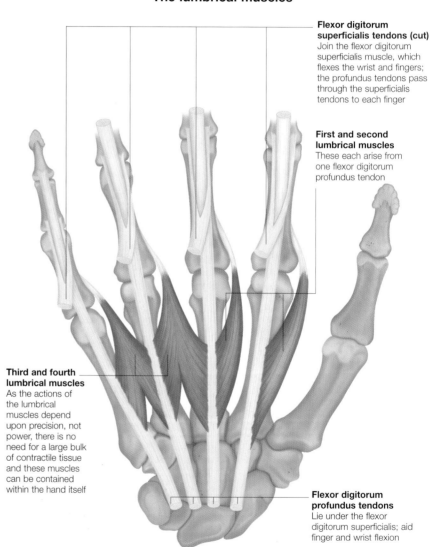

Flexor digitorum superficialis tendons (cut)
Join the flexor digitorum superficialis muscle, which flexes the wrist and fingers; the profundus tendons pass through the superficialis tendons to each finger

First and second lumbrical muscles
These each arise from one flexor digitorum profundus tendon

Third and fourth lumbrical muscles
As the actions of the lumbrical muscles depend upon precision, not power, there is no need for a large bulk of contractile tissue and these muscles can be contained within the hand itself

Flexor digitorum profundus tendons
Lie under the flexor digitorum superficialis; aid finger and wrist flexion

The lumbrical muscles act to flex (bend) the fingers at the knuckle joints and to extend (straighten) the fingers at the interphalangeal joints.

Moving the thumb and little finger

The muscles that move the thumb are contained in the thenar eminence, at the base of the thumb; those that move the little finger are found in the hypothenar eminence, between the little finger and the wrist.

The four small muscles of the thenar eminence act together to allow the thumb to move in the manner that is so important to humans. This action is known as 'opposition' and is the action whereby the tip of the thumb is brought around to touch the tip of any of the fingers.

The muscles of the thenar eminence that move the thumb include:

• Abductor pollicis brevis
Abductor pollicis brevis literally means the short muscle which abducts the thumb (lifts the thumb up away from the palm).
• Flexor pollicis brevis
Flexor pollicis brevis (which flexes the thumb) lies near to the centre of the palm.
• Opponens pollicis
Opponens pollicis (the muscle that opposes the thumb) originates in the flexor retinaculum and the trapezium bone of the wrist, and it inserts into the outer border of the first metacarpal bone.
• Adductor pollicis
Adductor pollicis brings the abducted thumb back in line with the palm. This is a deeply placed muscle which has two heads of origin, separated by the radial artery, which join to form a tendon. This tendon often contains a 'sesamoid' bone, a small bone which lies completely within the tendon and makes no connections with other bones.

Palmar view of the right hand

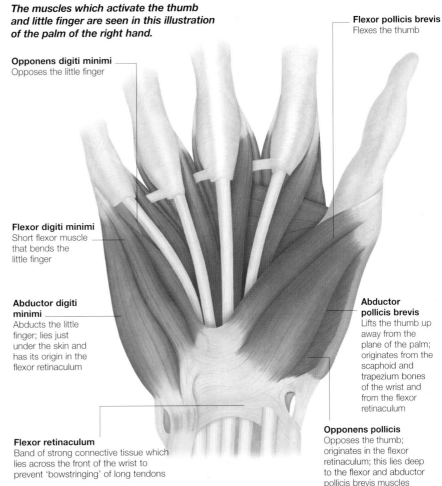

The muscles which activate the thumb and little finger are seen in this illustration of the palm of the right hand.

Opponens digiti minimi
Opposes the little finger

Flexor digiti minimi
Short flexor muscle that bends the little finger

Abductor digiti minimi
Abducts the little finger; lies just under the skin and has its origin in the flexor retinaculum

Flexor retinaculum
Band of strong connective tissue which lies across the front of the wrist to prevent 'bowstringing' of long tendons

Flexor pollicis brevis
Flexes the thumb

Abductor pollicis brevis
Lifts the thumb up away from the plane of the palm; originates from the scaphoid and trapezium bones of the wrist and from the flexor retinaculum

Opponens pollicis
Opposes the thumb; originates in the flexor retinaculum; this lies deep to the flexor and abductor pollicis brevis muscles

THE HYPOTHENAR EMINENCE

The muscles of the smaller hypothenar eminence form the swelling which lies between the little finger and the wrist. These muscles act together to move the little finger around towards the thumb in the action of cupping the hand, or when gripping the lid of a jar to twist it off.
• Abductor digiti minimi
This muscle lies just under the skin and has its origin in

the flexor retinaculum and the pisiform bone of the wrist. It inserts into the side of the base of the little finger.
• Flexor digiti minimi
This short flexor muscle lies alongside the previous muscle but nearer to the centre of the palm. It originates in the flexor retinaculum and the hamate bone of the wrist, and inserts into the base of the little finger.
• Opponens digiti minimi
This muscle, which opposes the little finger, lies

underneath the more superficial muscles of the hypothenar eminence.
• Palmaris brevis muscle
This short muscle has no attachments to bone, but originates in the palmar aponeurosis (connective tissue sheet which lies in the palm) and inserts into the skin overlying the hypothenar eminence. It acts to wrinkle the skin, which is believed to aid grip.

151

Nerves and blood vessels of the hand

The hand is supplied with numerous arteries and veins. These join to form networks of small, interconnecting blood vessels which ensure a good blood supply to all fingers, even if one artery is damaged.

Palmar view of arteries of the left hand

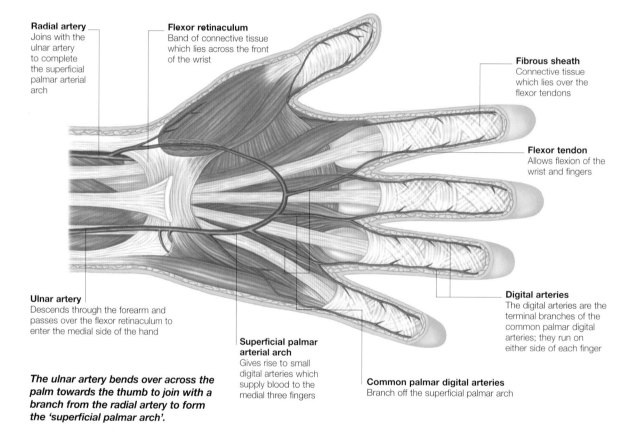

Radial artery
Joins with the ulnar artery to complete the superficial palmar arterial arch

Flexor retinaculum
Band of connective tissue which lies across the front of the wrist

Fibrous sheath
Connective tissue which lies over the flexor tendons

Flexor tendon
Allows flexion of the wrist and fingers

Ulnar artery
Descends through the forearm and passes over the flexor retinaculum to enter the medial side of the hand

Superficial palmar arterial arch
Gives rise to small digital arteries which supply blood to the medial three fingers

Common palmar digital arteries
Branch off the superficial palmar arch

Digital arteries
The digital arteries are the terminal branches of the common palmar digital arteries; they run on either side of each finger

The ulnar artery bends over across the palm towards the thumb to join with a branch from the radial artery to form the 'superficial palmar arch'.

The hand has a plentiful blood supply from the ulnar and radial arteries. These have many interconnections (anastomoses), maintaining the blood supply even if one artery is damaged.
• Superficial palmar arch
The ulnar artery enters the hand on the same side as the little finger and crosses the palm to join with the radial artery

to form the 'superficial palmar arch'. This gives off small digital arteries which supply blood to the little, ring and middle fingers.
• Deep palmar arch
The deep palmar arch is formed by a continuation of the radial artery. This enters the palm from below the base of the thumb and gives off small

arteries which supply the thumb and index finger as well as metacarpal branches which anastomose with the digital arteries.
• Back of the hand
An irregular network of small arteries lies over the back of the wrist, which supplies the back of the hand and the fingers.

Nerves of the hand

The structures of the hand receive their nerve supply from terminal branches of the three main nerves of the upper limb: the median, ulnar and radial nerves.

The median nerve enters the hand on the palmar side by passing under the flexor retinaculum (a restraining band of connective tissue) within the carpal tunnel.

In the hand the median nerve supplies:

- The three muscles of the thenar eminence – abductor pollicis brevis, flexor pollicis brevis and opponens pollicis. If the median nerve is damaged there will be loss of innervation to these muscles with corresponding loss of function of the thumb. This will include the inability to perform the important action of opposition of the thumb.
- The first and second lumbrical muscles.
- The skin of the palm and palmar surface of the first three-and-a-half digits as well as the dorsal surface (back) of the tips of those fingers. The branch of the median nerve which supplies the skin of the central palm arises before the median nerve enters the carpal tunnel and passes over, not under, the flexor retinaculum so the skin will continue to receive its nerve supply if the median nerve is damaged there.

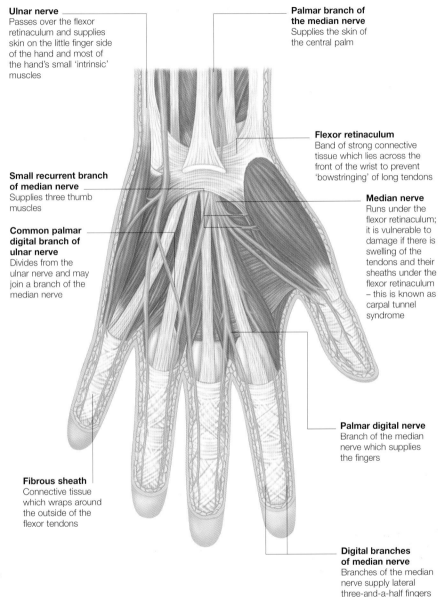

Ulnar nerve
Passes over the flexor retinaculum and supplies skin on the little finger side of the hand and most of the hand's small 'intrinsic' muscles

Palmar branch of the median nerve
Supplies the skin of the central palm

Flexor retinaculum
Band of strong connective tissue which lies across the front of the wrist to prevent 'bowstringing' of long tendons

Small recurrent branch of median nerve
Supplies three thumb muscles

Common palmar digital branch of ulnar nerve
Divides from the ulnar nerve and may join a branch of the median nerve

Median nerve
Runs under the flexor retinaculum; it is vulnerable to damage if there is swelling of the tendons and their sheaths under the flexor retinaculum – this is known as carpal tunnel syndrome

Fibrous sheath
Connective tissue which wraps around the outside of the flexor tendons

Palmar digital nerve
Branch of the median nerve which supplies the fingers

Digital branches of median nerve
Branches of the median nerve supply lateral three-and-a-half fingers

The hand is supplied by three major nerves: the median, ulnar and radial (not shown). Branches of these nerves supply all the muscles and skin of the hand.

Overview of the abdomen

The abdomen is the part of the trunk which lies between the thorax (above) and the pelvis (below). The contents of the abdominal cavity are supported by a bony framework and the abdominal wall.

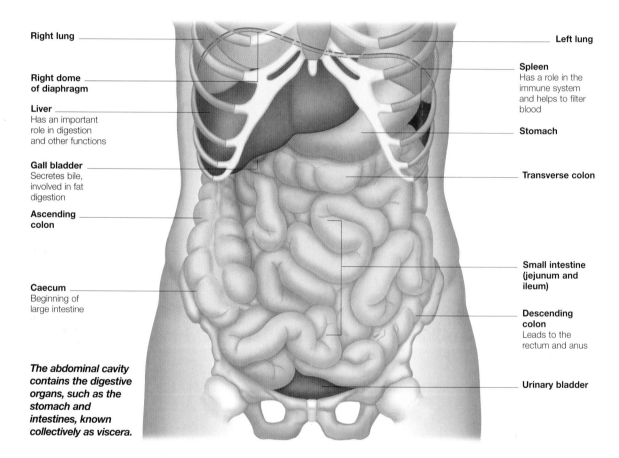

Right lung

Right dome of diaphragm

Liver
Has an important role in digestion and other functions

Gall bladder
Secretes bile, involved in fat digestion

Ascending colon

Caecum
Beginning of large intestine

Left lung

Spleen
Has a role in the immune system and helps to filter blood

Stomach

Transverse colon

Small intestine (jejunum and ileum)

Descending colon
Leads to the rectum and anus

Urinary bladder

The abdominal cavity contains the digestive organs, such as the stomach and intestines, known collectively as viscera.

The organs of the upper part of the abdominal cavity – the liver, gall bladder, stomach and spleen – lie under the domes of the diaphragm and are protected by the lower ribs.

The vertebrae and their associated muscles, form the back wall of the abdominal cavity, while the bones of the pelvis support it from beneath.

The abdomen is relatively unprotected by bone. This allows for mobility of the trunk and enables the abdomen to distend when necessary, such as during pregnancy or after a large meal.

ABDOMINAL CONTENTS
The contents of the abdominal cavity include:
• Much of the gastro-intestinal tract
• Liver
• Pancreas

• Spleen
• Kidneys.
As well as these viscera the abdominal cavity contains all the blood vessels, lymphatics and nerves which supply them, together with a variable amount of fatty tissue.

Planes and regions of the abdomen

In order to describe the position of organs or the site of abdominal pain, doctors find it useful to divide the abdomen into regions defined by imaginary vertical and horizontal planes. These areas help in the making of a clinical diagnosis.

The abdomen may be divided into nine regions for precise descriptions. These regions are delineated by two horizontal (subcostal and transtubercular) planes and two vertical (midclavicular) planes.

The nine regions are:
• Right hypochondrium
• Epigastrium
• Left hypochondrium
• Right flank (lumbar)
• Umbilical
• Left flank (lumbar)
• Right inguinal (groin)
• Suprapubic
• Left inguinal (groin).

FOUR QUADRANTS
For general clinical purposes it is usually sufficient to divide the abdomen into just four sections delineated by one horizontal (transumbilical) plane and one vertical (median) plane.

The four sections are known simply as the right and left upper quadrants and the right and left lower quadrants.

CLINICAL IMPORTANCE
It is important to know which of the abdominal contents lie in each of the regions. If an abnormality is found, or if the patient has an abdominal pain, these regions can be used to indicate the site in the patient's notes for reference.

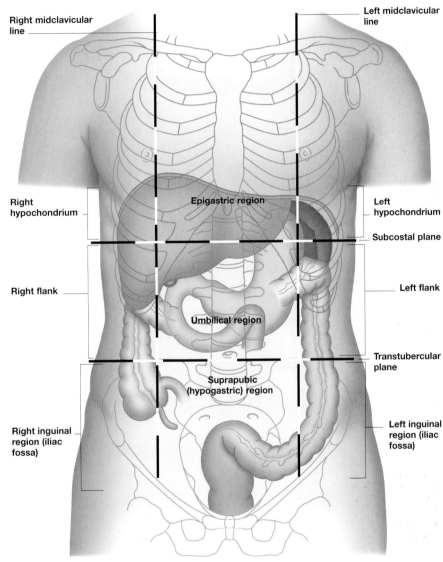

Right midclavicular line

Left midclavicular line

Right hypochondrium

Epigastric region

Left hypochondrium

Subcostal plane

Right flank

Left flank

Umbilical region

Transtubercular plane

Right inguinal region (iliac fossa)

Suprapubic (hypogastric) region

Left inguinal region (iliac fossa)

155

Abdominal wall

The abdominal cavity lies between the diaphragm and the pelvis. The abdominal wall at the front and sides of the body consists of different muscular layers, surrounding and supporting the cavity.

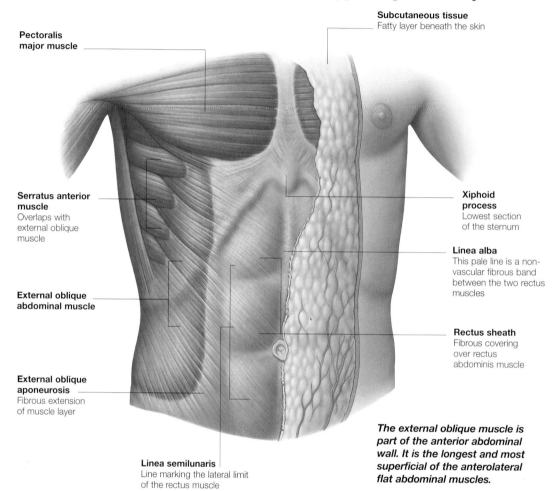

Subcutaneous tissue
Fatty layer beneath the skin

Pectoralis major muscle

Serratus anterior muscle
Overlaps with external oblique muscle

External oblique abdominal muscle

External oblique aponeurosis
Fibrous extension of muscle layer

Linea semilunaris
Line marking the lateral limit of the rectus muscle

Xiphoid process
Lowest section of the sternum

Linea alba
This pale line is a non-vascular fibrous band between the two rectus muscles

Rectus sheath
Fibrous covering over rectus abdominis muscle

The external oblique muscle is part of the anterior abdominal wall. It is the longest and most superficial of the anterolateral flat abdominal muscles.

The posterior (rear) abdominal wall is formed by the lower ribs, the spine and accompanying muscles, while the anterolateral (front and side) wall consists entirely of muscle and fibrous sheets (aponeuroses).

Under the skin and subcutaneous fat layer there lie the muscle layers of the abdominal wall. The muscles here lie in three broad sheets: the external oblique, the internal oblique and the transversus abdominis, which give the abdomen support in all directions. In addition, there is a wide band of muscle, the rectus abdominis, which runs vertically from the front of the ribcage down to the front of the pelvis.

EXTERNAL OBLIQUE
The external oblique muscle forms the most superficial layer of abdominal muscles. It is in the form of a broad, thin sheet whose fibres run down and inwards.

The muscle arises from the under-surfaces of the lower ribs. The fibres fan out into the wide sheet of tough connective tissue known as the external oblique aponeurosis. At the lower end the fibres insert into the top of the pubic bones.

Deeper muscles of the abdominal wall

Beneath the large external oblique muscle lie two more layers of sheet-like muscle, the internal oblique and the transversus abdominis. In addition, running vertically down the centre of the abdominal wall is the rectus abdominis.

The internal oblique muscle is a broad, thin sheet which lies deep to the external oblique. Its fibres run upwards and inwards, at approximately 90 degrees to those of the external oblique.

The fibres of the internal oblique originate from the lumbar fascia (a layer of connective tissue on either side of the spine), the iliac crest of the pelvis and the inguinal ligament (formed in the groin).

Like the external oblique, the internal oblique muscle inserts into a tough, broad aponeurosis which splits to enclose the rectus abdominis muscle (rectus sheath).

TRANSVERSUS ABDOMINIS

This is the innermost of the three sheets of muscle which support the abdominal contents. Its fibres run horizontally to insert into an aponeurosis which lies behind the rectus abdominis muscle for much of its length.

RECTUS ABDOMINIS

These two strap-like muscles run vertically down the front of the abdominal wall.

The upper part of each muscle is wider and thinner than the lower part. Lying between the muscles is a thin, tendinous band of tough connective tissue, the linea alba.

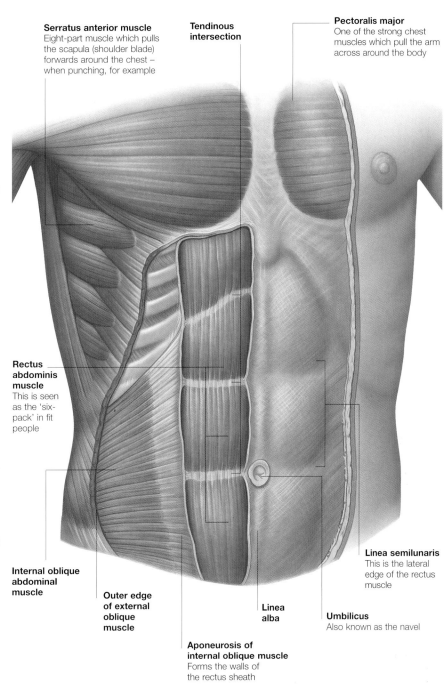

Serratus anterior muscle
Eight-part muscle which pulls the scapula (shoulder blade) forwards around the chest – when punching, for example

Tendinous intersection

Pectoralis major
One of the strong chest muscles which pull the arm across around the body

Rectus abdominis muscle
This is seen as the 'six-pack' in fit people

Internal oblique abdominal muscle

Outer edge of external oblique muscle

Aponeurosis of internal oblique muscle
Forms the walls of the rectus sheath

Linea alba

Umbilicus
Also known as the navel

Linea semilunaris
This is the lateral edge of the rectus muscle

Oesophagus

The oesophagus is the tubular connection between the pharynx in the neck and the stomach. It is used solely as a passage for food, and plays no part in digestion and absorption.

The oesophagus is the muscular tube for the passage of food from mouth to stomach.

SHAPE OF OESOPHAGUS

As it is soft and somewhat flexible, the contour and path of the oesophagus is not straight; rather it curves around, and is indented by, firmer structures such as the arch of the aorta and the left main bronchus.

PASSAGE OF FOOD

When no food is passing through the oesophagus, its inner lining lies in folds which fill the lumen, or central space. As a bolus (lump) of food is swallowed and passes down, it distends this lining and the oesophageal walls. Food is carried down the oesophagus by waves of contractions in a process known as peristalsis.

OESOPHAGEAL STRUCTURE

In cross-section, the oesophagus has four layers:
- Mucosa – the innermost layer lined by stratified squamous epithelium; it is resistant to abrasion by food
- Submucosa – composed of loose connective tissue; it also contains glands which secrete mucus to aid the passage of food
- Muscle layer – striated muscle lines the upper oesophagus; smooth muscle the lower part; and a combination in the mid-region.
- Adventitia – a covering layer of fibrous connective tissue.

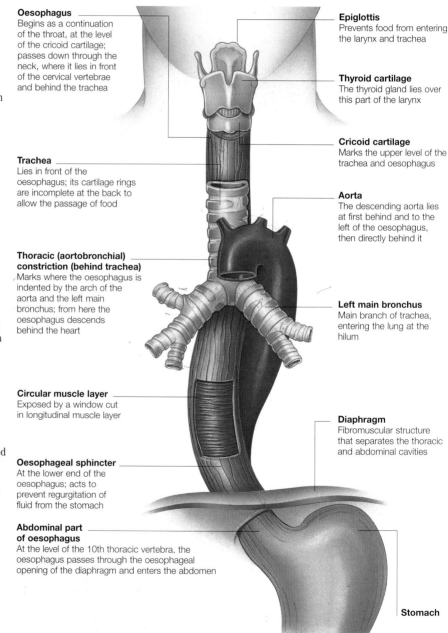

Oesophagus
Begins as a continuation of the throat, at the level of the cricoid cartilage; passes down through the neck, where it lies in front of the cervical vertebrae and behind the trachea

Trachea
Lies in front of the oesophagus; its cartilage rings are incomplete at the back to allow the passage of food

Thoracic (aortobronchial) constriction (behind trachea)
Marks where the oesophagus is indented by the arch of the aorta and the left main bronchus; from here the oesophagus descends behind the heart

Circular muscle layer
Exposed by a window cut in longitudinal muscle layer

Oesophageal sphincter
At the lower end of the oesophagus; acts to prevent regurgitation of fluid from the stomach

Abdominal part of oesophagus
At the level of the 10th thoracic vertebra, the oesophagus passes through the oesophageal opening of the diaphragm and enters the abdomen

Epiglottis
Prevents food from entering the larynx and trachea

Thyroid cartilage
The thyroid gland lies over this part of the larynx

Cricoid cartilage
Marks the upper level of the trachea and oesophagus

Aorta
The descending aorta lies at first behind and to the left of the oesophagus, then directly behind it

Left main bronchus
Main branch of trachea, entering the lung at the hilum

Diaphragm
Fibromuscular structure that separates the thoracic and abdominal cavities

Stomach

Blood vessels and nerves

The arterial supply of the oesophagus derives from branches of the aorta and subclavian artery. As with much of the body, the veins which drain blood from the oesophagus tend to run alongside the arteries.

Veins of the oesophagus

The oesophagus is surrounded by a network of veins. These drain either into the SVC via its tributaries or the portal vein via the left gastric vein.

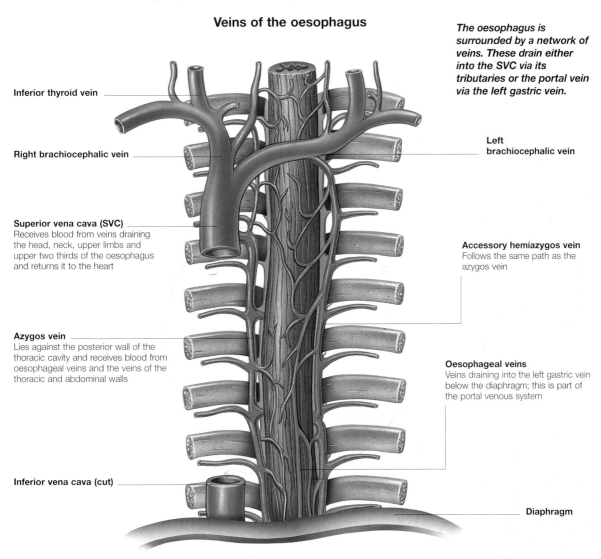

Inferior thyroid vein

Right brachiocephalic vein

Superior vena cava (SVC)
Receives blood from veins draining the head, neck, upper limbs and upper two thirds of the oesophagus and returns it to the heart

Azygos vein
Lies against the posterior wall of the thoracic cavity and receives blood from oesophageal veins and the veins of the thoracic and abdominal walls

Inferior vena cava (cut)

Left brachiocephalic vein

Accessory hemiazygos vein
Follows the same path as the azygos vein

Oesophageal veins
Veins draining into the left gastric vein below the diaphragm; this is part of the portal venous system

Diaphragm

A network of small veins surrounds and drains blood from the oesophagus.

UPPER OESOPHAGUS
The veins from the upper third of the oesophagus drain into the inferior thyroid veins. Blood from the middle third of the oesophagus is drained into the azygos venous system.

LOWER OESOPHAGUS
Blood from the lower third of the oesophagus may enter the left gastric vein – part of the portal venous system which drains blood via the liver. This is clinically important because increased pressure in the portal system, may cause blood to travel back up into the oesophageal veins. As a result, the veins in the lower oesophagus become distended (varices form) and may rupture.

Stomach

The stomach is the expanded part of the digestive tract that receives swallowed food from the oesophagus. Food is stored here before being propelled into the small intestine as digestion continues.

The stomach is a distendable muscular bag lined by mucous membrane. It is fixed at two points: the oesophageal opening at the top and at the beginning of the small intestine below. Between these points it is mobile and can vary in position.

STOMACH LINING

When empty, the stomach lining lies in numerous folds, or rugae, which run from one opening to the other.

The walls of the stomach are similar to other parts of the gut but with some modifications:

- The gastric epithelium – this is the layer of cells which lines the stomach; it contains many glands that secrete protective mucus, and others that produce enzymes and acid, which begin the process of digestion.
- The muscle layer – this has an inner oblique layer of muscle as well as the usual longitudinal and circular fibres. This arrangement helps the stomach to churn food thoroughly before propelling it on towards the small intestine.

REGIONS OF THE STOMACH

The stomach is said to have four parts, and two curvatures:

- The cardia
- The fundus
- The body
- The pyloric region – the outlet area of the stomach
- The lesser curvature
- The greater curvature.

Location and structure of the stomach

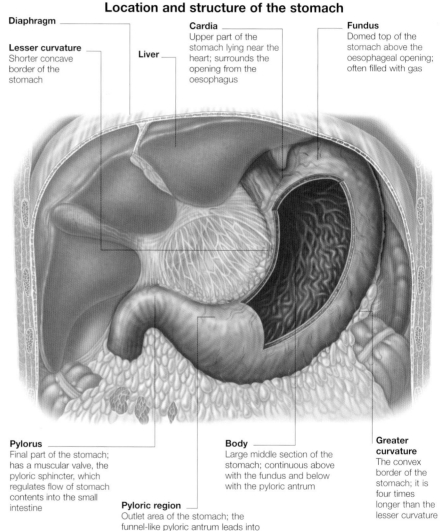

Diaphragm

Lesser curvature
Shorter concave border of the stomach

Liver

Cardia
Upper part of the stomach lying near the heart; surrounds the opening from the oesophagus

Fundus
Domed top of the stomach above the oesophageal opening; often filled with gas

Pylorus
Final part of the stomach; has a muscular valve, the pyloric sphincter, which regulates flow of stomach contents into the small intestine

Pyloric region
Outlet area of the stomach; the funnel-like pyloric antrum leads into the narrow pyloric canal; at the end of the canal lies the pylorus

Body
Large middle section of the stomach; continuous above with the fundus and below with the pyloric antrum

Greater curvature
The convex border of the stomach; it is four times longer than the lesser curvature

The stomach lies within the epigastric region of the abdomen, below the diaphragm. It lies to the right of the spleen and partly under the liver.

Blood supply of the stomach

The stomach has a profuse blood supply, which comes from the various branches of the coeliac trunk.

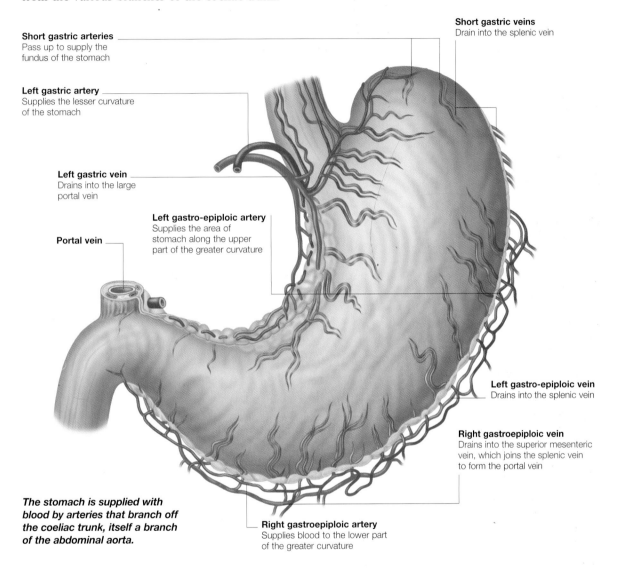

Short gastric arteries
Pass up to supply the fundus of the stomach

Left gastric artery
Supplies the lesser curvature of the stomach

Left gastric vein
Drains into the large portal vein

Portal vein

Left gastro-epiploic artery
Supplies the area of stomach along the upper part of the greater curvature

Short gastric veins
Drain into the splenic vein

Left gastro-epiploic vein
Drains into the splenic vein

Right gastroepiploic vein
Drains into the superior mesenteric vein, which joins the splenic vein to form the portal vein

The stomach is supplied with blood by arteries that branch off the coeliac trunk, itself a branch of the abdominal aorta.

Right gastroepiploic artery
Supplies blood to the lower part of the greater curvature

The vessels that supply the stomach are:
- Left gastric artery – a branch of the coeliac trunk
- Right gastric artery – usually arises from the hepatic artery (a branch of the coeliac trunk)
- Right gastroepiploic artery – arises from the gastroduodenal branch of the hepatic artery
- Left gastroepiploic artery – arises from the splenic artery
- Short gastric arteries – arise from the splenic artery.

VEINS AND LYMPHATICS
The gastric veins run alongside the various gastric arteries. Blood from the stomach is drained ultimately into the portal venous system, which takes blood through the liver before returning it back to the heart.

Lymph collected from the stomach walls drains through lymphatic vessels into the many lymph nodes which lie in groups along the lesser and greater curvature. It is then transported to the coeliac lymph nodes.

161

Small intestine

The small intestine extends from the stomach to the junction with the large intestine. It is made up of three parts, and is the main site in the body where food is digested and absorbed.

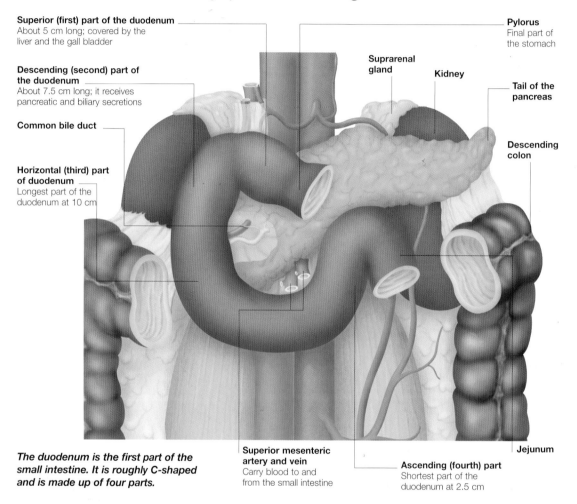

Superior (first) part of the duodenum
About 5 cm long; covered by the liver and the gall bladder

Descending (second) part of the duodenum
About 7.5 cm long; it receives pancreatic and biliary secretions

Common bile duct

Horizontal (third) part of duodenum
Longest part of the duodenum at 10 cm

Suprarenal gland

Kidney

Pylorus
Final part of the stomach

Tail of the pancreas

Descending colon

The duodenum is the first part of the small intestine. It is roughly C-shaped and is made up of four parts.

Superior mesenteric artery and vein
Carry blood to and from the small intestine

Ascending (fourth) part
Shortest part of the duodenum at 2.5 cm

Jejunum

The small intestine is the main site of digestion and absorption of food. It is about seven metres in length in adults and extends from the stomach to the junction with the large intestine. It is divided into three parts: the duodenum, the jejunum and the ileum.

THE DUODENUM
The duodenum is the first part of the small intestine and the shortest. It receives the contents of the stomach with each wave of contraction of the stomach walls. In the duodenum the contents are mixed with secretions from the duodenal walls, pancreas and gall bladder.

The duodenum cannot move, but it is fixed in place behind the peritoneum, the sheet of connective tissue that lines the abdominal cavity.

BLOOD SUPPLY
OF THE DUODENUM
The duodenum receives arterial blood from various

branches off the aorta. These, in turn, give off small branches that provide each part of the duodenum with a rich supply of blood. Venous blood supply mirrors the arterial pattern, returning blood to the hepatic portal venous system.

The jejunum and ileum

The jejunum and the ileum together form the longest part of the small intestine. Unlike the duodenum, they are able to move within the abdomen.

The jejunum and the ileum comprise the longest part of the small intestine. They are surrounded and supported by a fan-shaped fold of the peritoneum – the mesentery – which allows them to move within the abdominal cavity. The mesentery is 15 cm long.

BLOOD SUPPLY

The jejunum and the ileum receive their arterial blood supply from 15–18 branches of the superior mesenteric artery. These branches anastomose (join) to form arches, called arterial arcades. Straight arteries pass out from the arterial arcades to supply all parts of the small intestine. Venous blood from the jejunum and ileum enters the superior mesenteric vein. This vein lies alongside the superior mesenteric artery and drains into the hepatic portal venous system.

ROLE OF LYMPH IN DIGESTION

Fat is absorbed from the contents of the small intestine into specialized lymphatic vessels, known as lacteals, which are found within the mucosa. The milky lymphatic fluid produced by this absorption enters lymphatic plexuses (networks of lymphatic vessels) within the walls of the intestine. The fluid is then carried to special nodes called mesenteric lymph nodes.

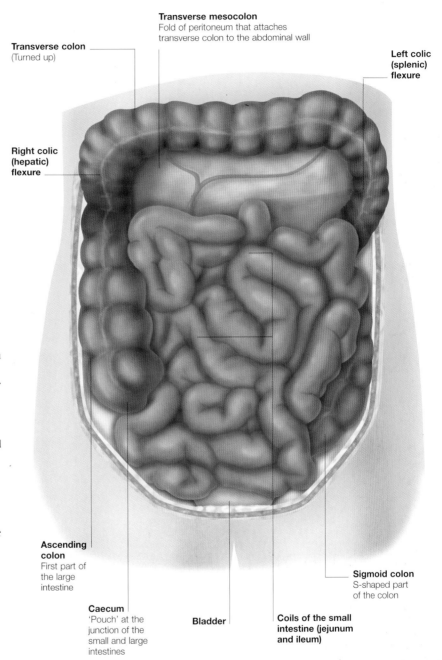

Transverse colon
(Turned up)

Transverse mesocolon
Fold of peritoneum that attaches transverse colon to the abdominal wall

Left colic (splenic) flexure

Right colic (hepatic) flexure

Ascending colon
First part of the large intestine

Caecum
'Pouch' at the junction of the small and large intestines

Bladder

Coils of the small intestine (jejunum and ileum)

Sigmoid colon
S-shaped part of the colon

The coils of the jejunum and ileum take up a large amount of central space in the abdominal cavity. They are not fixed and are able to move within the cavity.

163

Liver and biliary system

The liver is the largest abdominal organ, weighing about 1.5 kg in adult men. It plays an important role in digestion, and also produces bile, which is secreted into the duodenum.

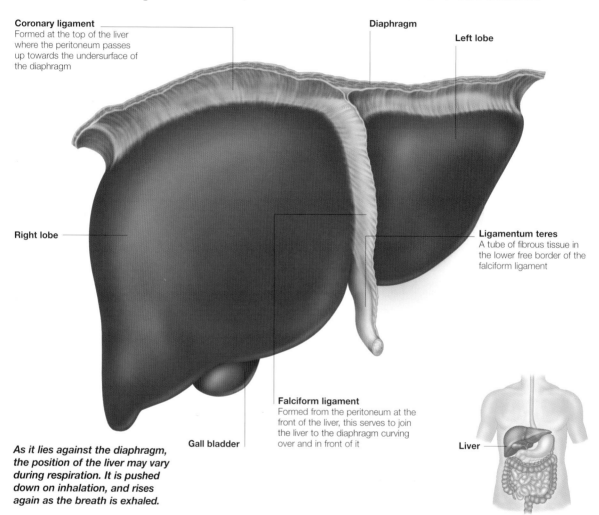

Coronary ligament
Formed at the top of the liver where the peritoneum passes up towards the undersurface of the diaphragm

Diaphragm

Left lobe

Right lobe

Ligamentum teres
A tube of fibrous tissue in the lower free border of the falciform ligament

As it lies against the diaphragm, the position of the liver may vary during respiration. It is pushed down on inhalation, and rises again as the breath is exhaled.

Gall bladder

Falciform ligament
Formed from the peritoneum at the front of the liver, this serves to join the liver to the diaphragm curving over and in front of it

Liver

The liver lies under the diaphragm in the abdominal cavity, on the right side, protected largely by the ribcage.

The tissue of the liver is soft and pliable, and reddish brown in colour. It has a rich blood supply from both the portal vein and the hepatic artery and so will bleed profusely if cut or damaged.

LOBES OF THE LIVER
Although it has four lobes, functionally, the liver is divided into two parts, right and left, each receiving its own separate blood supply. The two smaller lobes, the caudate and the quadrate, can only be seen on the underside of the liver.

PERITONEAL COVERINGS
The greater part of the liver is covered with the peritoneum, a sheet of connective tissue which lines the walls and structures of the abdomen. Folds of the peritoneum form the various ligaments of the liver.

Visceral surface of the liver

The underside of the liver is known as the visceral surface as it lies against the abdominal organs, or viscera. The impressions of adjacent organs, the related vessels and the positions of the inferior vena cava and gall bladder can be seen.

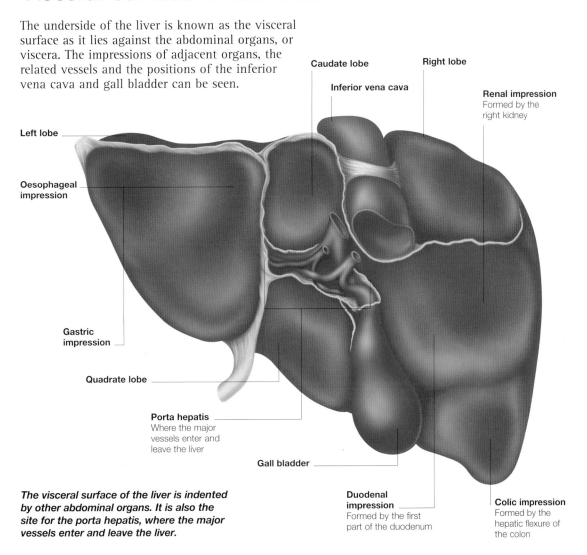

Caudate lobe

Inferior vena cava

Right lobe

Renal impression
Formed by the right kidney

Left lobe

Oesophageal impression

Gastric impression

Quadrate lobe

Porta hepatis
Where the major vessels enter and leave the liver

Gall bladder

Duodenal impression
Formed by the first part of the duodenum

Colic impression
Formed by the hepatic flexure of the colon

The visceral surface of the liver is indented by other abdominal organs. It is also the site for the porta hepatis, where the major vessels enter and leave the liver.

The liver lies closely against many other organs in the abdomen. Because the tissue of the liver is soft and pliable, these surrounding structures may leave impressions on its surface. The largest and most obvious impressions are seen on the surfaces of the right and left lobes.

PORTA HEPATIS
The porta hepatis is an area which is similar to the hilum of the lungs, in that major vessels enter and leave the liver together clothed in a sleeve of connective tissue, in this case peritoneum.

Structures which pass through the porta hepatis include the portal vein, the hepatic artery, the bile ducts, lymphatic vessels and nerves.

BLOOD SUPPLY
The liver is unusual in that it receives blood from two sources:

• The hepatic artery. Conveys 30 per cent of the liver's blood supply. It arises from the common hepatic artery and carries fresh oxygenated blood. On entering the liver it divides into right and left branches. The right branch supplies the right lobe and the left branch supplies the caudate, quadrate and left lobes.
• The hepatic portal vein. Conveys 70 per cent of the liver's blood supply. This large vein drains blood from the gastro-intestinal tract, from the stomach to the rectum. Portal blood is rich in nutrients which have been absorbed after digestion in the gut. Like the hepatic artery, it divides into right and left branches with similar distributions. Venous blood from the liver is returned to the heart via the hepatic vein.

Caecum and appendix

The caecum and appendix lie at the junction of the large and small intestine, an area also known as the ileocaecal region. The caecum, from which the appendix arises, receives food from the small intestine.

The caecum is the first part of the large intestine. Food is passed from the terminal ileum, part of the small intestine, into the large intestine through the ileocaecal valve; the caecum lies below this valve. The caecum is a blind-ending pouch, about 7.5 cm in length and breadth, which continues above as the ascending colon, the next part of the large intestine. The appendix, a long, thin pouch of intestine arises from the caecum.

MUSCLE FIBRES

The muscular 'coat' of the small intestine is continued in the walls of the large intestine, but here it becomes separated into three strips of muscle, the taeniae coli. As food passes through the intestine, the caecum may become distended with faeces or gas. It may then be palpable through the abdominal wall.

BLOOD SUPPLY

The arterial blood supply to the caecum is from the anterior and posterior caecal arteries, which arise from the ileocolic artery. Venous blood returns through a similar layout of veins, ultimately draining into the superior mesenteric vein.

The ileocaecal region is the area surrounding the junction where the small intestine meets the large intestine. It consists of the caecum and the appendix.

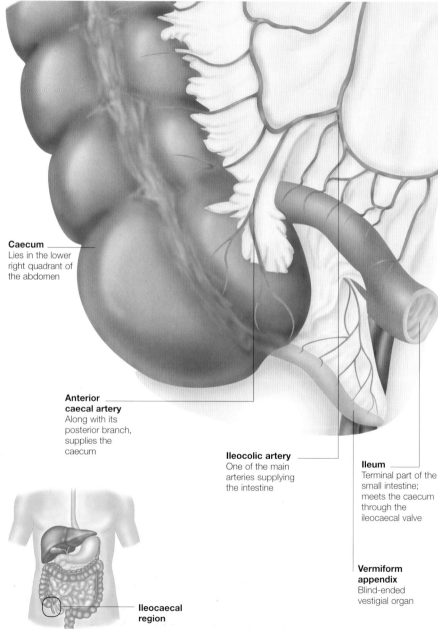

Caecum
Lies in the lower right quadrant of the abdomen

Anterior caecal artery
Along with its posterior branch, supplies the caecum

Ileocolic artery
One of the main arteries supplying the intestine

Ileum
Terminal part of the small intestine; meets the caecum through the ileocaecal valve

Vermiform appendix
Blind-ended vestigial organ

Ileocaecal region

The appendix

The appendix is a narrow, muscular outpouching of the caecum. It is usually between six and 10 cm in length, although it may be much longer or shorter. It arises from the back of the caecum, its lower end being free and mobile.

The vermiform (or 'wormlike') appendix is attached to the caecum at the beginning of the large intestine. The walls of the appendix contain lymphoid tissue. The lymphoid tissue of the appendix and that within the walls of the small intestine, protects the body from micro-organisms within the gut.

MUSCLE LAYER
Whereas the longitudinal muscle in the walls of the rest of the large intestine is present only in three strips – the taeniae coli – the appendix has a complete muscle layer. This is because the three taeniae coli converge on the base of the appendix and their fibres join to cover its entire surface.

PERITONEUM
The appendix is enclosed within a covering of peritoneum which forms a fold between the ileum, the caecum and the first part of the appendix. This fold is known as the mesoappendix.

BASE OF THE APPENDIX
The base of the appendix, where it arises from the caecum, is usually in a fixed position. The corresponding area on the surface of the abdomen is known as McBurney's point.

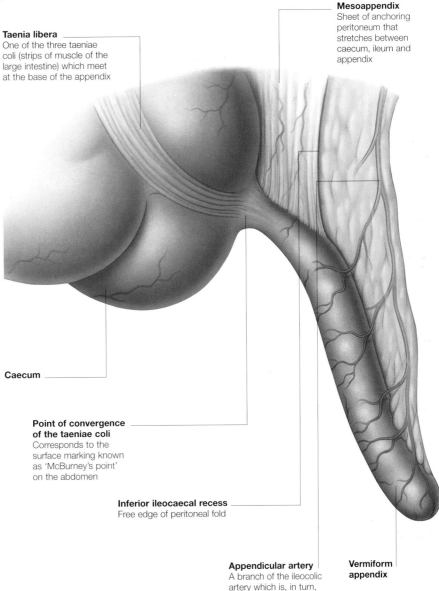

Taenia libera
One of the three taeniae coli (strips of muscle of the large intestine) which meet at the base of the appendix

Mesoappendix
Sheet of anchoring peritoneum that stretches between caecum, ileum and appendix

Caecum

Point of convergence of the taeniae coli
Corresponds to the surface marking known as 'McBurney's point' on the abdomen

Inferior ileocaecal recess
Free edge of peritoneal fold

Appendicular artery
A branch of the ileocolic artery which is, in turn, a branch of the superior mesenteric artery

Vermiform appendix

The appendix arises from the caecum, below the ileocaecal junction. As a vestigial organ, it has become functionless in the course of evolution, and has no role in the process of digestion.

Colon

The colon forms the main part of the large intestine. Although a continuous tube, the colon has four parts: the ascending colon, the transverse colon, the descending colon and the sigmoid colon.

The colon receives the liquefied contents of the small intestine and reabsorbs the water to form semi-solid waste, which is then expelled through the rectum and anal canal as faeces. There are two sharp bends, or flexures, in the colon known as the right colic (hepatic) flexure and the left colic (splenic) flexure.

ASCENDING COLON

The ascending colon runs from the ileocaecal valve up to the right colic flexure, where it becomes the transverse colon. It is about 12 cm long and lies against the posterior (back) abdominal wall, being covered on the front and sides only by the peritoneum, the thin sheet of connective tissue that lines the abdominal organs.

TRANSVERSE COLON

The transverse colon begins at the right colic flexure, under the right lobe of the liver, and runs across the body towards the left colic flexure next to the spleen.

With a length of about 45 cm, the transverse colon is the longest and the most mobile part of the large intestine, as it hangs down suspended within a fold of peritoneum (or mesentery).

THE DESCENDING COLON

The descending colon runs from the left colic flexure down to the brim of the pelvis where it becomes the sigmoid colon. As the left colic flexure is higher than the right, the descending colon is longer than the ascending colon.

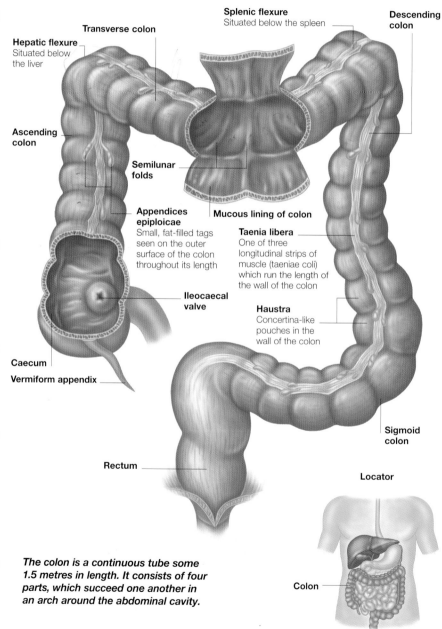

Splenic flexure
Situated below the spleen

Descending colon

Transverse colon

Hepatic flexure
Situated below the liver

Ascending colon

Semilunar folds

Appendices epiploicae
Small, fat-filled tags seen on the outer surface of the colon throughout its length

Mucous lining of colon

Taenia libera
One of three longitudinal strips of muscle (taeniae coli) which run the length of the wall of the colon

Ileocaecal valve

Haustra
Concertina-like pouches in the wall of the colon

Caecum

Vermiform appendix

Sigmoid colon

Rectum

Locator

Colon

The colon is a continuous tube some 1.5 metres in length. It consists of four parts, which succeed one another in an arch around the abdominal cavity.

Blood supply and drainage of the colon

Like the rest of the intestine, each of the parts of the colon is readily supplied with blood from a network of arteries.

Venous blood draining from the colon passes through the hepatic portal system, for treatment by the liver, before re-entering the general circulation.

ARTERIAL SUPPLY OF THE COLON

The arterial supply to the colon comes from the superior and inferior mesenteric branches of the aorta, the large central artery of the abdomen.

The ascending colon and first two thirds of the transverse colon are supplied by the superior mesenteric artery, while the last third of the transverse colon, the descending colon and the sigmoid colon are supplied by the inferior mesenteric.

PATTERN OF THE ARTERIES

As in other parts of the gastro-intestinal tract, there are anastomoses, or connections, between the branches of these two arteries.

The superior mesenteric artery gives off the ileocolic, right colic and middle colic arteries which anastomose with each other and with the left colic and sigmoid branches of the inferior mesenteric artery.

In this way an 'arcade' of arteries is formed around the wall of the colon, supplying all parts with arterial blood.

Arterial system of the colon

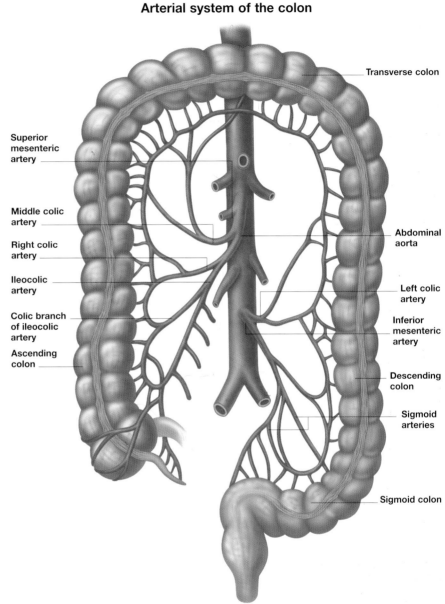

Superior mesenteric artery

Middle colic artery

Right colic artery

Ileocolic artery

Colic branch of ileocolic artery

Ascending colon

Transverse colon

Abdominal aorta

Left colic artery

Inferior mesenteric artery

Descending colon

Sigmoid arteries

Sigmoid colon

The arteries of the ascending colon and greater part of the transverse colon are supplied by the superior mesenteric artery. The inferior mesenteric artery supplies the descending colon and the left part of the transverse colon.

Rectum and anal canal

The rectum and anal canal together form the last part of the gastro-intestinal tract. They receive waste matter in the form of faeces and allow it to be passed out of the body.

The rectum continues on from the sigmoid colon, which lies at the level of the third sacral vertebra. Rectum means 'straight' but in fact the rectum follows the curve of the sacrum and coccyx, which form the back of the bony pelvis.

The lower end of the rectum joins to the anal canal with an 80–90 degree change in direction. This anorectal flexure prevents faeces passing into the anal canal until required.

The longitudinal muscle of the rectum is in two broad bands, which run down the front and the back surfaces. There are three horizontal folds in the wall of the rectum, known as the superior (upper), middle and inferior (lower) transverse folds. Below the inferior fold the rectum widens into the ampulla.

THE ANAL CANAL

The anal canal runs from the anorectal flexure down to the anus. Except during defecation the canal is empty and closed.

The lining of the anal canal changes along its length. The upper part carries longitudinal ridges called anal columns which begin above at the anorectal junction and end below at the pectinate line.

At the lower end of the anal columns are the anal sinuses and anal valves. The anal sinuses produce mucus when faeces are being passed, which acts as a lubricant. The valves help to prevent the passage of mucus out of the anal canal at other times.

Coronal section through the rectum and anal canal

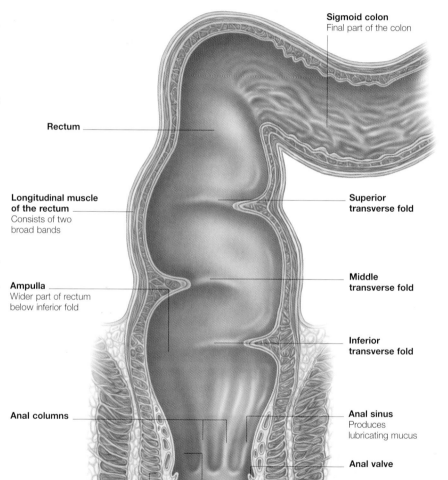

Sigmoid colon
Final part of the colon

Rectum

Longitudinal muscle of the rectum
Consists of two broad bands

Superior transverse fold

Ampulla
Wider part of rectum below inferior fold

Middle transverse fold

Inferior transverse fold

Anal columns

Anal sinus
Produces lubricating mucus

Anal valve

Anal canal
Remains closed, with walls touching, until defecation

Pectinate line
Anal canal above it has a different blood supply, lymphatic drainage and nerve supply from that below

Anorectal flexure
Prevents faeces passing into anal canal until required

Vessels of the rectum and anus

The rectum and anal canal have a rich blood supply. A network of veins drains blood from this area.

Beneath the lining of the rectum and anal canal lies a network of small veins, the rectal venous plexus. This is in two parts:
- The internal rectal venous plexus – lies just under the lining
- The external rectal venous plexus – lies outside the muscle layer.

These receive blood from the tissues and carry it to the larger veins that drain the area. These larger veins are the superior, middle and inferior rectal veins which drain the corresponding parts of the rectum.

The internal venous plexus of the anal canal drains blood in two directions on either side of the pectinate line region. Above this level blood drains mainly into the superior rectal vein while from below it drains into the inferior rectal vein.

ARTERIAL BLOOD SUPPLY
The rectum receives its blood supply from three sources. The upper part is supplied by the superior rectal artery, the lower portion is supplied by the middle rectal arteries while the anorectal junction receives blood from the inferior rectal arteries.

Within the anal canal the superior rectal artery travels down to provide blood above the pectinate line. The two inferior rectal arteries, branches of the pudendal, supply the anal canal below the pectinate line. as well as the surrounding muscle and skin around the anus.

Venous drainage system of the rectum and anus

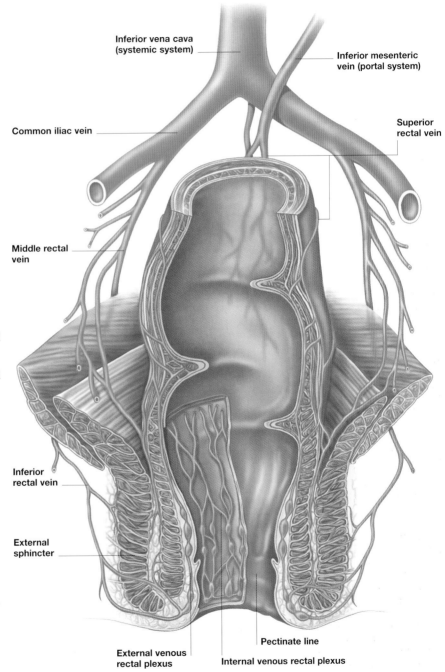

Inferior vena cava (systemic system)

Inferior mesenteric vein (portal system)

Common iliac vein

Superior rectal vein

Middle rectal vein

Inferior rectal vein

External sphincter

Pectinate line

External venous rectal plexus

Internal venous rectal plexus

Pancreas and spleen

The pancreas is a large gland that produces both enzymes and hormones. It lies in the upper abdomen behind the stomach, one end in the curve of the duodenum and the other end touching the spleen.

The pancreas secretes enzymes into the duodenum, the first part of the small intestine, to aid the digestion of food. It also produces the hormones insulin and glucagon, which regulate the use of glucose by cells. Lying across the posterior wall of the abdomen, the pancreas is said to have four parts:

- The head – which lies within the C-shaped curve of the duodenum. It is attached to the inner side of the duodenum; a small, hook-like projection, the uncinate process, projects towards the midline
- The neck – is narrower than the head, due to the large hepatic portal vein behind; it lies over the superior mesenteric blood vessels
- The body – which is triangular in cross-section and lies in front of the aorta; it passes up and to the left to merge with the tail
- The tail – which comes to a tapering end within the concavity of the spleen.

BLOOD SUPPLY
The pancreas has a very rich blood supply. The pancreatic head is supplied from two arterial arcades which are formed from the superior and inferior pancreaticoduodenal arteries. The body and tail of the pancreas are supplied with blood by branches of the splenic artery.

Venous blood from the pancreas travels to the liver via the portal venous system, the veins lying in an arrangement which mirrors the arterial supply.

Location of pancreas

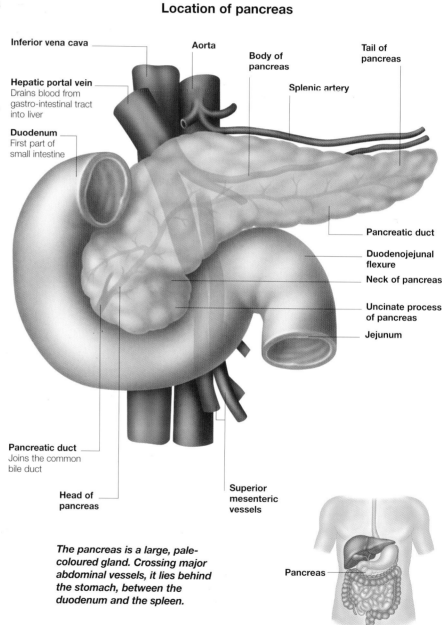

Inferior vena cava

Aorta

Body of pancreas

Tail of pancreas

Hepatic portal vein
Drains blood from gastro-intestinal tract into liver

Splenic artery

Duodenum
First part of small intestine

Pancreatic duct

Duodenojejunal flexure

Neck of pancreas

Uncinate process of pancreas

Jejunum

Pancreatic duct
Joins the common bile duct

Head of pancreas

Superior mesenteric vessels

The pancreas is a large, pale-coloured gland. Crossing major abdominal vessels, it lies behind the stomach, between the duodenum and the spleen.

Pancreas

The spleen

The spleen is the largest of the lymphatic organs. It is dark purple in colour and lies under the lower ribs on the left side of the upper abdomen.

The dimensions of the spleen can vary greatly, but is usually about the size of a clenched fist. In old age, the spleen naturally atrophies and reduces in size.

The hilum of the spleen contains its blood vessels (the splenic artery and vein) and some lymphatic vessels. The hilum also contains lymph nodes and the tail of the pancreas, all enclosed within the lienorenal ligament – a fold of peritoneum.

SURFACE OF THE SPLEEN

The spleen shows indentations of the organs which surround it. The surface which lies against the diaphragm is curved smoothly, while the visceral surface carries the impressions of the stomach, the left kidney and the splenic flexure of the colon.

SPLEEN COVERINGS

The spleen is surrounded and protected by a thin capsule, which is composed of irregular fibro-elastic connective tissue. Contained within the tissue of the capsule are muscle fibres that allow the spleen to contract periodically. These contractions expel the blood the spleen has filtered back into the circulation.

Outside the capsule the spleen is completely enclosed by the peritoneum, the thin sheet of connective tissue which lines the abdominal cavity and covers the organs within it.

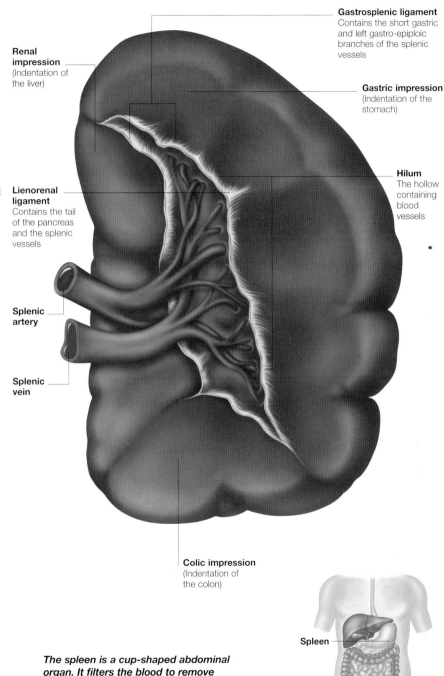

Gastrosplenic ligament
Contains the short gastric and left gastro-epiploic branches of the splenic vessels

Renal impression
(Indentation of the liver)

Gastric impression
(Indentation of the stomach)

Hilum
The hollow containing blood vessels

Lienorenal ligament
Contains the tail of the pancreas and the splenic vessels

Splenic artery

Splenic vein

Colic impression
(Indentation of the colon)

Spleen

The spleen is a cup-shaped abdominal organ. It filters the blood to remove damaged red blood cells; and produces lymphatic cells and antibodies in the fight against infection.

Inguinal region

The inguinal region, commonly known as the groin, is the site of inguinal hernias. The abdominal wall has an area of weakness, which may allow the abdominal contents to protrude through it.

The bilateral areas of weakness in the inguinal region are due to the presence of the inguinal canals, tubes through which pass the spermatic cords in males, and the round ligaments in females.

INGUINAL CANAL

The design of the inguinal canal minimizes the likelihood of herniation (protrusion) of abdominal contents. It passes down towards the midline of the body from its origin, the deep inguinal ring (the entrance to the inguinal canal), to emerge at the superficial inguinal ring (the exit from the canal).

WALLS OF THE CANAL

The inguinal canal has a roof, a floor and two walls:
- Roof – formed by the arching fibres of the internal oblique and the transversus abdominis muscle
- Floor – a shallow gutter, formed by the inguinal ligament
- Anterior wall – formed mainly by the strong aponeurosis of the external oblique muscle, with a contribution from the internal oblique at the outer edges
- Posterior wall – formed by the transversalis fascia with the medial part of the wall being reinforced by the conjoint tendon.

Inguinal region in a male

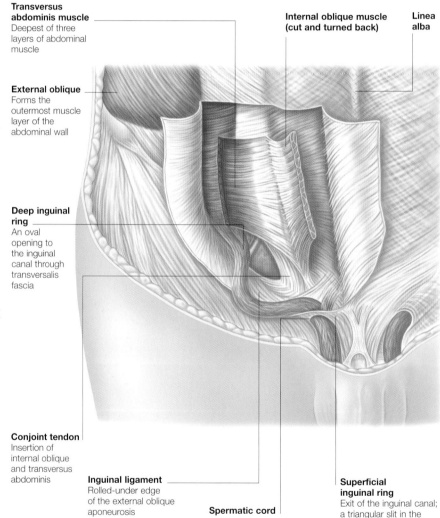

Transversus abdominis muscle
Deepest of three layers of abdominal muscle

External oblique
Forms the outermost muscle layer of the abdominal wall

Deep inguinal ring
An oval opening to the inguinal canal through transversalis fascia

Internal oblique muscle (cut and turned back)

Linea alba

Conjoint tendon
Insertion of internal oblique and transversus abdominis

Inguinal ligament
Rolled-under edge of the external oblique aponeurosis

Spermatic cord

Superficial inguinal ring
Exit of the inguinal canal; a triangular slit in the external oblique muscle

The inguinal canal passes through the layers of the lower abdominal wall. In adults, it is about 4 cm long, although in babies it is much shorter.

Behind the inguinal ligament

The inguinal ligament encloses behind it a number of vital structures including blood vessels and nerves serving the lower limb and two groups of lymph nodes (deep and superficial).

There are two major blood vessels which pass behind the inguinal ligament:
- Femoral artery – the main vessel supplying blood to the lower limb
- Femoral vein – lies medially to the femoral artery (on the inner side).

FEMORAL NERVE
Lateral to these vessels, on the outer side, lies the femoral nerve which is the largest branch of the lumbar plexus, a network of nerves within the abdomen.

FEMORAL SHEATH
The femoral blood vessels are enclosed within a thin, funnel-shaped sheet of connective tissue, known as the femoral sheath. This allows the femoral vessels to glide harmlessly against the inguinal ligament during movements of the hip.

INGUINAL LYMPH NODES
Within the groin lie two groups of lymph nodes:
- Superficial inguinal nodes – these lie just under the skin in a horizontal and a vertical group, and drain areas which include the buttocks, external genitalia and superficial layers of the lower limb
- Deep inguinal nodes – these lie around the femoral artery and vein as they pass under the inguinal ligament, and drain lymph from the lower limb.

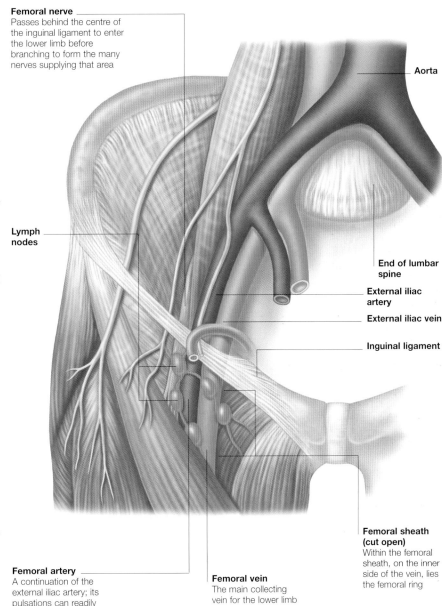

Femoral nerve
Passes behind the centre of the inguinal ligament to enter the lower limb before branching to form the many nerves supplying that area

Aorta

Lymph nodes

End of lumbar spine

External iliac artery

External iliac vein

Inguinal ligament

Femoral sheath (cut open)
Within the femoral sheath, on the inner side of the vein, lies the femoral ring

Femoral artery
A continuation of the external iliac artery; its pulsations can readily be felt with the fingers

Femoral vein
The main collecting vein for the lower limb

Overview of the urinary tract

The urinary tract consists of the kidneys, ureters, urinary bladder and urethra. Together, these organs are responsible for the production of urine and its expulsion from the body.

The paired kidneys filter the blood to remove waste chemicals and excess fluid, which they excrete as urine. Urine passes down through the narrow ureters to the bladder, which stores it temporarily before it is expelled through the urethra.

• Kidneys

The bean-shaped kidneys lie within the abdomen, against the posterior abdominal wall behind the intestines

• Ureters

From the hilus, or 'stalk' of each of the two kidneys, emerge the right and left ureters. These are narrow tubes which receive the urine produced continuously by the kidney

• Bladder

Urine is received and stored temporarily in the urinary bladder, a collapsible, balloon-like structure which lies within the pelvis

• Urethra

When appropriate, the bladder contracts to expel its contents through the urethra, a thin-walled muscular tube.

The urinary tract consists of the structures involved in the, production, storage and expulsion of urine. It extends from the abdomen into the pelvis.

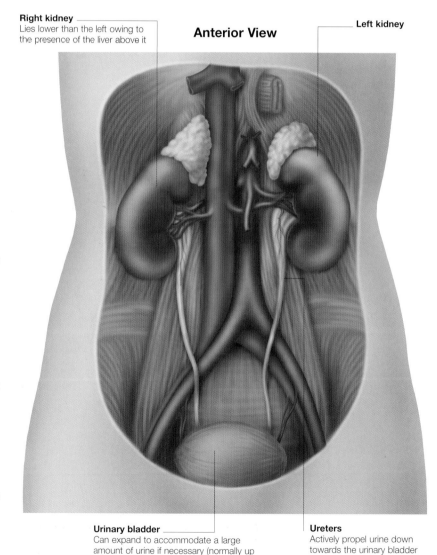

Right kidney
Lies lower than the left owing to the presence of the liver above it

Anterior View

Left kidney

Urinary bladder
Can expand to accommodate a large amount of urine if necessary (normally up to about one litre); expels urine through the urethra

Ureters
Actively propel urine down towards the urinary bladder by contractions of their walls

The adrenal glands

The adrenal glands are situated above the kidneys, but are not part of the urinary tract. Each one consists of two separate parts: a medulla surrounded by a cortex.

Lying on top of the kidneys are the paired adrenal glands, also known as the suprarenal glands. Although they are physically close to the kidneys they play no part in the urinary system: they are endocrine glands which produce hormones vital to the healthy functioning of the body.

SURROUNDING TISSUES
The yellowish adrenal glands lie above the kidneys and under the diaphragm. They are surrounded by a thick layer of fatty tissue and are enclosed by renal fascia although they are separated from the kidneys themselves by fibrous tissue.

This separation is what allows a kidney to be surgically removed without damaging these delicate and important glands.

GLAND DIFFERENCES
Due to the position of the surrounding structures the soft adrenal glands differ in appearance:
• The right adrenal gland
The right adrenal gland is pyramid-shaped and sits on the upper pole of the right kidney. It lies in contact with the diaphragm, the liver and the inferior vena cava, the main vein of the abdomen
• The left adrenal gland
The left adrenal gland has the shape of a half moon and lies along the upper surface of the left kidney from the pole down to the hilus. It lies in contact with the spleen, the

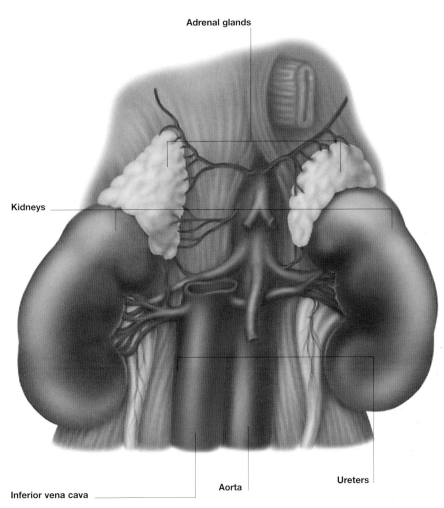

Adrenal glands

Kidneys

Inferior vena cava

Aorta

Ureters

Although the adrenal glands sit on top of the kidneys, they are entirely unrelated to the urinary tract. Instead, they are endocrine glands that secrete hormones into the bloodstream.

stomach, the pancreas and the diaphragm.

BLOOD SUPPLY
As with other endocrine glands, which secrete their hormones directly into the bloodstream, the adrenal glands have a very rich blood supply. They receive

arterial blood from three sources – the superior, middle and inferior adrenal arteries – which arise from the inferior phrenic artery, the aorta and the renal artery respectively.

Near the adrenal glands these arteries branch repeatedly, numerous tiny

arteries entering the glands over their entire surface.

A single vein leaves the adrenal glands on each side to drain blood into the inferior vena cava on the right and the renal vein on the left.

Kidneys

The kidneys are a pair of solid organs situated at the back of the abdomen. They act as filtering units for blood and maintain the balance and composition of fluids within the body.

The paired kidneys lie within the abdominal cavity against the posterior abdominal wall. Each kidney is about 10 cm in length, reddish brown in colour and has the characteristic shape, after which the 'kidney bean' is named. On the medial, or inward facing, surface lies the hilus of the kidney from which the blood vessels enter and leave. The hilus is also the site of exit for the right and left ureters, via which urine leaves the kidney and is transported to the bladder.

REGIONS OF THE KIDNEY
The kidney has three regions, each of which plays a role in the production or collection of urine:
- The renal cortex – the most superficial layer; it is quite pale and has a granular appearance
- The renal medulla – composed of dark reddish tissue, it lies within the cortex in the form of 'pyramids'
- The renal pelvis – the central, funnel-like area of the kidney which collects the urine and is continuous with the ureters at the hilus.

OUTER LAYERS
Each kidney is covered by a tough, fibrous capsule. Outside the kidney lies a protective layer of fat which is contained within the renal fascia – a dense connective tissue that anchors the kidneys and adrenal glands, to surrounding structures.

Cross-section through kidney

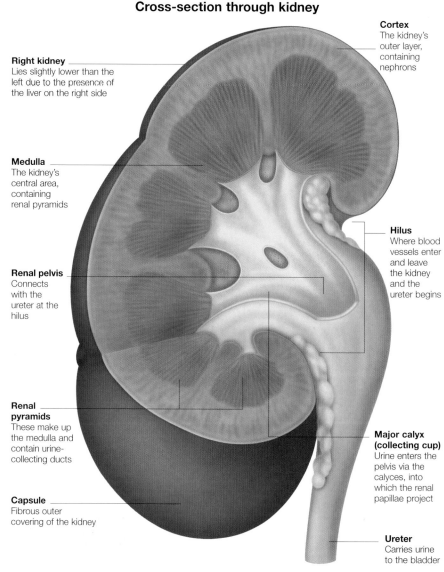

Cortex
The kidney's outer layer, containing nephrons

Right kidney
Lies slightly lower than the left due to the presence of the liver on the right side

Medulla
The kidney's central area, containing renal pyramids

Hilus
Where blood vessels enter and leave the kidney and the ureter begins

Renal pelvis
Connects with the ureter at the hilus

Renal pyramids
These make up the medulla and contain urine-collecting ducts

Major calyx (collecting cup)
Urine enters the pelvis via the calyces, into which the renal papillae project

Capsule
Fibrous outer covering of the kidney

Ureter
Carries urine to the bladder

The kidneys are responsible for the excretion of waste from the blood. Each has three regions: cortex, medulla and renal pelvis.

Blood supply to the kidneys

The function of the kidneys is to filter blood, for which they receive an exceedingly rich blood supply. As with other parts of the body, the pattern of drainage of venous blood mirrors the pattern of arterial supply.

Arterial blood is carried to the kidneys by the right and left renal arteries, which arise directly from the main artery of the body, the aorta. The right renal artery is longer than the left as the aorta lies slightly to the left of the midline. One in three people have an additional, accessory, renal artery.

RENAL ARTERIES
The renal artery enters the kidney at the hilus and divides into between three and five segmental arteries, each of which further divides into lobar arteries. There are no connections between branches of neighbouring segmental arteries.

The interlobar arteries pass between the renal pyramids and branch to form the arcuate arteries, which run along the junction of cortex and medulla. Numerous interlobular arteries pass into the tissue of the renal cortex to carry blood to the glomeruli of the nephrons, where it is filtered to remove excess fluid and waste products.

VENOUS DRAINAGE
Blood enters the interlobular, arcuate and then interlobar veins before being collected by the renal vein and returned to the inferior vena cava, the main collecting vein of the abdomen.

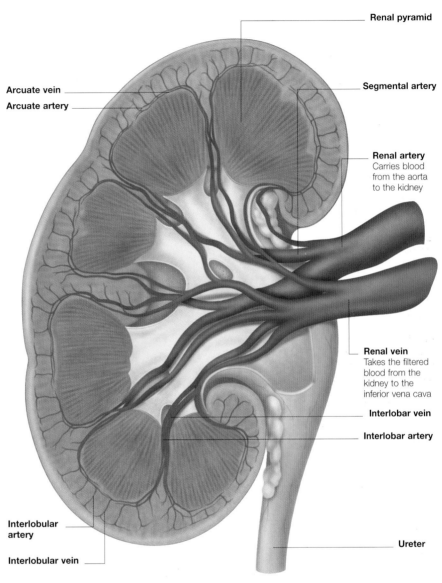

Arcuate vein

Arcuate artery

Renal pyramid

Segmental artery

Renal artery
Carries blood from the aorta to the kidney

Renal vein
Takes the filtered blood from the kidney to the inferior vena cava

Interlobar vein

Interlobar artery

Interlobular artery

Interlobular vein

Ureter

Each day, the kidneys process about 1,700 litres of blood. The renal arteries that supply the blood arise from the aorta – the main blood vessel in the body.

Bladder and ureters

The ureters channel urine produced by the kidneys down their length and into the urinary bladder. Urine is stored in the bladder until it is expelled from the body via the urethra.

Urine is continuously produced by the kidneys and is carried down to the urinary bladder by two muscular tubes, the ureters.

THE BLADDER

The bladder stores urine until it is passed out via the urethra. When the bladder is empty, it is pyramidal in shape, its walls thrown into folds, or rugae, which flatten out on filling. The position of the bladder varies:

- In adults the empty bladder lies low within the pelvis, rising up into the abdomen as it fills
- In infants the bladder is higher, being within the abdomen even when empty
- The walls contain many muscle fibres, collectively known as the detrusor muscle, which allow the bladder to contract and expel its contents.

TRIGONE

The trigone is a triangular area of the bladder wall at the base of the structure. The wall here contains muscle fibres which act to prevent urine from ascending the ureters when the bladder contracts. A muscular sphincter around the urethral opening keeps it closed until urine is passed out of the body.

Coronal section of female bladder and urethra

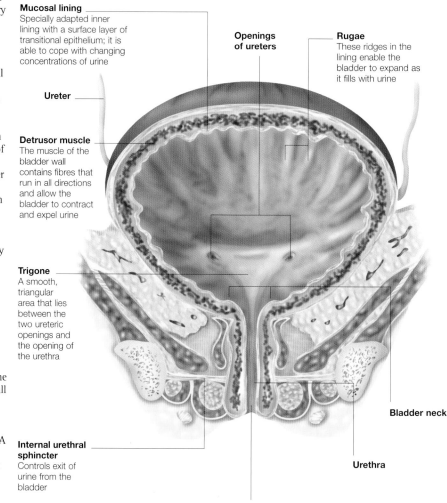

Mucosal lining
Specially adapted inner lining with a surface layer of transitional epithelium; it is able to cope with changing concentrations of urine

Ureter

Detrusor muscle
The muscle of the bladder wall contains fibres that run in all directions and allow the bladder to contract and expel urine

Trigone
A smooth, triangular area that lies between the two ureteric openings and the opening of the urethra

Internal urethral sphincter
Controls exit of urine from the bladder

Openings of ureters

Rugae
These ridges in the lining enable the bladder to expand as it fills with urine

Bladder neck

Urethra

Urethral orifice
In females, situated just in front of the entrance to the vagina

The urinary bladder is flexible enough to expand as it fills. It is made from strong muscle fibres that facilitate the expulsion of urine when necessary.

The ureters

The ureters are tubular and propel the urine towards the bladder. Each ureter squeezes and contracts its muscles to encourage the free flow of urine.

The ureters are narrow, thin-walled muscular tubes which carry urine from the kidneys to the urinary bladder.

Each of the two ureters is 25–30 cm in length and about 3 mm wide. They originate at the kidney and pass down the posterior abdominal wall to cross the bony brim of the pelvis and enter the bladder by piercing its posterior wall.

PARTS OF THE URETER
Each ureter consists of three anatomically distinct parts:

• Renal pelvis
This is the first part of the ureter, which lies within the hilum of the kidney. It is funnel-shaped as it receives urine from the major calyces and then tapers to form the narrow ureteric tube.
The junction of this part of the ureter with the next is one of the narrowest parts of the whole structure.
• Abdominal ureter
The ureter passes downwards through the abdomen and then towards the midline until it reaches the pelvic brim and enters the pelvis. During its course through the abdomen the ureter runs behind the peritoneum, the membranous lining of the abdominal cavity.
• Pelvic ureter
The ureter enters the pelvis just in front of the division of the large common iliac artery. It runs down the back wall of the pelvis before turning to enter the posterior wall of the bladder.

View of the ureters and bladder from behind

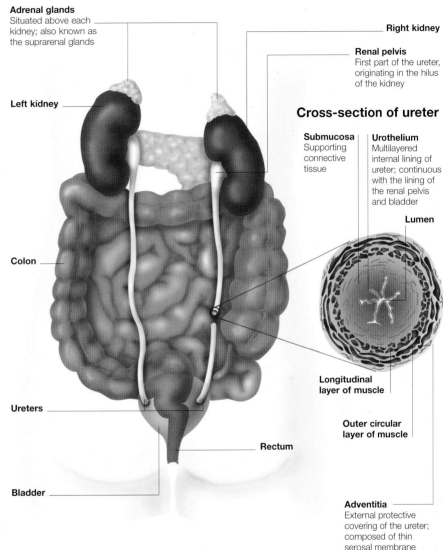

Adrenal glands
Situated above each kidney; also known as the suprarenal glands

Left kidney

Colon

Ureters

Bladder

Rectum

Right kidney

Renal pelvis
First part of the ureter, originating in the hilus of the kidney

Cross-section of ureter

Submucosa
Supporting connective tissue

Urothelium
Multilayered internal lining of ureter; continuous with the lining of the renal pelvis and bladder

Lumen

Longitudinal layer of muscle

Outer circular layer of muscle

Adventitia
External protective covering of the ureter; composed of thin serosal membrane

Urine is actively propelled along the ureters to the bladder by contraction of the muscular walls. This is the action known as 'peristalsis'.

Male reproductive system

The male reproductive system includes the penis, scrotum and the two testes (contained within the scrotum). The internal structures of the reproductive system are contained within the pelvis.

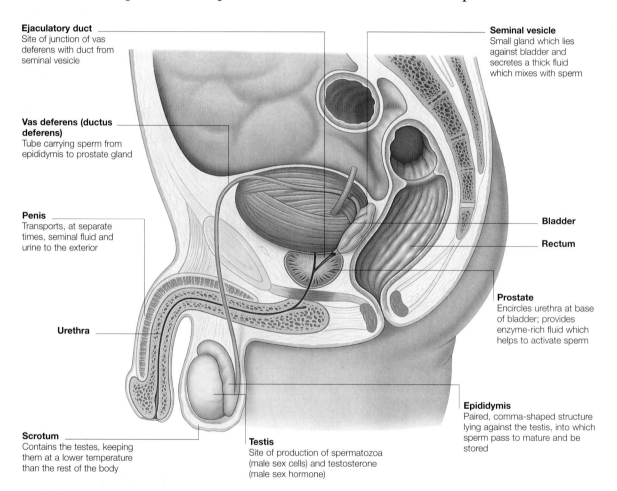

Ejaculatory duct
Site of junction of vas deferens with duct from seminal vesicle

Seminal vesicle
Small gland which lies against bladder and secretes a thick fluid which mixes with sperm

Vas deferens (ductus deferens)
Tube carrying sperm from epididymis to prostate gland

Penis
Transports, at separate times, seminal fluid and urine to the exterior

Bladder

Rectum

Urethra

Prostate
Encircles urethra at base of bladder; provides enzyme-rich fluid which helps to activate sperm

Scrotum
Contains the testes, keeping them at a lower temperature than the rest of the body

Testis
Site of production of spermatozoa (male sex cells) and testosterone (male sex hormone)

Epididymis
Paired, comma-shaped structure lying against the testis, into which sperm pass to mature and be stored

The structures constituting the male reproductive tract are responsible for the production of sperm and seminal fluid and their carriage out of the body. Unlike other organs it is not until the male reaches puberty that they develop and become fully functional.

CONSTITUENT PARTS
The male reproductive system consists of a number of interrelated parts:
- Testis – the paired testes lie suspended in the scrotum. Sperm are carried away from the testes through tubes or ducts, the first of which is the epididymis
- Epididymis – on ejaculation sperm leave the epididymis and enter the vas deferens
- Vas deferens – sperm are carried along this muscular tube en route to the prostate gland
- Seminal vesicle – on leaving the vas deferens sperm mix with fluid from the seminal vesicle gland in a combined 'ejaculatory' duct
- Prostate – the ejaculatory duct empties into the urethra within the prostate gland
- Penis – on leaving the prostate gland, the urethra then becomes the central core of the penis.

Prostate gland

The prostate gland forms a vital part of the male reproductive system, providing enzyme-rich fluid, and produces up to a third of the total volume of the seminal fluid.

About 3 cm in length, the prostate gland lies just under the bladder and encircles the first part of the urethra. Its base lies closely attached to the base of the bladder, its rounded anterior (front) surface lying just behind the pubic bone.

CAPSULE
The prostate is covered by a tough capsule made up of dense fibrous connective tissue. Outside this true capsule is a further layer of fibrous connective tissue, which is known as the prostatic sheath.

INTERNAL STRUCTURE
The urethra, the outflow tract from the bladder, runs vertically through the centre of the prostate gland, where it is known as the prostatic urethra. The ejaculatory ducts open into the prostatic urethra on a raised ridge, the seminal colliculus.

The prostate is said to be divided into lobes, although they are not as distinct as they may be in other organs:
- Anterior lobe – lies in front of the urethra and contains mainly fibromuscular tissue
- Posterior lobe – lies behind the urethra and beneath the ejaculatory ducts
- Lateral lobes – these two lobes, lying on either side of the urethra, form the main part of the gland
- Median lobe – this lies between the urethra and the ejaculatory ducts.

Location of the prostate gland

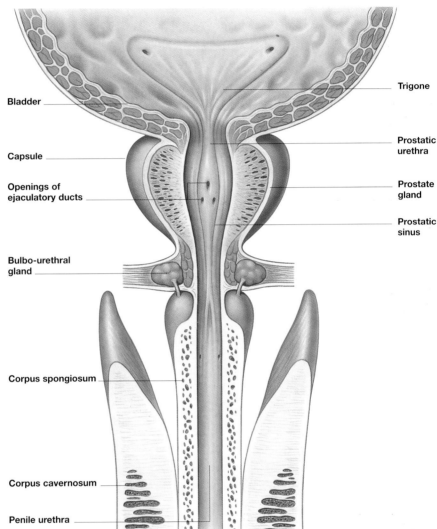

Bladder

Capsule

Openings of ejaculatory ducts

Bulbo-urethral gland

Corpus spongiosum

Corpus cavernosum

Penile urethra

Trigone

Prostatic urethra

Prostate gland

Prostatic sinus

The prostate is located at the base of the bladder, surrounding the urethra. It is a firm, smooth organ, approximately the size of a walnut.

Cross-section of prostate

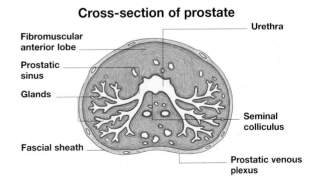

Fibromuscular anterior lobe

Prostatic sinus

Glands

Fascial sheath

Urethra

Seminal colliculus

Prostatic venous plexus

183

Testes, scrotum and epididymis

The testes, which lie suspended within the scrotum, are the sites of sperm production. The scrotum also contains the two epididymides – long, coiled tubes, which connect to the vas deferens.

The paired testes are firm, mobile, oval-shaped structures about 4 cm in length and 2.5 cm in width. The testes lie within the scrotum, a bag formed as an outpouching of the anterior abdominal wall, and are attached above to the spermatic cord, from which they hang.

TEMPERATURE CONTROL
Normal sperm can only be produced if the temperature of the testes is about three degrees lower than the internal body temperature. Muscle fibres within the spermatic cord and walls of the scrotum help to regulate the scrotal temperature by lifting the testes up towards the body when it is cold, and relaxing when the ambient temperature is higher.

EPIDIDYMIS
Each epididymis is a firm, comma-shaped structure which lies closely attached to the upper pole of the testis, running down its posterior surface. The epididymis receives the sperm made in the testis and is composed of a highly coiled tube which, if extended, would be six metres in length.

From the tail of the epididymis emerges the vas deferens. This tube will carry the sperm back up the spermatic cord and into the pelvic cavity on the next stage of the journey.

Sagittal section of the contents of the scrotum

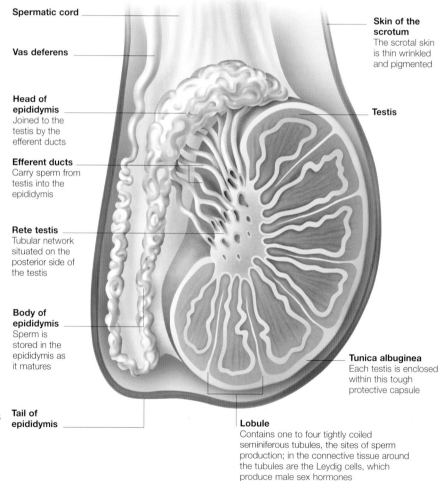

Spermatic cord

Vas deferens

Head of epididymis
Joined to the testis by the efferent ducts

Efferent ducts
Carry sperm from testis into the epididymis

Rete testis
Tubular network situated on the posterior side of the testis

Body of epididymis
Sperm is stored in the epididymis as it matures

Tail of epididymis

Skin of the scrotum
The scrotal skin is thin wrinkled and pigmented

Testis

Tunica albuginea
Each testis is enclosed within this tough protective capsule

Lobule
Contains one to four tightly coiled seminiferous tubules, the sites of sperm production; in the connective tissue around the tubules are the Leydig cells, which produce male sex hormones

The paired testes are the male sex organs that produce sperm. The testis and the epididymis on each side lie within the soft scrotal sac.

Blood supply of the testes

The arterial blood supply of the testes arises from the abdominal aorta, and descends to the scrotum. Venous drainage follows the same route in reverse.

During embryonic life, the testes develop within the abdomen; it is only at birth that they descend into their final position within the scrotum. Because of this the blood supply of the testes arises from the abdominal aorta, and travels down with the descending testis to the scrotum.

TESTICULAR ARTERIES

The paired testicular arteries are long and narrow and arise from the abdominal aorta. They then pass down on the posterior abdominal wall, crossing the ureters as they go, until they reach the deep inguinal rings and enter the inguinal canal.

As part of the spermatic cord they leave the inguinal canal and enter the scrotum where they supply the testis, also forming inter-connections with the artery to the vas deferens.

TESTICULAR VEINS

Testicular veins arise from the testis and epididymis on each side. Their course differs from that of the testicular arteries within the spermatic cord where, instead of a single vein, there is a network of veins, known as the pampiniform plexus.

Further up in the abdomen, the right testicular vein drains into the large inferior vena cava, while the left normally drains into the left renal vein.

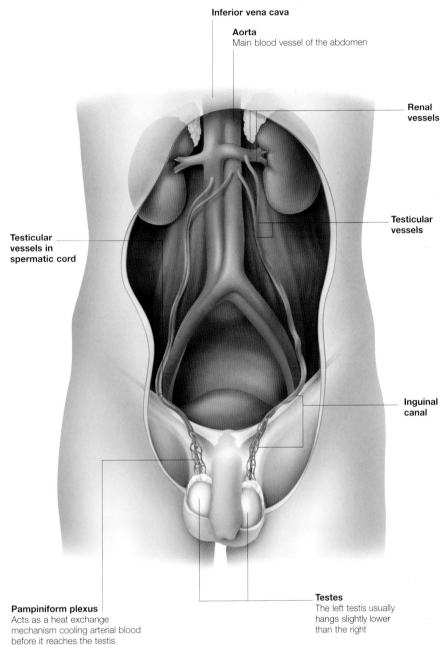

Inferior vena cava

Aorta
Main blood vessel of the abdomen

Renal vessels

Testicular vessels in spermatic cord

Testicular vessels

Inguinal canal

Pampiniform plexus
Acts as a heat exchange mechanism cooling arterial blood before it reaches the testis

Testes
The left testis usually hangs slightly lower than the right

The blood supply to the testes originates from high up in the abdominal blood vessels. These resulting long vessels allow for the testes' descent in early life.

Penis

The penis is the male copulatory organ, which, when erect, conveys sperm into the vagina during sexual intercourse. To enable this, the penis is largely composed of erectile tissue.

The penis is mostly composed of three columns of sponge-like erectile tissue, the two corpora cavernosa and the corpus spongiosum. These are able to fill and become engorged with blood, causing an erection.

STRUCTURE OF THE PENIS

There is only a small amount of muscular tissue associated with the penis, and what there is lies in its root. The shaft and glans have no muscle fibres.

The main components of the penis are:

- Root – this first part of the penis is fixed in position and is made up of the expanded bases of the three columns of erectile tissue covered by muscle fibres
- Shaft – this hangs down in the flaccid condition and is made up of erectile tissue, connective tissue, and blood and lymphatic vessels
- Glans – the tip of the penis, this is formed from the expanded end of the corpus spongiosum and carries the outlet of the urethra, the external urethral orifice
- Skin – this is continuous with that of the scrotum and is thin, dark and hairless. It is attached only loosely to the underlying fascia and lies in wrinkles when the penis is flaccid.

At the tip of the penis the skin extends as a double layer which covers the glans; this is known as the prepuce, or foreskin.

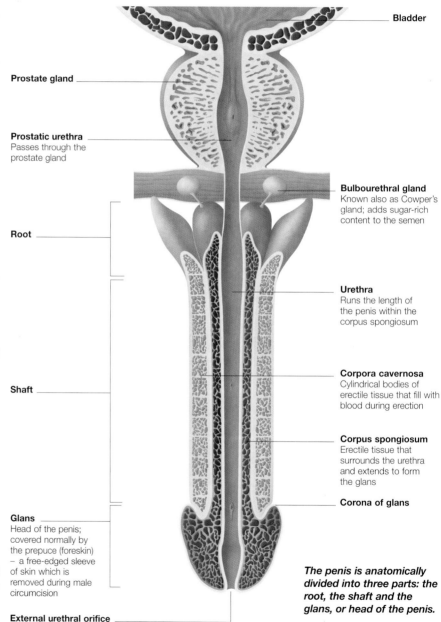

Bladder

Prostate gland

Prostatic urethra
Passes through the prostate gland

Bulbourethral gland
Known also as Cowper's gland; adds sugar-rich content to the semen

Root

Urethra
Runs the length of the penis within the corpus spongiosum

Corpora cavernosa
Cylindrical bodies of erectile tissue that fill with blood during erection

Shaft

Corpus spongiosum
Erectile tissue that surrounds the urethra and extends to form the glans

Corona of glans

Glans
Head of the penis; covered normally by the prepuce (foreskin) – a free-edged sleeve of skin which is removed during male circumcision

The penis is anatomically divided into three parts: the root, the shaft and the glans, or head of the penis.

External urethral orifice

Muscles associated with the penis

Several muscles are associated with the penis. Their fibres are confined to the root and structures around the penis, rather than to the shaft or glans.

These muscles are known collectively as the superficial perineal muscles, due to the fact that they lie in the perineum, the area around the anus and external genitalia.

There are three main muscles in this area:

• Superficial transverse perineal muscle
This narrow, paired muscle lies just under the skin in front of the anus. It runs from the ischial tuberosity of the pelvic bone on each side right across to the midline of the body.

• Bulbospongiosus
This muscle acts to compress the base of the corpus spongiosum, and thus the urethra, to help expel its contents. It originates in a central tendon or raphe, which unites the two sides and passes round to encircle the root of the penis.

• Ischiocavernosus
This muscle originates from the ischial tuberosity of the pelvic bone to surround the crura or bases of the corpora cavernosa on each side. Contraction of this muscle helps to maintain erection of the penis.

The muscles near the penis are known as the superficial perineal muscles. They surround the base of the penis and help to maintain an erection.

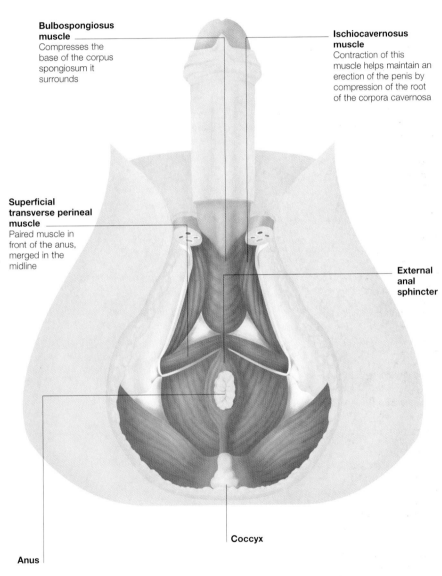

Bulbospongiosus muscle
Compresses the base of the corpus spongiosum it surrounds

Ischiocavernosus muscle
Contraction of this muscle helps maintain an erection of the penis by compression of the root of the corpora cavernosa

Superficial transverse perineal muscle
Paired muscle in front of the anus, merged in the midline

External anal sphincter

Coccyx

Anus

Female reproductive system

The role of the female reproductive tract is twofold. The ovaries produce eggs for fertilization, and the uterus nurtures and protects any resulting fetus for its nine-month gestation.

The female reproductive tract is composed of the internal genitalia – the ovaries, uterine (Fallopian) tubes, uterus and vagina – and the external genitalia (the vulva).

INTERNAL GENITALIA

The almond-shaped ovaries lie on either side of the uterus, suspended by ligaments. Above the ovaries are the paired uterine tubes, each of which provides a site for fertilization of the oocyte (egg), which then travels down the tube to the uterus.

The uterus lies within the pelvic cavity and rises into the lower abdominal cavity as a pregnancy progresses. The vagina, which connects the cervix to the vulva, can be distended greatly, as occurs during childbirth when it forms much of the birth canal.

EXTERNAL GENITALIA

The female external genitalia, or vulva, is where the reproductive tract opens to the exterior. The vaginal opening lies behind the opening of the urethra in an area known as the vestibule. This is covered by two folds of skin on each side, the labia minora and labia majora, in front of which lies the raised clitoris.

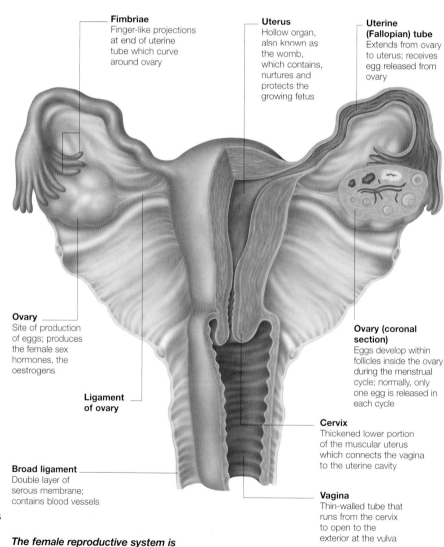

Fimbriae
Finger-like projections at end of uterine tube which curve around ovary

Uterus
Hollow organ, also known as the womb, which contains, nurtures and protects the growing fetus

Uterine (Fallopian) tube
Extends from ovary to uterus; receives egg released from ovary

Ovary
Site of production of eggs; produces the female sex hormones, the oestrogens

Ligament of ovary

Broad ligament
Double layer of serous membrane; contains blood vessels

Ovary (coronal section)
Eggs develop within follicles inside the ovary during the menstrual cycle; normally, only one egg is released in each cycle

Cervix
Thickened lower portion of the muscular uterus which connects the vagina to the uterine cavity

Vagina
Thin-walled tube that runs from the cervix to open to the exterior at the vulva

The female reproductive system is composed of internal and external organs. The internal genitalia are T-shaped and lie within the pelvic cavity.

Blood supply of the internal genitalia

The female reproductive tract receives a rich blood supply via an interconnecting network of arteries. Venous blood is drained by a network of veins.

The four principal arteries of the female genitalia are:

- Ovarian artery – this runs from the abdominal aorta to the ovary.
  Branches from the ovarian artery on each side pass through the mesovarium, the fold of peritoneum where the ovary lies, to supply the ovary and uterine (Fallopian) tubes. The ovarian artery in the tissue of the mesovarium connects with the uterine artery
- Uterine artery – this is a branch of the large internal iliac artery of the pelvis. The uterine artery approaches the uterus at the level of the cervix, which is anchored in place by cervical ligaments.
  The uterine artery connects with the ovarian artery above, while a branch connects with the arteries below to supply the cervix and vagina
- Vaginal artery – this is also a branch of the internal iliac artery. Together with blood from the uterine artery, its branches supply blood to the vaginal walls
- Internal pudendal artery – this contributes to the blood supply of the lower third of the vagina and anus.

VEINS
A plexus, or network, of small veins lies within the walls of the uterus and vagina. Blood received into these vessels drains into the internal iliac veins via the uterine vein.

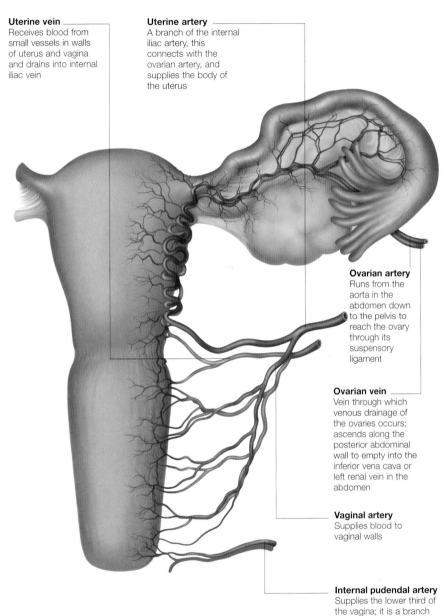

Uterine vein
Receives blood from small vessels in walls of uterus and vagina and drains into internal iliac vein

Uterine artery
A branch of the internal iliac artery, this connects with the ovarian artery, and supplies the body of the uterus

Ovarian artery
Runs from the aorta in the abdomen down to the pelvis to reach the ovary through its suspensory ligament

Ovarian vein
Vein through which venous drainage of the ovaries occurs; ascends along the posterior abdominal wall to empty into the inferior vena cava or left renal vein in the abdomen

Vaginal artery
Supplies blood to vaginal walls

Internal pudendal artery
Supplies the lower third of the vagina; it is a branch of the internal iliac artery

In this illustration, the surface layer of the female pelvic organs has been removed. This reveals the vasculature beneath.

189

Uterus

The uterus, or womb, is the part of the female reproductive tract that nurtures and protects the fetus during pregnancy. It lies within the pelvic cavity and is a hollow, muscular organ.

During a woman's reproductive years, in the non-pregnant state, the uterus is about 7.5 cm long and 5 cm across at its widest point. However, it can expand enormously to accommodate the fetus during pregnancy.

STRUCTURE

The uterus is said to be made up of two parts:

- The body, forming the upper part of the uterus – this is fairly mobile as it must expand during pregnancy. The central triangular space, or cavity, of the body receives the openings of the paired uterine (Fallopian) tubes
- The cervix, the lower part of the uterus – this is a thick, muscular canal, which is anchored to the surrounding pelvic structures for stability.

UTERINE WALLS

The main part of the uterus, the body, has a thick wall composed of three layers:

- Perimetrium – the thin outer coat which is continuous with the pelvic peritoneum
- Myometrium – forming the great bulk of the uterine wall
- Endometrium – the delicate lining, which is specialized to allow implantation of an embryo should fertilization occur.

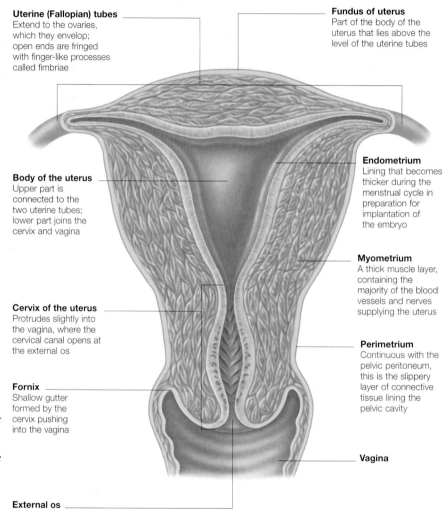

Uterine (Fallopian) tubes
Extend to the ovaries, which they envelop; open ends are fringed with finger-like processes called fimbriae

Fundus of uterus
Part of the body of the uterus that lies above the level of the uterine tubes

Body of the uterus
Upper part is connected to the two uterine tubes; lower part joins the cervix and vagina

Endometrium
Lining that becomes thicker during the menstrual cycle in preparation for implantation of the embryo

Cervix of the uterus
Protrudes slightly into the vagina, where the cervical canal opens at the external os

Myometrium
A thick muscle layer, containing the majority of the blood vessels and nerves supplying the uterus

Fornix
Shallow gutter formed by the cervix pushing into the vagina

Perimetrium
Continuous with the pelvic peritoneum, this is the slippery layer of connective tissue lining the pelvic cavity

External os

Vagina

The uterus resembles an inverted pear in shape. It is suspended in the pelvic cavity by peritoneal folds or ligaments.

The uterus in pregnancy

In pregnancy the uterus must enlarge to hold the growing fetus. From being a small pelvic organ, it increases in size to take up much of the space of the abdominal cavity.

Pressure of the enlarged uterus on the abdominal organs pushes them up against the diaphragm, encroaching on the thoracic cavity and causing the ribs to flare out to compensate. Organs such as the stomach and bladder are compressed to such an extent in late pregnancy that their capacity is greatly diminished and they become full sooner.

After pregnancy, the uterus will rapidly decrease in size again although it will always remain slightly larger than one which has never been pregnant.

HEIGHT OF FUNDUS
During pregnancy the enlarging uterus can be accommodated within the pelvis for the first 12 weeks, at which time the uppermost part, the fundus, can just be palpated in the lower abdomen. By 20 weeks, the fundus will have reached the region of the umbilicus, and by late pregnancy it may have reached the xiphisternum, the lowest part of the breastbone.

WEIGHT OF UTERUS
In the final stages of pregnancy the uterus will have increased in weight from a pre-pregnant 45 g to around 900 g. The myometrium (muscle layer) grows as the individual fibres increase in size (hypertrophy). In addition, the fibres increase in number (hyperplasia).

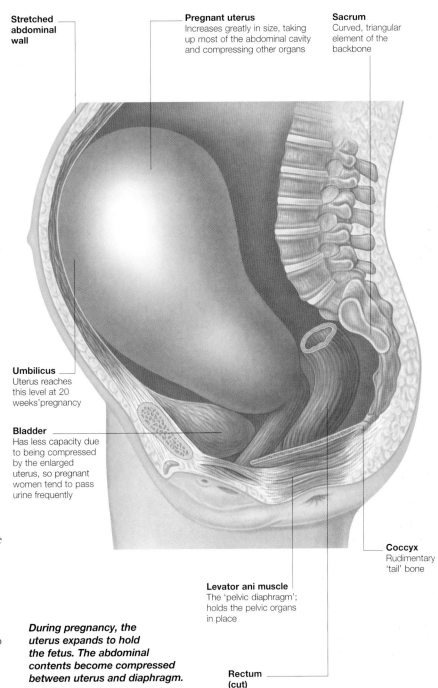

Stretched abdominal wall

Pregnant uterus
Increases greatly in size, taking up most of the abdominal cavity and compressing other organs

Sacrum
Curved, triangular element of the backbone

Umbilicus
Uterus reaches this level at 20 weeks' pregnancy

Bladder
Has less capacity due to being compressed by the enlarged uterus, so pregnant women tend to pass urine frequently

Coccyx
Rudimentary 'tail' bone

Levator ani muscle
The 'pelvic diaphragm'; holds the pelvic organs in place

Rectum (cut)

During pregnancy, the uterus expands to hold the fetus. The abdominal contents become compressed between uterus and diaphragm.

Vagina and cervix

The vagina is the thin-walled muscular tube that extends from the cervix of the uterus to the external genitalia. The vagina is closed at rest but is designed to stretch during intercourse or childbirth.

The vagina is approximately 8 cm in length and lies between the bladder and the rectum. It forms the main part of the birth canal and receives the penis during sexual intercourse.

STRUCTURE OF THE VAGINA

The front and back walls of the vagina normally lie in contact with one another, closing the lumen (central space), although the vagina can expand greatly, as occurs in childbirth.

The cervix, the lower end of the uterus, projects down into the lumen of the vagina at its upper end. Where the vagina arches up to meet the cervix, it forms recesses known as the vaginal fornices. These are divided into anterior, posterior, right and left fornices, although they form a complete ring.

The thin wall of the vagina has three layers:
- Adventitia – outer layer composed of fibroelastic connective tissue which allows distension when necessary
- Muscularis – the central muscular layer of the vaginal wall
- Mucosa – the inner layer of the vagina; this is thrown into many rugae (deep folds), and has a layered, stratified squamous (skin-like) epithelium (cell lining), which helps to resist abrasion during intercourse.

Coronal section through the vagina

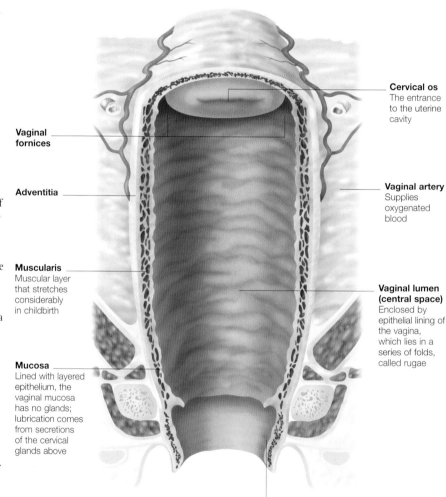

Vaginal fornices

Adventitia

Muscularis
Muscular layer that stretches considerably in childbirth

Mucosa
Lined with layered epithelium, the vaginal mucosa has no glands; lubrication comes from secretions of the cervical glands above

Cervical os
The entrance to the uterine cavity

Vaginal artery
Supplies oxygenated blood

Vaginal lumen (central space)
Enclosed by epithelial lining of the vagina, which lies in a series of folds, called rugae

Hymenal caruncle
Remains of the hymen – a fold of mucosa covering the entrance to the vagina at birth; it divides the vagina from the vestibule

The vagina is a muscular, tubular organ designed to expand during sexual intercourse and childbirth. It is approximately 8 cm in length.

The cervix

The cervix, or neck of the uterus, is the narrowed, lower part of the uterus which projects down into the upper vagina.

The cervix is fixed in position by the cervical ligaments, and so anchors the relatively mobile uterine body above.

CERVICAL STRUCTURE

The cervix has a narrow canal which is approximately 2.5 cm long in adult women. The walls of the cervix are tough, containing much fibrous tissue as well as muscle, unlike the body of the uterus, which is mainly muscular.

The central canal of the cervix is the downwards continuation of the uterine cavity which opens at its lower end, the external os, into the vagina. The canal is widest at its central point, constricting slightly at the internal os at the upper end and the external os below.

LINING OF THE CERVIX

The epithelium, or lining, of the cervix is of two types:
• Endocervix – this is the lining of the cervical canal, inside the cervix. The epithelium is a simple, single layer of columnar cells which overlies a surface thrown into many folds containing glands.
• Ectocervix – this covers the portion of the cervix which projects down into the vagina; it is composed of squamous epithelium and has many layers.

The cervix is located at the lower end of the uterus. It contains less muscle tissue than the uterus and is lined with two different types of epithelial cell.

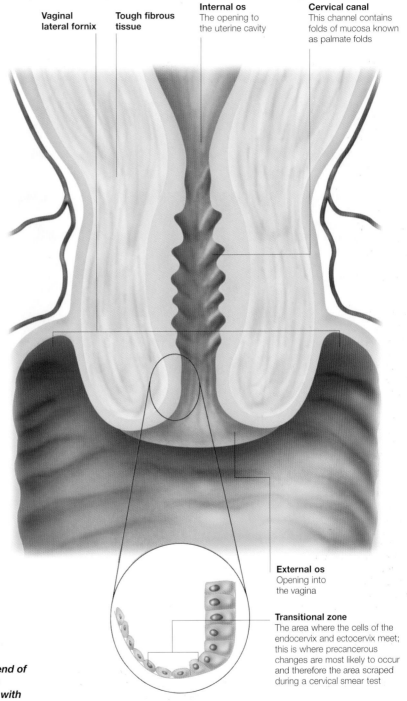

Vaginal lateral fornix

Tough fibrous tissue

Internal os
The opening to the uterine cavity

Cervical canal
This channel contains folds of mucosa known as palmate folds

External os
Opening into the vagina

Transitional zone
The area where the cells of the endocervix and ectocervix meet; this is where precancerous changes are most likely to occur and therefore the area scraped during a cervical smear test

Ovaries and uterine tubes

The ovaries are the site of production of oocytes, or eggs, which are fertilized by sperm to produce embryos. The uterine (or Fallopian) tubes conduct the oocytes from the ovaries to the uterus.

The paired ovaries are situated in the lower abdomen and lie on either side of the uterus. Their position may be variable, especially after childbirth, when the supporting ligaments have been stretched.

Each ovary consists of:
- Tunica albuginea – a protective layer of fibrous tissue
- Medulla – a central region with blood vessels and nerves
- Cortex – within which the oocytes develop
- Surface layer – smooth before puberty but becoming more pitted in the reproductive years.

BLOOD SUPPLY
The arterial supply to the ovaries comes via the ovarian arteries, which arise from the abdominal aorta. After supplying the uterine tubes also, the ovarian arteries overlap with the uterine arteries.

Blood from the ovaries enters a network of tiny veins, the pampiniform plexus, within the broad ligament, from which it enters the right and left ovarian veins. These ascend into the abdomen to drain ultimately into the large inferior vena cava and the renal vein respectively.

Cross-section of an ovary

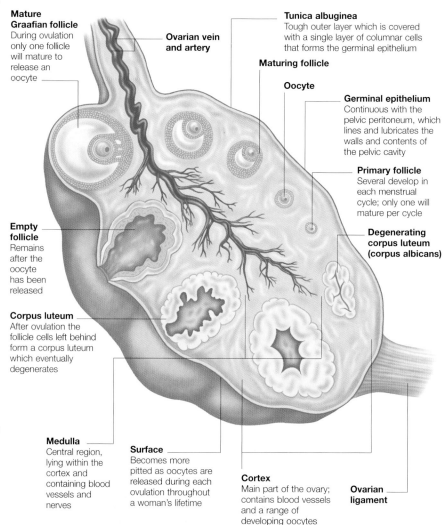

Mature Graafian follicle
During ovulation only one follicle will mature to release an oocyte

Ovarian vein and artery

Tunica albuginea
Tough outer layer which is covered with a single layer of columnar cells that forms the germinal epithelium

Maturing follicle

Oocyte

Germinal epithelium
Continuous with the pelvic peritoneum, which lines and lubricates the walls and contents of the pelvic cavity

Primary follicle
Several develop in each menstrual cycle; only one will mature per cycle

Degenerating corpus luteum (corpus albicans)

Empty follicle
Remains after the oocyte has been released

Corpus luteum
After ovulation the follicle cells left behind form a corpus luteum which eventually degenerates

Medulla
Central region, lying within the cortex and containing blood vessels and nerves

Surface
Becomes more pitted as oocytes are released during each ovulation throughout a woman's lifetime

Cortex
Main part of the ovary; contains blood vessels and a range of developing oocytes

Ovarian ligament

This cross-section shows the follicles situated in the cortex of the ovary. Each follicle contains an oocyte at a different stage of development.

The uterine tubes

The uterine, or Fallopian, tubes collect the oocytes released from the ovaries and transport them to the uterus. They also provide a site for fertilization of the oocyte by a sperm to take place.

Major parts of a uterine tube

The uterine tubes lie on either side of the body. The outer part of each tube lies near the ovary, its end opening there into the abdominal cavity.

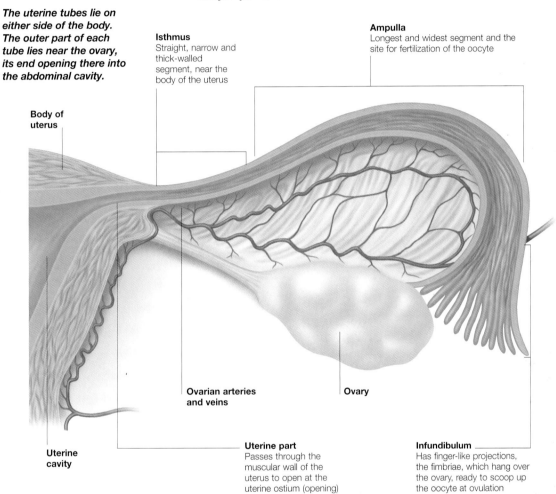

Isthmus
Straight, narrow and thick-walled segment, near the body of the uterus

Ampulla
Longest and widest segment and the site for fertilization of the oocyte

Body of uterus

Ovarian arteries and veins

Ovary

Uterine cavity

Uterine part
Passes through the muscular wall of the uterus to open at the uterine ostium (opening)

Infundibulum
Has finger-like projections, the fimbriae, which hang over the ovary, ready to scoop up the oocyte at ovulation

Each uterine tube is about 10 cm long and extends outwards from the upper part of the body of the uterus towards the lateral wall of the pelvic cavity.

The tubes run within the upper edge of the broad ligament and open into the peritoneal cavity in the region of the ovary.

STRUCTURE
The tubes are divided anatomically into four parts which, from outer to inner are:

- Infundibulum – the funnel-shaped outer end of the uterine tubes which opens into the peritoneal cavity
- Ampulla – the longest and widest part and the most usual site for fertilization of the oocyte
- Isthmus – a constricted region with thick walls
- Uterine part – this is the shortest part of the tube.

BLOOD SUPPLY
The uterine tubes have a very rich blood supply which comes from both the ovarian and the uterine arteries; these overlap to form an arterial arcade.

Venous blood drains from the tubes in a pattern which mirrors the arterial supply.

Bones of the pelvis

The basin-like pelvis is formed by the hip bones, sacrum and coccyx. The pelvic bones provide sites of attachment for many important muscles, and also help to protect the vital pelvic organs.

Adult female pelvis from the front

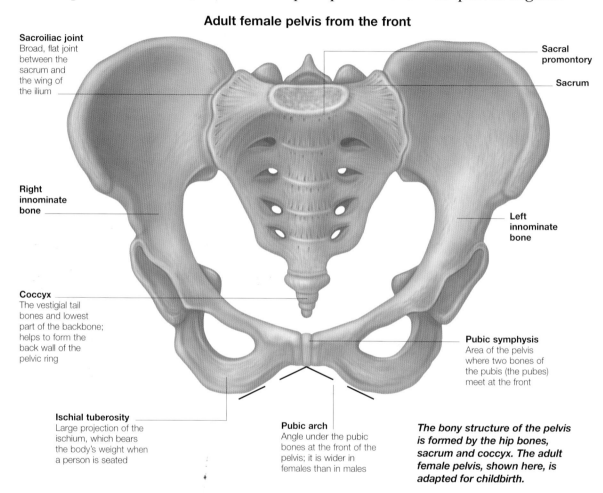

Sacroiliac joint
Broad, flat joint between the sacrum and the wing of the ilium

Sacral promontory

Sacrum

Right innominate bone

Left innominate bone

Coccyx
The vestigial tail bones and lowest part of the backbone; helps to form the back wall of the pelvic ring

Pubic symphysis
Area of the pelvis where two bones of the pubis (the pubes) meet at the front

Ischial tuberosity
Large projection of the ischium, which bears the body's weight when a person is seated

Pubic arch
Angle under the pubic bones at the front of the pelvis; it is wider in females than in males

The bony structure of the pelvis is formed by the hip bones, sacrum and coccyx. The adult female pelvis, shown here, is adapted for childbirth.

The bones of the pelvis form a ring which connects the spine to the lower limbs and protects the pelvic contents, including the reproductive organs and bladder.

The pelvic bones, to which many powerful muscles are attached, allow the weight of the body to be transferred to the legs with great stability.

STRUCTURE OF THE PELVIS
The basin-like pelvis consists of the innominate (hip) bones, the sacrum and the coccyx. The innominate bones meet at the pubic symphysis anteriorly. Posteriorly, these two bones are joined to the sacrum. Extending down from the sacrum at the back of the pelvis is the coccyx.

FALSE AND TRUE PELVIS
The pelvis can be said to be divided into two parts by an imaginary plane passing through the sacral promontory and the pubic symphysis:
• Above the sacral promontory, the false pelvis flares out and supports the lower abdominal contents
• Below this plane lies the true pelvis lies; in females, it forms the constricted birth canal through which the baby passes.

The hip bone

The two hip bones are fused together at the front and join with the sacrum at the back. They each consist of three bones – the ilium, ischium and pubis.

The two innominate (hip) bones constitute the greater part of the pelvis, joining with each other at the front and with the sacrum at the back.

STRUCTURE

The hip bone is large and strong, due to its function of transmitting the forces between the legs and the spine. As with most bones, it has areas which are raised or roughened by the attachments of muscle or ligaments.

The hip bone is formed by the fusion of three separate bones: the ilium, the ischium and the pubis. In children, these three bones are joined only by cartilage. At puberty, they fuse to form the single innominate, or hip, bone on each side.

FEATURES

The upper margin of the hip bone is formed by the widened iliac crest. Further down the hip bone is the ischial tuberosity, a projection of the ischium.

The obturator foramen lies below and slightly in front of the acetabulum, the latter receiving the head of the femur (thigh bone).

This lateral view of the hip bone clearly shows its constituent parts of ilium, ischium and pubis. These three bones fuse together at puberty.

Right hip bone, lateral view

Ilium
Bone which forms the upper part of each hip bone

Iliac crest
Convex upper margin of the hip bone which passes back from the prominent anterior superior iliac spine

Anterior superior iliac spine
Bony projection of ilium; provides attachment for the sartorius muscle at the front of the thigh and the inguinal ligament in the groin

Ischium
Bone which forms the lower posterior part of each hip bone

Ischial spine

Ischial tuberosity
Large projection of the ischium; part of the pelvis which bears the weight when sitting

Obturator foramen
Large opening in lower hip bone which is almost completely covered by a fibrous connective tissue sheet

Acetabulum
Cup-like depression which receives head of the femur to form the hip joint

Pubis
Bone which forms the lower, anterior part of each hip bone

Pelvic floor muscles

The muscles of the pelvic floor play a vital role in supporting the abdominal and pelvic organs. They also help to regulate the processes of defecation and urination.

Female pelvic diaphragm from above

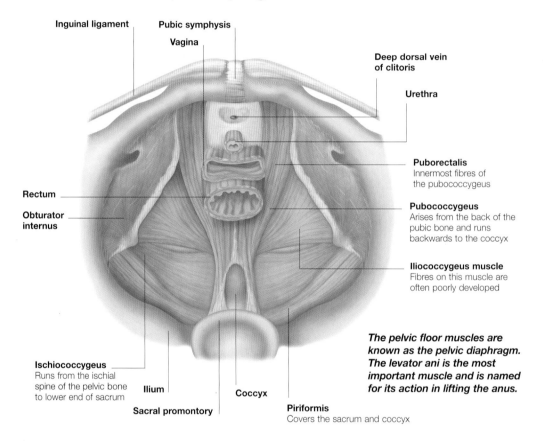

Inguinal ligament

Pubic symphysis

Vagina

Deep dorsal vein of clitoris

Urethra

Puborectalis
Innermost fibres of the pubococcygeus

Rectum

Obturator internus

Pubococcygeus
Arises from the back of the pubic bone and runs backwards to the coccyx

Iliococcygeus muscle
Fibres on this muscle are often poorly developed

The pelvic floor muscles are known as the pelvic diaphragm. The levator ani is the most important muscle and is named for its action in lifting the anus.

Ischiococcygeus
Runs from the ischial spine of the pelvic bone to lower end of sacrum

Ilium

Coccyx

Sacral promontory

Piriformis
Covers the sacrum and coccyx

The pelvic floor muscles play an important role in supporting the abdominal and pelvic organs. In pregnancy, they help to carry the growing weight of the uterus, and in childbirth they support the baby's head as the cervix dilates.

MUSCLES
The muscles of the pelvic floor are attached to the inside of the ring of bone that makes up the pelvic skeleton, and slope downwards to form a rough funnel shape.

The levator ani is the largest muscle of the pelvic floor. It is a wide, thin sheet made up of three parts:
- Pubococcygeus – the main part of the levator ani muscle
- Puborectalis – joins with its counterpart on the other side to form a U-shaped sling around the rectum
- Iliococcygeus – the posterior fibres of the levator ani. A second muscle, the coccygeus (or ischiococcygeus), lies behind the levator ani.

PELVIC WALLS
The pelvic cavity is described as having an anterior, a posterior and two lateral walls.

The anterior wall is formed by the pubic bones and their connection, the pubic symphysis. The posterior wall is formed by the sacrum and coccyx and the neighbouring parts of the iliac bones. The two lateral walls are formed by the obturator internus muscles overlying the hip bones.

Openings of the pelvic floor

The pelvic floor resembles the diaphragm in the chest in that it forms a nearly continuous sheet, but does have openings to allow important structures to pass through it. There are two important openings situated in the pelvic floor region.

From below, the pelvic floor can be seen to assume a funnel shape. The muscles of the pelvic floor are so arranged that there are two main openings:

- Anorectal hiatus – this opening, or hiatus, allows the rectum and anal canal to pass through the sheet of pelvic floor muscles to reach the anus beneath. The U-shaped fibres of the puborectalis muscle form the posterior edge of this hiatus
- Urogenital hiatus – lying in front of the anorectal hiatus there is an opening in the pelvic floor for the urethra, which carries urine from the bladder out of the body. In females, the vagina also passes through the pelvic diaphragm within this opening, just behind the urethra.

FUNCTIONS OF THE PELVIC FLOOR MUSCLES

- Supporting internal organs of the abdomen and pelvis
- Helping to resist rises in pressure within the abdomen, such as during coughing and sneezing, which would cause the bladder/bowel to empty
- Assisting in the control of defecation and urination
- Helping to brace the trunk during forceful movements of the upper limbs, such as weight-lifting.

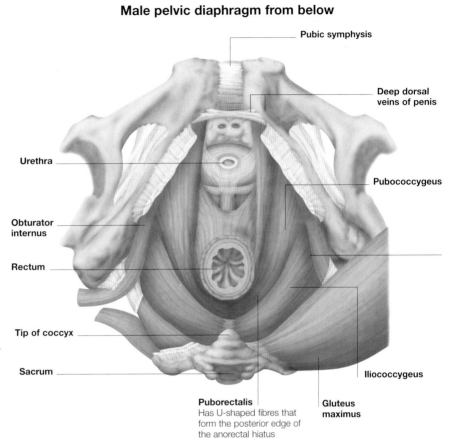

Male pelvic diaphragm from below

Pubic symphysis

Deep dorsal veins of penis

Urethra

Pubococcygeus

Obturator internus

Rectum

Tip of coccyx

Sacrum

Iliococcygeus

Puborectalis
Has U-shaped fibres that form the posterior edge of the anorectal hiatus

Gluteus maximus

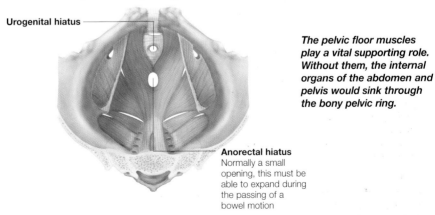

Male pelvic diaphragm from above

Urogenital hiatus

The pelvic floor muscles play a vital supporting role. Without them, the internal organs of the abdomen and pelvis would sink through the bony pelvic ring.

Anorectal hiatus
Normally a small opening, this must be able to expand during the passing of a bowel motion

Muscles of the gluteal region

The gluteus maximus is the largest and heaviest of all the gluteal muscles and is situated in the buttock region. This strong, thick muscle plays an important part in enabling humans to stand.

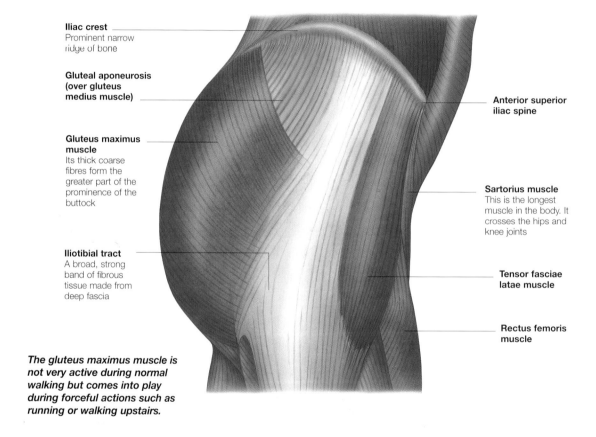

Iliac crest
Prominent narrow ridge of bone

Gluteal aponeurosis (over gluteus medius muscle)

Gluteus maximus muscle
Its thick coarse fibres form the greater part of the prominence of the buttock

Iliotibial tract
A broad, strong band of fibrous tissue made from deep fascia

Anterior superior iliac spine

Sartorius muscle
This is the longest muscle in the body. It crosses the hips and knee joints

Tensor fasciae latae muscle

Rectus femoris muscle

The gluteus maximus muscle is not very active during normal walking but comes into play during forceful actions such as running or walking upstairs.

The gluteal, or buttock, region lies behind the pelvis. The shape is formed by a number of large muscles which help to stabilize and move the hip joint. A layer of fat covers these muscles.

GLUTEUS MAXIMUS
This is one of the largest muscles in the body. It covers the other gluteal muscles with the exception of about one-third of the smaller gluteus medius. The gluteus maximus arises from the ilium (a part of the bony pelvis), the back of the sacrum and the coccyx. Its fibres run down and outwards at a 45 degree angle towards the femur.

Most of the fibres then insert into a band (the iliotibial tract).

ACTIONS
The main function of the gluteus maximus is to extend (straighten) the leg as in standing from a sitting position. When the leg is extended, as in standing, the gluteus maximus covers the bony ischial tuberosity. This bears the weight of the body when sitting. However, we never sit on the gluteus maximus muscle itself as it moves up and away from the ischial tuberosity when the leg is flexed (bent forward).

Deeper muscles of the gluteal region

The muscles that lie deep to the gluteus maximus region play an important part in walking. They keep the pelvis level as each foot is lifted off the ground.

Quadratus femoris
This rectangular-shaped muscle extends laterally from the pelvis

Gluteus medius muscle
This is a thick muscle largely covered by the gluteus maximus

Gluteus minimus muscle
The smallest and deepest of the gluteal muscles

Ischial tuberosity
The ischial tuberosities are the strongest parts of the hip bones

Piriformis muscle

Superior gemellus muscle

Obturator internus

Gluteus maximus

Inferior gemellus muscle

Greater trochanter

The deep muscles of this group rotate the thigh laterally and stabilize the hip joint. The main muscles are the gluteus medius and gluteus minimus.

Beneath the gluteus maximus lie a number of other muscles which act to stabilise the hip joint and move the lower limb.

GLUTEUS MEDIUS AND MINIMUS

The gluteus medius and gluteus minimus muscles lie deep to the gluteus maximus. They are both fan-shaped muscles with fibres that run in the same direction.

Gluteus medius lies directly beneath gluteus maximus, with only about one-third of it not covered by this larger

muscle. Its fibres originate from the external surface of the ilium (part of the pelvis) and insert into the greater trochanter, a protuberance of the femur.

Gluteus minimus lies directly beneath gluteus medius and is of a similar fan-like shape. Its fibres also originate from the ilium and insert into the greater trochanter.

ESSENTIAL ROLE

The gluteus medius and gluteus minimus together have an essential role in the

action of walking. These muscles act to hold the pelvis level when one foot is lifted from the ground, rather than letting it sag to that side. This allows the non-weight-bearing foot to clear the ground before being swung further forward.
A number of other muscles lie within this region, acting mainly to help certain movements of the lower limb at the hip. These include:
• Piriformis – this muscle, which is named for its pear shape, lies below gluteus minimus. It acts to rotate

the thigh laterally, a movement which results in the foot turning outwards
• Obturator internus, superior and inferior gemelli – these three muscles together form a composite three-headed muscle which lies below the piriformis muscle. These muscles rotate the thigh laterally and stabilize the hip joint
• Quadratus femoris – this short, thick muscle rotates the thigh laterally and helps to stabilize the hip joint.

201

Hip joint

The hip joint is the strong ball-and-socket joint that connects the lower limb to the pelvis. Of all the body's joints, the hip is second only to the shoulder in the variety of movements it allows.

In the hip joint, the head of the femur (thigh bone) is the 'ball' that fits tightly into the 'socket' formed by the cup-like acetabulum of the hip bone of the pelvis.

The articular surfaces – the parts of the bone which come into contact with each other – are covered by a protective layer of hyaline cartilage, which is very smooth and slippery. The hip joint is a synovial joint, which means that movement is further lubricated by a thin layer of synovial fluid, which lies between these articular surfaces within the synovial cavity. The fluid is secreted by the synovial membrane.

ACETABULAR LABRUM
The depth of the socket formed by the acetabulum is increased by the presence of the acetabular labrum. This structure brings greater stability to the joint, allowing the almost spherical femoral head to rest deep within the joint.

The cartilage-covered articular surface of the acetabulum is not a continuous cup, or even a ring, but is horseshoe-shaped. There is a gap, the acetabular notch, at the lowest point, which is bridged by the complete ring of the acetabular labrum. The open centre of the 'horseshoe' is filled with a cushioning pad of fat.

Cross-section of the right hip joint

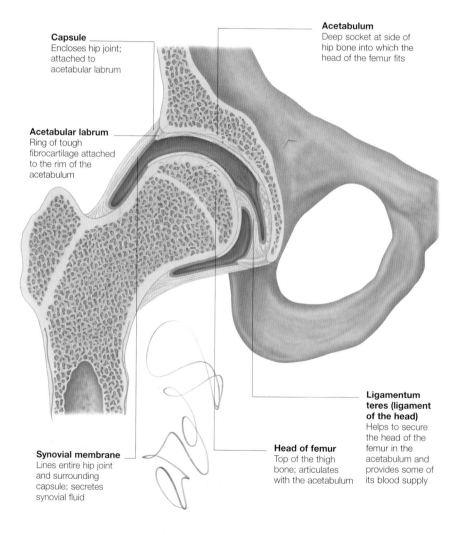

Capsule
Encloses hip joint; attached to acetabular labrum

Acetabular labrum
Ring of tough fibrocartilage attached to the rim of the acetabulum

Acetabulum
Deep socket at side of hip bone into which the head of the femur fits

Ligamentum teres (ligament of the head)
Helps to secure the head of the femur in the acetabulum and provides some of its blood supply

Head of femur
Top of the thigh bone; articulates with the acetabulum

Synovial membrane
Lines entire hip joint and surrounding capsule; secretes synovial fluid

The hip joint is the ball-and-socket joint between the head of the femur and the hip bone. The joint is capable of a wide range of movement.

Ligaments of the hip joint

The hip joint is enclosed and protected by a thick, fibrous capsule. The capsule is flexible enough to allow the joint a wide range of movements but is strengthened by a number of tough ligaments.

The ligaments of the hip joint are thickened parts of the joint capsule, which extends from the rim of the acetabulum down to the neck of the femur. These ligaments, which generally follow a spiral path from the hip bone to the femur, are named according to the parts of the bone to which they attach:
• Iliofemoral ligament
• Pubofemoral ligament
• Ischiofemoral ligament.

MOVEMENT AND STABILITY
The ball-and-socket nature of the hip joint allows it great mobility, second only to the shoulder joint in its range of movement. Unlike the shoulder, however, it needs to be very stable as it is a major weight-bearing joint. It is capable of the following movements:

• Flexion (bending forward, the knee coming up)
• Extension (bending the leg back behind the body)
• Abduction (moving the leg out to the side)
• Adduction (bringing the leg back to the midline)
• Rotation, which is greatest when the leg is flexed.

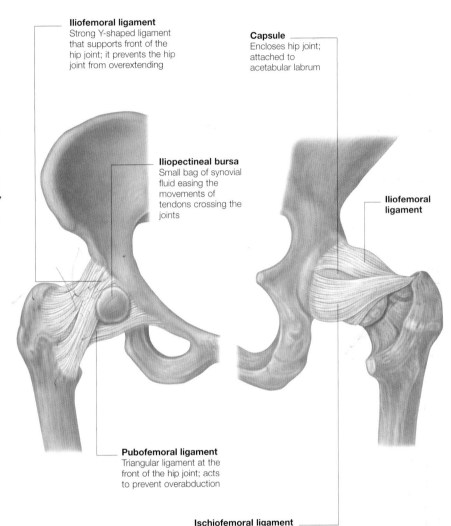

Anterior view of right hip

Iliofemoral ligament
Strong Y-shaped ligament that supports front of the hip joint; it prevents the hip joint from overextending

Iliopectineal bursa
Small bag of synovial fluid easing the movements of tendons crossing the joints

Pubofemoral ligament
Triangular ligament at the front of the hip joint; acts to prevent overabduction

Posterior view of right hip

Capsule
Encloses hip joint; attached to acetabular labrum

Iliofemoral ligament

Ischiofemoral ligament
Large spiral ligament lying at back of hip joint; prevents overextension

A fibrous capsule encloses the hip joint. This capsule is reinforced by a number of ligaments which spiral down from the hip bone to the femur.

Femur

The femur, or thigh bone, is the longest and heaviest bone in the body. Measuring approximately 45 cm in length in adult males, the femur makes up about one quarter of a person's total height.

The femur has a long, thick shaft with two expanded ends. The upper end articulates with the pelvis to form the hip joint while the lower end articulates with the tibia and patella to form the knee joint.

UPPER END
The femur upper end includes:
- Head – this is the near-spherical projection which forms the 'ball' of the ball and socket hip joint
- Neck – this is the narrowed area which connects the head to the body of the femur
- Greater and lesser trochanters – projections of bone allowing the attachment of muscles.

SHAFT
The long central shaft of the femur is slightly bowed, being concave on its posterior surface. For much of its length the femur appears cylindrical, with a circular cross-section.

LOWER END
The lower end of the femur is made up of two enlarged bony processes, the medial and lateral femoral condyles. These carry the smooth, curved surfaces which articulate with the tibia and patella to form the knee joint. The shape of the femoral condyles is outlined when the leg is viewed with the knee bent.

The femur is the thigh bone, running from the hip joint to the knee. It is the longest bone in the body and is very strong.

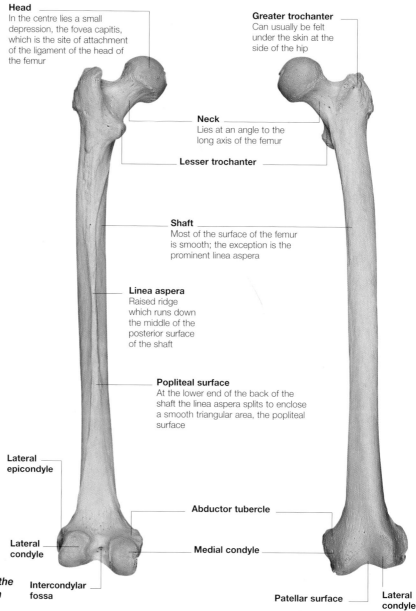

Posterior view (left leg)

Head
In the centre lies a small depression, the fovea capitis, which is the site of attachment of the ligament of the head of the femur

Neck
Lies at an angle to the long axis of the femur

Lesser trochanter

Shaft
Most of the surface of the femur is smooth; the exception is the prominent linea aspera

Linea aspera
Raised ridge which runs down the middle of the posterior surface of the shaft

Popliteal surface
At the lower end of the back of the shaft the linea aspera splits to enclose a smooth triangular area, the popliteal surface

Lateral epicondyle

Lateral condyle

Intercondylar fossa

Anterior view (left leg)

Greater trochanter
Can usually be felt under the skin at the side of the hip

Abductor tubercle

Medial condyle

Patellar surface

Lateral condyle

Muscle attachments of the femur

The femur is a very strong bone which provides sites of attachment for many of the muscles of locomotion in the hip joint and legs.

MUSCLE ORIGINS

Some muscles, such as the powerful gluteus muscles, have their origins on the pelvic bones and so cross the hip joint to insert into the femur. When these muscles contract they cause the hip joint to move, allowing the leg to bend, straighten or move sideways.

Other muscles originate on the femur itself and pass down across the knee joint to insert on the tibia or the fibula, the two bones of the lower leg. These muscles allow the knee to bend or straighten.

Together these muscles bring about movements of the legs such as in climbing, or rising from a sitting position.

BONY PROCESSES

Where muscle is attached to bone it causes a projection, or bony process, to arise. If the muscle is powerful, or if a number of muscles attach at the same site, the bony process can be pronounced. This is the case in the femur. The surface of the bone at the site of muscle attachment can also become quite roughened, unlike the smooth bone surface in between.

The surface of the femur is roughened by projections to which muscles are attached. These muscles bring about movements in the legs and hips.

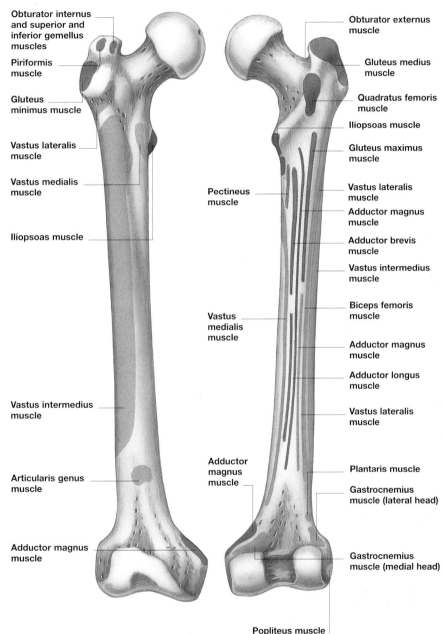

Anterior view (right leg)

Obturator internus and superior and inferior gemellus muscles

Piriformis muscle

Gluteus minimus muscle

Vastus lateralis muscle

Vastus medialis muscle

Iliopsoas muscle

Vastus intermedius muscle

Articularis genus muscle

Adductor magnus muscle

Posterior view (right leg)

Obturator externus muscle

Gluteus medius muscle

Quadratus femoris muscle

Iliopsoas muscle

Gluteus maximus muscle

Pectineus muscle

Vastus lateralis muscle

Adductor magnus muscle

Adductor brevis muscle

Vastus intermedius muscle

Biceps femoris muscle

Vastus medialis muscle

Adductor magnus muscle

Adductor longus muscle

Vastus lateralis muscle

Adductor magnus muscle

Plantaris muscle

Gastrocnemius muscle (lateral head)

Gastrocnemius muscle (medial head)

Popliteus muscle

Tibia and fibula

The tibia and fibula together form the skeleton of the lower leg. The tibia is much larger and stronger than the fibula as it must bear the weight of the body.

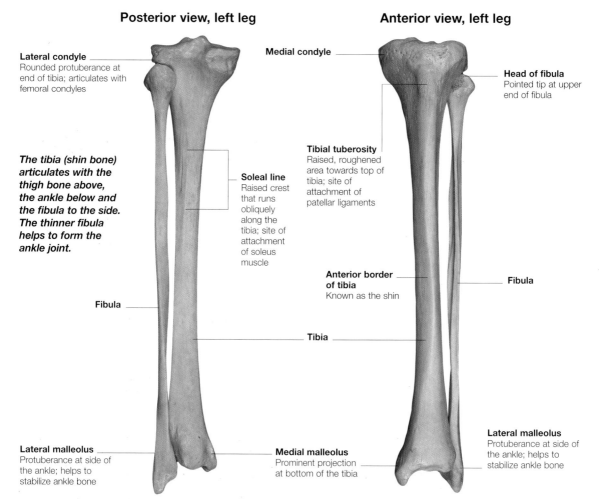

Posterior view, left leg

Anterior view, left leg

Lateral condyle
Rounded protuberance at end of tibia; articulates with femoral condyles

Medial condyle

Head of fibula
Pointed tip at upper end of fibula

The tibia (shin bone) articulates with the thigh bone above, the ankle below and the fibula to the side. The thinner fibula helps to form the ankle joint.

Soleal line
Raised crest that runs obliquely along the tibia; site of attachment of soleus muscle

Tibial tuberosity
Raised, roughened area towards top of tibia; site of attachment of patellar ligaments

Anterior border of tibia
Known as the shin

Fibula

Fibula

Tibia

Lateral malleolus
Protuberance at side of the ankle; helps to stabilize ankle bone

Medial malleolus
Prominent projection at bottom of the tibia

Lateral malleolus
Protuberance at side of the ankle; helps to stabilize ankle bone

Second only to the femur (thigh bone) in size, the tibia (shin bone) has the shape of a typical long bone, with an elongated shaft and two expanded ends. The tibia lies alongside the fibula, on the medial (inner) side, and articulates with the fibula at its upper and lower ends.

TIBIAL CONDYLES
The upper end of the tibia is expanded to form the medial and lateral tibial condyles, which articulate with the femoral condyles at the knee joint. The lower end of the tibia is less pronounced than the upper end. It articulates with both the talus (ankle bone) and with the lower end of the fibula.

FIBULA
The fibula is a long, narrow bone which has none of the strength of the tibia. It lies next to the tibia, on its lateral (outer) side, and articulates with that bone. The fibula plays no part in the knee joint, but is an important support for the ankle.
The shaft of the fibula is narrow and bears the grooves and ridges associated with its major role as a site of attachment for leg muscles.

Ligaments of the tibia and fibula

The ligaments which surround the tibia and fibula bind the two bones to each other and to the other leg bones with which they articulate.

Anterior view of left leg with ligament attachments

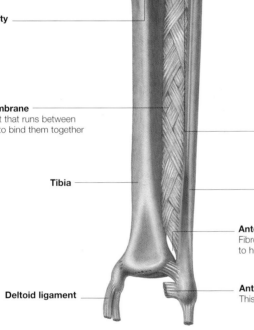

Posterior cruciate ligament

Anterior cruciate ligament

Fibular collateral ligament

Tendon of biceps femoris muscle

Tibial collateral ligament

Patellar ligament

Tibial tuberosity

Anterior ligament of fibular head
Ligament that runs from the front of the fibular head across to the front of the lateral tibial condyle in broad, flat bands

Interosseous membrane
Dense fibrous sheet that runs between the tibia and fibula to bind them together

Oval-shaped opening
This opening in the membrane allows blood vessels to pass through to reach the front of the leg

Tibia

Fibula

The tibia and fibula of the lower leg are surrounded by a number of ligaments. Ligaments are tough fibrous bands of connective tissue that bind bones together where they articulate at a joint.

Anterior tibiofibular ligament
Fibrous ligament that binds the tibia to the fibula to help maintain the stability of the ankle joint

Deltoid ligament

Anterior talofibular ligament
This is often injured in a sprained ankle

Ligaments are the strong fibrous bands which bind bones together. There are a number of ligaments which surround the tibia and the fibula; they bind the two bones to each other and to other bones of the leg.

PROXIMAL (UPPER) END
Just under the knee is the upper joint between the head of the fibula and the underside of the lateral tibial condyle. The joint is surrounded and protected by a fibrous joint capsule, which

is strengthened by the anterior and posterior tibiofibular ligaments.

The anterior ligament of the head of the fibula runs from the front of the fibular head across to the front of the lateral tibial condyle. The posterior ligament of the head of the fibula runs in a similar fashion behind the fibular head.

Other ligaments bind the bones of the lower leg to the femur. The strongest of these are the medial and lateral collateral ligaments of the

knee joint, which run vertically down from the femur to the corresponding bone (tibia or fibula) beneath.

DISTAL (LOWER) END
The joint between the lower ends of the tibia and fibula allows no movement of one bone upon the other. Rather, the fibula is bound tightly to the tibia by fibrous ligaments in order to maintain the stability of the ankle joint. The main ligaments concerned are the anterior and posterior inferior (lower)

tibiofibular ligaments. Other ligaments around the ankle bind the tibia and fibula to the bones of the foot.

INTEROSSEOUS MEMBRANE
The fibres of the dense interosseous membrane run obliquely from the sharp interosseous border of the tibia across to the front of the fibula, binding the two bones together.

207

Knee joint and patella

The knee is the joint between the end of the thigh bone and the top of the tibia. In front of the knee is the patella (kneecap), the convex surface of which can readily be felt under the skin.

The knee is the joint between the lower end of the femur (thigh bone) and the upper end of the tibia (the largest bone of the lower leg). The fibula (the smaller of the two lower leg bones) plays no part in the joint.

STRUCTURE

The knee is a synovial joint – one in which movement is lubricated by synovial fluid which is secreted by a membrane lining the joint cavity.

Although the knee tends to be thought of as a single joint it is, in fact, the most complex joint in the body, being made up of three joints which share a common joint cavity. These three joints are:

- The joint between the patella (kneecap) and the lower end of the femur. Classified as a plane joint, this allows one bone to slide upon the other
- A joint on either side between the femoral condyles (the large bulbous ends of the femur) and the corresponding part of the upper tibia. These are said to be hinge joints as the movement they allow is akin to the movement of a door on its hinges.

STABILITY OF THE KNEE

Considering that there is not a very good 'fit' between the femoral condyles and the upper end of the tibia, the knee is actually a reasonably stable joint. It relies heavily on the surrounding muscles and ligaments for its stability.

Sagittal section of knee

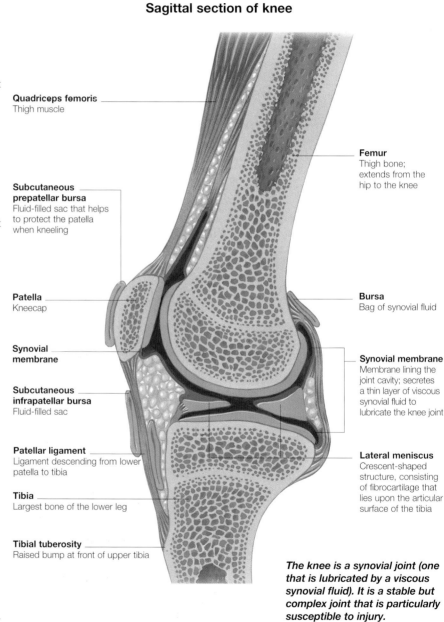

Quadriceps femoris
Thigh muscle

Subcutaneous prepatellar bursa
Fluid-filled sac that helps to protect the patella when kneeling

Patella
Kneecap

Synovial membrane

Subcutaneous infrapatellar bursa
Fluid-filled sac

Patellar ligament
Ligament descending from lower patella to tibia

Tibia
Largest bone of the lower leg

Tibial tuberosity
Raised bump at front of upper tibia

Femur
Thigh bone; extends from the hip to the knee

Bursa
Bag of synovial fluid

Synovial membrane
Membrane lining the joint cavity; secretes a thin layer of viscous synovial fluid to lubricate the knee joint

Lateral meniscus
Crescent-shaped structure, consisting of fibrocartilage that lies upon the articular surface of the tibia

The knee is a synovial joint (one that is lubricated by a viscous synovial fluid). It is a stable but complex joint that is particularly susceptible to injury.

Inside the knee – the menisci

The menisci are crescent-shaped plates of tough fibrocartilage lying on the articular surface of the tibia. They act as 'shock absorbers' within the knee and prevent sideways movement of the femur.

Superior view of knee (tibial plateau)

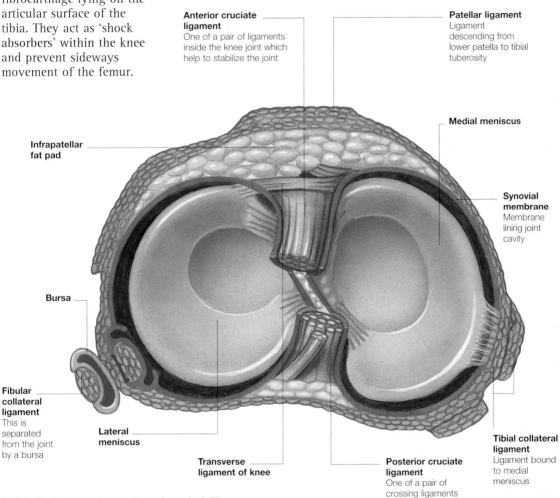

Anterior cruciate ligament
One of a pair of ligaments inside the knee joint which help to stabilize the joint

Patellar ligament
Ligament descending from lower patella to tibial tuberosity

Medial meniscus

Infrapatellar fat pad

Synovial membrane
Membrane lining joint cavity

Bursa

Fibular collateral ligament
This is separated from the joint by a bursa

Lateral meniscus

Transverse ligament of knee

Posterior cruciate ligament
One of a pair of crossing ligaments inside the knee joint

Tibial collateral ligament
Ligament bound to medial meniscus

Inside the knee are two c-shaped menisci. These are plates of tough fibrocartilage into which fit the femoral condyles (ends of the femur).

Looking down on the upper surface of the tibia within the opened knee, the two c-shaped menisci can clearly be seen. Named after the Greek word for 'crescent', the menisci are plates of tough fibrocartilage which lie upon the articular surface of the tibia, deepening the depression into which the femoral condyles fit.

SHOCK ABSORBERS
The menisci also have the function of acting as 'shock absorbers' within the knee and help to prevent the side-to-side rocking of the joint.

STRUCTURE OF THE MENISCI
The two menisci are wedge-shaped in cross-section, their external margins being widest. Centrally, they taper to a thin, unattached edge. Anteriorly, the two menisci are attached to each other by the transverse ligament of the knee, while the outer edges of the menisci are firmly attached to the joint capsule.

ATTACHMENT
Attachment of the medial meniscus to the tibial collateral ligament is of great clinical significance as the meniscus can be damaged when this ligament is injured during contact sports.

Ligaments and bursae of the knee

The knee joint is only partially enclosed in a capsule and relies on ligaments for its stability. Bursae are situated around the knee and allow smooth movement to take place.

The bones of the knee do not fit together in a particularly stable fashion. For this reason the stability of the knee joint depends upon the ligaments and muscles that surround it.

The joint cavity of the knee is enclosed within a fibrous capsule. The ligaments that support the knee can be divided into two groups, depending on their relationship to this capsule.

EXTRACAPSULAR LIGAMENTS

The extracapsular ligaments lie outside the capsule and act to prevent the lower leg bending forward at the knee, or hyperextending. They include:

- Quadriceps tendon – extends from tendon of the quadriceps femoris muscle. This supports the front of the knee (not shown)
- Fibular (or lateral) collateral ligament – a strong cord which binds the lower end of the outer femur to the head of the fibula
- Tibial (or medial) collateral ligament – a strong flat band, which runs from the lower end of the inner femur down to the tibia. It is weaker than the fibular collateral ligament
  - Oblique popliteal ligament – this strengthens the capsule at the back
  - Arcuate popliteal ligament – also adds strength to the back of the knee (not shown).

Anterior view of flexed left knee

Posterior cruciate ligament

Medial condyle of femur

Medial meniscus

Deep portion of tibial collateral ligament
Its attachment to the meniscus makes it prone to injury on twisting

Transverse ligament of the knee

Tibial collateral ligament
Prevents medial movement when the knee is extended

Tibia

Tibial tuberosity
Site of insertion of the patellar ligament – the distal end of the quadriceps tendon

Anterior cruciate ligament

Lateral condyle of femur

Lateral meniscus

Popliteus tendon
Strengthens the back of the knee

Fibular collateral ligament
A strong cord that prevents lateral movement when the knee is extended

Head of fibula

Extracapsular ligaments are responsible for preventing hyperextension of the knee. These ligaments are stretched when extending the knee.

Bursae of the knee

The bursae of the knee are small sacs filled with synovial fluid. They act to protect the structures inside the knee, reducing friction, as they slide over each other when the joint is moving.

Bursae are small fluid-filled sacs found between two structures, usually bone and tendon, that regularly move against each other. The bursae protect the structures from wear and tear.

There are a number of bursae around the knee that protect the tendons during movement or allow easy movement of the skin across the patella.

SUPRAPATELLAR BURSA
Some of the bursae around the knee joint are continuous with the joint cavity, the fluid-filled space between the articular surfaces. The suprapatellar bursa lies above the joint cavity between the lower end of the femur and the powerful quadriceps femoris muscle.

PREPATELLAR AND INFRAPATELLAR BURSAE
These surround the patella and the patellar ligament. The prepatellar bursa allows the skin to move freely over the patella during movement. The superficial and deep infrapatellar bursae lie around the lower end of the patellar ligament where it attaches to the tibial tuberosity.

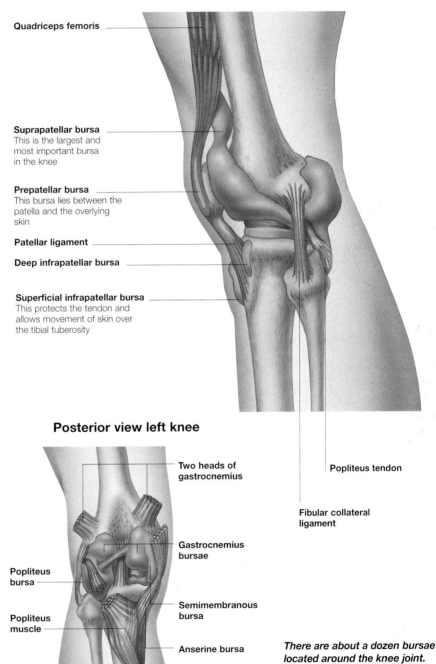

Lateral view left knee

Quadriceps femoris

Suprapatellar bursa
This is the largest and most important bursa in the knee

Prepatellar bursa
This bursa lies between the patella and the overlying skin

Patellar ligament

Deep infrapatellar bursa

Superficial infrapatellar bursa
This protects the tendon and allows movement of skin over the tibial tuberosity

Popliteus tendon

Fibular collateral ligament

Posterior view left knee

Two heads of gastrocnemius

Gastrocnemius bursae

Popliteus bursa

Semimembranous bursa

Popliteus muscle

Anserine bursa

There are about a dozen bursae located around the knee joint. They allow the structures of the knee to move freely over one another, reducing friction.

211

Muscles of the thigh

The thigh is composed mainly of groups of large muscles which act to move the hip and the knee joint. Muscles that effect the movements of the thigh are among the strongest in the body.

The thigh muscles are divided into three basic groups; the anterior muscles lie in front of the femur, the posterior muscles lie behind, and the medial muscles (adductors) run between the inner femur (thigh) and the pelvis.

ANTERIOR MUSCLES

The muscles of the anterior compartment of the thigh flex or bend the hip, and extend or straighten the knee. These are the actions associated with lifting the leg up and bringing it forward during walking.

The muscles of this group include:

- Iliopsoas. This large muscle arises partly from the inside of the pelvis and partly from the lumbar (lower) vertebrae. Its fibres insert into the projection of the upper femur known as the lesser trochanter. The iliopsoas is the most powerful of the muscles that flex the thigh, bringing the knee up and forwards
- Tensor fasciae latae. This muscle inserts into the strong band of connective tissue that runs down the outside of the leg to the tibia below the knee
- Sartorius. The longest muscle in the body, the sartorius runs as a flat strap across the thigh from the anterior superior iliac spine of the pelvis. It crosses both the hip and the knee joint before it inserts into the inner side of the top of the tibia.
- Quadriceps femoris. A large four-headed muscle.

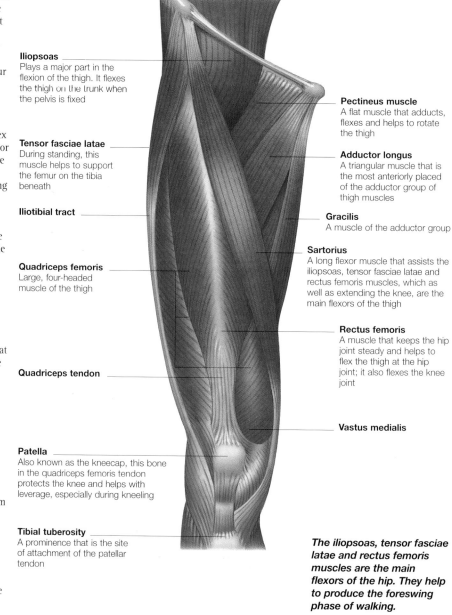

Iliopsoas
Plays a major part in the flexion of the thigh. It flexes the thigh on the trunk when the pelvis is fixed

Tensor fasciae latae
During standing, this muscle helps to support the femur on the tibia beneath

Iliotibial tract

Quadriceps femoris
Large, four-headed muscle of the thigh

Quadriceps tendon

Patella
Also known as the kneecap, this bone in the quadriceps femoris tendon protects the knee and helps with leverage, especially during kneeling

Tibial tuberosity
A prominence that is the site of attachment of the patellar tendon

Pectineus muscle
A flat muscle that adducts, flexes and helps to rotate the thigh

Adductor longus
A triangular muscle that is the most anteriorly placed of the adductor group of thigh muscles

Gracilis
A muscle of the adductor group

Sartorius
A long flexor muscle that assists the iliopsoas, tensor fasciae latae and rectus femoris muscles, which as well as extending the knee, are the main flexors of the thigh

Rectus femoris
A muscle that keeps the hip joint steady and helps to flex the thigh at the hip joint; it also flexes the knee joint

Vastus medialis

The iliopsoas, tensor fasciae latae and rectus femoris muscles are the main flexors of the hip. They help to produce the foreswing phase of walking.

Posterior thigh muscles

The three large muscles of the posterior thigh are commonly known as the hamstrings. These three muscles are the biceps femoris, semitendinosus and semimembranosus.

The hamstring muscles can both extend the hip and flex the knee. However, they cannot do both fully at the same time.

BICEPS FEMORIS
The biceps femoris has two heads. The long head arises from the ischial tuberosity of the pelvis and the short head arises from the back of the femur. The rounded tendon of the biceps femoris can easily be felt and seen behind the outer side of the knee, especially if the knee is flexed against resistance.

SEMITENDINOSUS
Like the biceps femoris muscle, the semitendinosus arises from the ischial tuberosity of the pelvis. It is named for its long tendon, which begins about two-thirds of the way down its course. This tendon attaches to the inner side of the upper tibia.

SEMIMEMBRANOSUS
This muscle arises from a flattened, membranous attachment to the ischial tuberosity of the pelvis. The muscle runs down the back of the thigh, deep to the semitendinosus. It inserts into the inner side of the upper tibia.

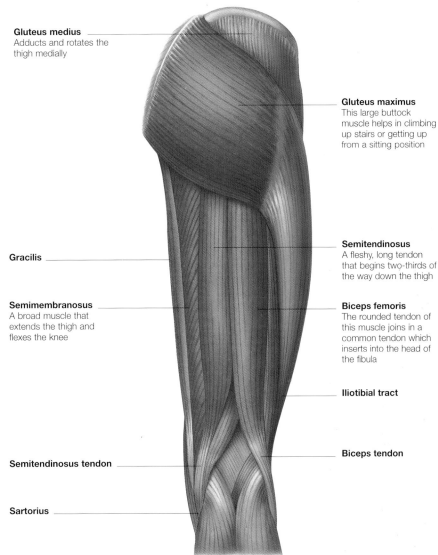

Gluteus medius
Adducts and rotates the thigh medially

Gluteus maximus
This large buttock muscle helps in climbing up stairs or getting up from a sitting position

Gracilis

Semimembranosus
A broad muscle that extends the thigh and flexes the knee

Semitendinosus
A fleshy, long tendon that begins two-thirds of the way down the thigh

Biceps femoris
The rounded tendon of this muscle joins in a common tendon which inserts into the head of the fibula

Iliotibial tract

Semitendinosus tendon

Biceps tendon

Sartorius

The hamstrings consist of three fleshy muscles, situated in the posterior thigh. They extend from the pelvis to the back of the tibia.

213

Muscles of the lower leg

There are three groups of muscles in the lower leg.
Depending where they lie, they support and flex the ankle and foot,
extend the toes and assist in lifting the body weight at the heel.

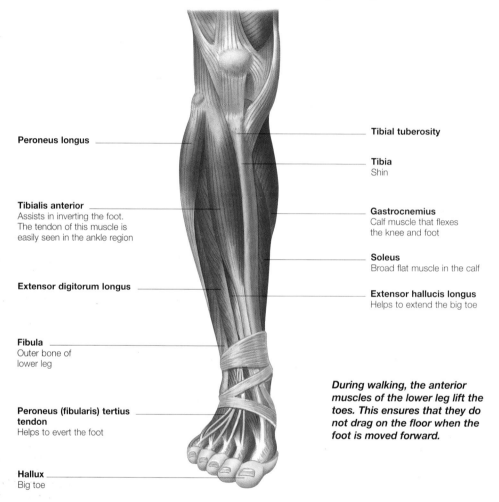

Peroneus longus

Tibialis anterior
Assists in inverting the foot.
The tendon of this muscle is
easily seen in the ankle region

Extensor digitorum longus

Fibula
Outer bone of
lower leg

Peroneus (fibularis) tertius
tendon
Helps to evert the foot

Hallux
Big toe

Tibial tuberosity

Tibia
Shin

Gastrocnemius
Calf muscle that flexes
the knee and foot

Soleus
Broad flat muscle in the calf

Extensor hallucis longus
Helps to extend the big toe

*During walking, the anterior
muscles of the lower leg lift the
toes. This ensures that they do
not drag on the floor when the
foot is moved forward.*

The muscles of the lower leg can be divided into three groups: the anterior group which lie in front of the tibia, the lateral group which lie on the outer side of the lower leg and the posterior group.

ANTERIOR MUSCLES
The anterior muscles of the lower leg include:

- Tibialis anterior – this muscle can be felt under the skin alongside the edge of the tibia
- Extensor digitorum longus – this muscle lies under the tibialis anterior and attaches to the outer four toes
- Peroneus (fibularis) tertius – this muscle is not always present but, when it is, it may join the extensor digitorum longus muscle. It inserts into the fifth metatarsal bone near the little toe
- Extensor hallucis longus – this thin muscle runs down to insert into the end of the hallux (big toe).

ACTION OF THE ANTERIOR MUSCLES
These muscles all have a similar action in that they are dorsiflexors of the foot. This means that when they contract they bend the ankle, bringing the toes up and the heel down.

Posterior muscles of the lower leg

The posterior group of muscles of the lower leg form the mound of the calf. Together, these muscles are strong and heavy, enabling them to work together to flex the foot and to support the weight of the body.

The muscles that lie within the posterior compartment form the largest group of the lower leg. Also known as the 'calf muscles', this group can be further divided into superficial and deep layers.

SUPERFICIAL CALF MUSCLES

The superficial group of posterior muscles forms the bulk of the rounded calf, a feature peculiar to humans and which is due to our upright posture. These muscles include:

- Gastrocnemius – this large fleshy muscle is the most superficial. It has a distinctive shape with two heads which arise from the medial and lateral condyles of the femur. Its fibres run mainly vertically, which allows for the rapid and strong contractions needed in running and jumping
- Soleus – this is a large and powerful muscle which lies under the gastrocnemius. It takes its name from its shape, being flat like a sole (a type of flatfish). Contraction of the soleus muscle is important for maintaining balance when standing
- Plantaris – this muscle is sometimes absent, and when present it is small and thin. Due to its relative unimportance in the lower leg, it is sometimes used to

replace damaged tendons in the hand.

ACTIONS OF THE SUPERFICIAL MUSCLES

These muscles have the job of plantar flexing the foot, which means lifting the heel

and pointing the toes downwards. Strong muscles are needed for this job because, during walking, running and jumping, the heel needs to be lifted against the whole weight of the body.

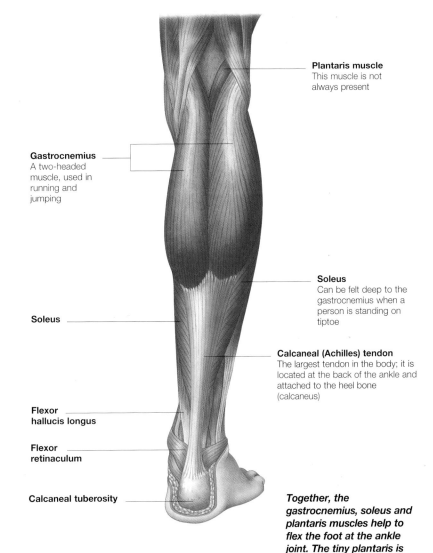

Plantaris muscle
This muscle is not always present

Gastrocnemius
A two-headed muscle, used in running and jumping

Soleus

Flexor hallucis longus

Flexor retinaculum

Calcaneal tuberosity

Soleus
Can be felt deep to the gastrocnemius when a person is standing on tiptoe

Calcaneal (Achilles) tendon
The largest tendon in the body; it is located at the back of the ankle and attached to the heel bone (calcaneus)

Together, the gastrocnemius, soleus and plantaris muscles help to flex the foot at the ankle joint. The tiny plantaris is the weakest of the three muscles.

Gastrocnemius and soleus have a single, common tendon that is known as the large and powerful Achilles tendon, which runs down from the lower edge of the calf to the heel.

Arteries of the leg

The lower limb is supplied by a series of arteries which arise from the external iliac artery of the pelvis. These arteries pass down the leg, branching to reach muscles, bones, joints and skin.

A network of arteries supply the tissues of the lower limb with nutrients. The main arteries give off important and smaller branches to provide nourishment to various joints and muscles.

THE ARTERIES

- The femoral artery – the main artery of the leg. Its main branch is the profunda femoris (deep femoral) artery. Small branches supply nearby muscles before the artery enters a gap in the adductor magnus muscle, the 'adductor hiatus', to enter the popliteal fossa (behind the knee)
- Profunda femoris (deep femoral artery) – the main artery of the thigh. It gives off several branches including the medial and lateral circumflex femoral arteries and the four perforating arteries
- Popliteal artery – a continuation of the femoral artery. It runs down the back of the knee, giving off small branches to nourish that joint, before dividing into the anterior and posterior tibial arteries
- Anterior tibial artery – this supplies the structures within the anterior (front) compartment of the lower leg. It runs downwards to the foot and becomes the dorsalis pedis artery
- Posterior tibial artery – this artery remains at the back of the lower leg and, together with the peroneal (fibular) artery, it supplies the structures of the back and outer compartments.

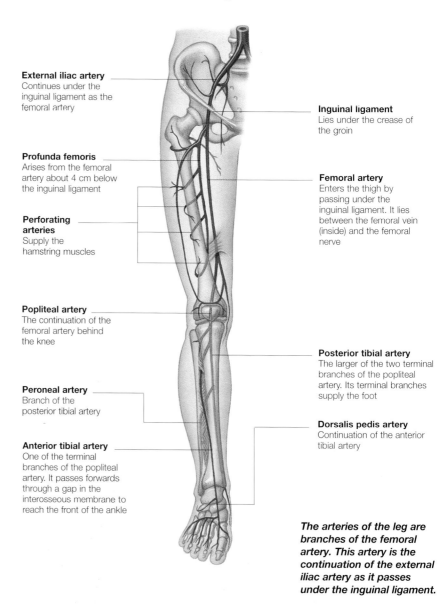

External iliac artery
Continues under the inguinal ligament as the femoral artery

Profunda femoris
Arises from the femoral artery about 4 cm below the inguinal ligament

Perforating arteries
Supply the hamstring muscles

Popliteal artery
The continuation of the femoral artery behind the knee

Peroneal artery
Branch of the posterior tibial artery

Anterior tibial artery
One of the terminal branches of the popliteal artery. It passes forwards through a gap in the interosseous membrane to reach the front of the ankle

Inguinal ligament
Lies under the crease of the groin

Femoral artery
Enters the thigh by passing under the inguinal ligament. It lies between the femoral vein (inside) and the femoral nerve

Posterior tibial artery
The larger of the two terminal branches of the popliteal artery. Its terminal branches supply the foot

Dorsalis pedis artery
Continuation of the anterior tibial artery

The arteries of the leg are branches of the femoral artery. This artery is the continuation of the external iliac artery as it passes under the inguinal ligament.

Arteries of the foot

In a pattern similar to that in the hand, the small arteries of the foot form arches which interconnect, giving off branches to each side of the toes. Branches of the arteries give the sole of the foot a particularly rich blood supply.

Plantar aspect of foot (sole)

Plantar digital arteries

Plantar metatarsal arteries

Medial plantar artery

Deep plantar artery

Plantar arch

Perforating branches

Superficial arch

Medial plantar artery

Lateral plantar artery

Posterior tibial artery

Calcanean branch

Dorsum of foot (top)

Dorsal digital arteries

First dorsal metatarsal artery

Perforating branches of deep plantar arch

Deep plantar artery

Arcuate artery

Medial tarsal arteries

Lateral tarsal artery

Lateral malleolar artery

Medial malleolar artery

Perforating branch of fibular artery

Anterior tibial artery

Dorsalis pedis artery

The arteries supplying the feet branch in a similar fashion to those of the hands. The sole of the foot has a particularly rich blood supply.

The arterial supply of the foot is provided by the terminal branches of the anterior and posterior tibial arteries.

TOP OF THE FOOT
As the anterior tibial artery passes down in front of the ankle it becomes the 'dorsalis pedis' artery. This then runs down across the top of the foot towards the space between the first and second toes, where it gives off a deep branch that joins the arteries on the sole of the foot.

Branches of the dorsalis pedis on the top of the foot join to form an arch which gives off branches to the toes.

The dorsalis pedis pulse can be felt by an examining doctor on the top of the foot next to the tendon of tibialis anterior. As the artery lies just under the skin the pulse should be fairly easy to feel if the blood vessels are healthy.

SOLE OF THE FOOT
The sole of the foot has a rich blood supply which is

provided by branches of the posterior tibial artery. As the artery enters the sole, it divides into two parts to form the medial and lateral plantar arteries.
- The medial plantar artery – this is the smaller of the two branches of the posterior tibial artery. It provides blood for the muscles of the big toe and sends tiny branches to the other toes.
- The lateral plantar artery – this artery is much

larger than the medial plantar artery and curves around under the metatarsal bones to form the deep plantar arch.

The deep branch of the dorsalis pedis artery joins the inner end of this arch so making a connection between the arterial supply of the top of the foot and the sole.

217

Veins of the leg

The lower limb is drained by a series of veins which can be divided into two groups, superficial and deep. The perforating veins connect the two groups of veins.

Lying within the subcutaneous (beneath the skin) tissue, there are two main superficial veins of the leg, the great and small saphenous veins.

GREAT SAPHENOUS VEIN
The great saphenous vein is the longest vein in the body and is sometimes used during surgical procedures to replace damaged or diseased arteries in areas such as the heart. It arises from the medial (inner) end of the dorsal venous arch of the foot and runs up the leg towards the groin.

On its journey, the great saphenous vein passes in front of the medial malleolus (inner ankle bone), tucks behind the medial condyle of the femur at the knee and passes through the saphenous opening in the groin to drain into the large femoral vein.

SMALL SAPHENOUS VEIN
This smaller superficial vein arises from the lateral (outer) end of the dorsal venous arch and passes behind the lateral malleolus (outer ankle bone) and up the centre of the back of the calf. As it approaches the knee, the small saphenous vein empties into the deep popliteal vein.

TRIBUTARIES
The great and small saphenous veins receive blood along the way from many smaller veins and also intercommunicate freely, or 'anastomose', with each other.

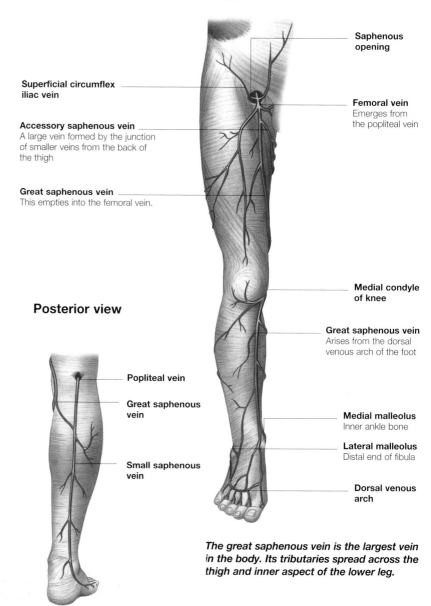

Superficial veins of leg, anterior view

Superficial circumflex iliac vein

Accessory saphenous vein
A large vein formed by the junction of smaller veins from the back of the thigh

Great saphenous vein
This empties into the femoral vein.

Saphenous opening

Femoral vein
Emerges from the popliteal vein

Medial condyle of knee

Great saphenous vein
Arises from the dorsal venous arch of the foot

Medial malleolus
Inner ankle bone

Lateral malleolus
Distal end of fibula

Dorsal venous arch

Posterior view

Popliteal vein

Great saphenous vein

Small saphenous vein

The great saphenous vein is the largest vein in the body. Its tributaries spread across the thigh and inner aspect of the lower leg.

Deep veins of the leg

The deep veins of the leg follow the pattern of the arteries, which they accompany along their length. As well as draining venous blood from the tissues of the leg, the deep veins receive blood from the superficial veins via the perforating veins.

Although the deep leg veins are referred to and illustrated as single veins they are usually, in fact, paired veins which lie either side of the artery. These veins are known as venae comitantes and are common throughout the body.

DEEP VEINS
The main deep veins comprise:
• The posterior tibial vein – this is formed by the joining together of the small medial and lateral plantar veins of the sole of the foot. As it approaches the knee it is joined from its lateral side by the fibular (peroneal) vein before joining with the anterior tibial vein to form the large popliteal vein
• The anterior tibial vein – this is the continuation of the dorsalis pedis vein on the top of the foot. It passes up the front of the lower leg
• The popliteal vein – this lies behind the knee and receives blood from the small veins which surround the knee joint
• The femoral vein – this is the continuation of the popliteal vein as it passes up the thigh. The large femoral vein receives blood from the superficial veins and continues up into the groin to become the external iliac vein of the pelvis.

Deep veins of leg, anterior view

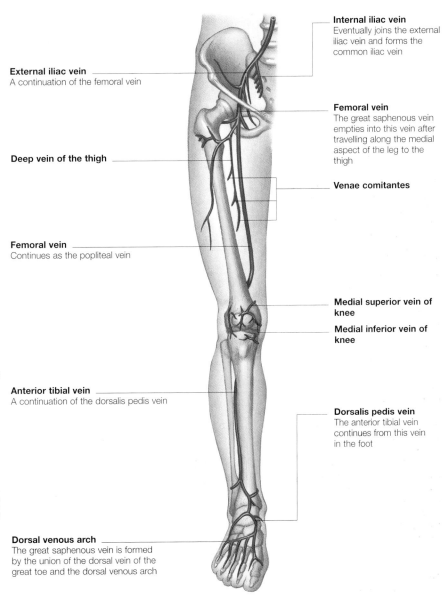

External iliac vein
A continuation of the femoral vein

Deep vein of the thigh

Femoral vein
Continues as the popliteal vein

Anterior tibial vein
A continuation of the dorsalis pedis vein

Dorsal venous arch
The great saphenous vein is formed by the union of the dorsal vein of the great toe and the dorsal venous arch

Internal iliac vein
Eventually joins the external iliac vein and forms the common iliac vein

Femoral vein
The great saphenous vein empties into this vein after travelling along the medial aspect of the leg to the thigh

Venae comitantes

Medial superior vein of knee

Medial inferior vein of knee

Dorsalis pedis vein
The anterior tibial vein continues from this vein in the foot

In the leg, the deep veins have more valves than the superficial veins. The deep veins are usually in pairs and they often follow the same route as the arteries.

Nerves of the leg

The main nerve of the leg – the sciatic nerve – is the largest
nerve in the body. Its branches supply the muscles of the hip, many of the
thigh and all of the muscles of the lower leg and foot.

The sciatic nerve is made up
of two nerves, the tibial nerve
and the common peroneal (or
fibular) nerve. These are
bound together by connective
tissue to form a wide band
that runs the full length of
the back of the thigh.

ORIGIN AND COURSE

The sciatic nerve arises from
a network of nerves at the
base of the spine, called the
sacral plexus. From here, it
passes out through the
greater sciatic foramen and
then curves downwards
through the gluteal region
under the gluteus maximus
muscle (midway between the
bony landmarks of the
greater trochanter of the
femur and the ischial
tuberosity of the pelvis).

The sciatic nerve leaves the
gluteal region by passing
under the long head of the
biceps femoris muscle to enter
the thigh and runs down the
centre of the back of the
thigh, branching off into the
hamstring muscles (a
collective name for the biceps
femoris, semitendinosus and
semimembranosus muscles). It
then divides to form two
branches, the tibial nerve and
the common peroneal nerve
just above the knee.

HIGHER DIVISION

In a few cases the sciatic
nerve divides into two at a
much higher level. In this
situation the common
peroneal nerve may pass
above or even through the
piriformis muscle.

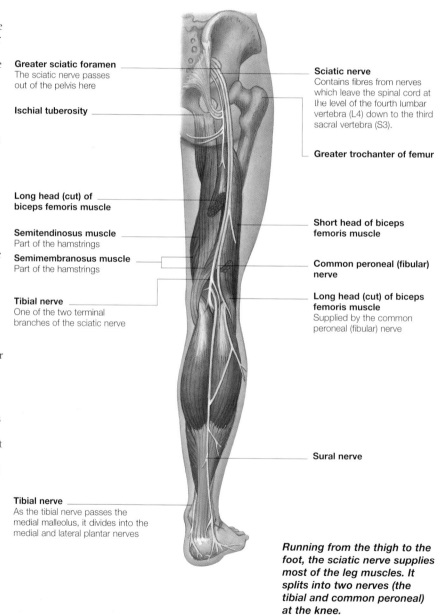

Greater sciatic foramen
The sciatic nerve passes
out of the pelvis here

Ischial tuberosity

**Long head (cut) of
biceps femoris muscle**

Semitendinosus muscle
Part of the hamstrings

Semimembranosus muscle
Part of the hamstrings

Tibial nerve
One of the two terminal
branches of the sciatic nerve

Tibial nerve
As the tibial nerve passes the
medial malleolus, it divides into the
medial and lateral plantar nerves

Sciatic nerve
Contains fibres from nerves
which leave the spinal cord at
the level of the fourth lumbar
vertebra (L4) down to the third
sacral vertebra (S3).

Greater trochanter of femur

**Short head of biceps
femoris muscle**

**Common peroneal (fibular)
nerve**

**Long head (cut) of biceps
femoris muscle**
Supplied by the common
peroneal (fibular) nerve

Sural nerve

*Running from the thigh to the
foot, the sciatic nerve supplies
most of the leg muscles. It
splits into two nerves (the
tibial and common peroneal)
at the knee.*

Terminal branches of the sciatic nerve

The sciatic nerve divides into two terminal branches: the common peroneal (fibular) nerve and the tibial nerve. The common peroneal nerve supplies the front of the leg, while the tibial nerve supplies the muscles and skin at the back.

The common peroneal nerve leaves the sciatic nerve in the lower third of the thigh and runs down around the outer side of the lower leg before dividing into two just below the knee.

NERVE BRANCHES
The two branches of the peroneal nerve comprise:
- The superficial branch of the peroneal nerve – supplies the lateral (outer) compartment of the lower leg in which it lies. It sub-divides into smaller branches to supply the muscles around it
- The deep peroneal nerve – runs in front of the interosseous membrane between the tibia and the fibula, and then passes over the ankle into the foot.
These two branches also supply the knee joint and the skin over the outer side of the calf and the top of the foot.

DAMAGE
As the common peroneal nerve passes around the outer side of the lower leg, it lies just under the skin and very close to the head of the fibula. It is very vulnerable to damage, especially if the fibula suffers a fracture, and is the most commonly damaged nerve in the leg.

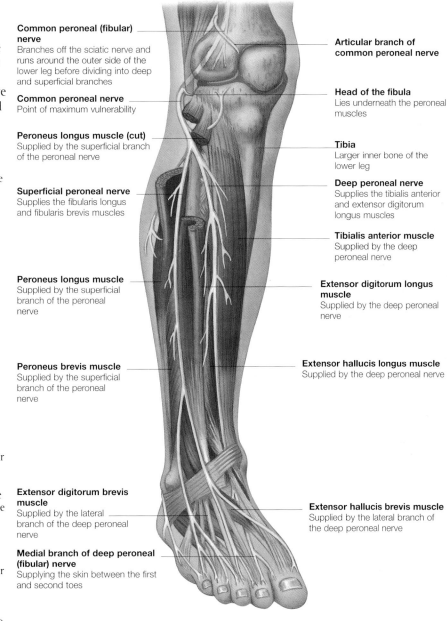

Common peroneal (fibular) nerve
Branches off the sciatic nerve and runs around the outer side of the lower leg before dividing into deep and superficial branches

Common peroneal nerve
Point of maximum vulnerability

Peroneus longus muscle (cut)
Supplied by the superficial branch of the peroneal nerve

Superficial peroneal nerve
Supplies the fibularis longus and fibularis brevis muscles

Peroneus longus muscle
Supplied by the superficial branch of the peroneal nerve

Peroneus brevis muscle
Supplied by the superficial branch of the peroneal nerve

Extensor digitorum brevis muscle
Supplied by the lateral branch of the deep peroneal nerve

Medial branch of deep peroneal (fibular) nerve
Supplying the skin between the first and second toes

Articular branch of common peroneal nerve

Head of the fibula
Lies underneath the peroneal muscles

Tibia
Larger inner bone of the lower leg

Deep peroneal nerve
Supplies the tibialis anterior and extensor digitorum longus muscles

Tibialis anterior muscle
Supplied by the deep peroneal nerve

Extensor digitorum longus muscle
Supplied by the deep peroneal nerve

Extensor hallucis longus muscle
Supplied by the deep peroneal nerve

Extensor hallucis brevis muscle
Supplied by the lateral branch of the deep peroneal nerve

The common peroneal nerve splits into two branches to supply the inner and outer lower leg. Close to the skin at points, it is vulnerable to damage.

221

Ankle

The ankle is the joint between the lower ends of
the tibia and fibula, and the upper surface of the large foot
bone, the talus. It is an example of a hinge joint.

At the ankle, a deep socket is
formed by the lower ends of
the tibia and fibula, the bones
of the lower leg. Into this
socket fits the pulley-shaped
upper surface of the talus. The
shape of the bones and the
presence of strong supporting
ligaments mean that the ankle
is very stable. This is an
important feature for such a
major weight-bearing joint.

THE JOINT
The articular surfaces of the
ankle joint – those parts of
the bone which move against
each other – are covered with
a layer of smooth hyaline
cartilage. This cartilage is
surrounded by a thin synovial
membrane that secretes a
viscous fluid and helps to
lubricate the joint.

The articular surfaces of the
ankle joint consist of the:
- Inside of the lateral
  malleolus, the expanded
  lower end of the fibula. This
  carries a facet (depression)
  that articulates with the
  outer side of the upper
  surface of the talus
- Undersurface of the lower
  end of the tibia. This forms
  the roof of the socket, which
  articulates with the talus
- Inside of the medial
  malleolus, the projection at
  the lower end of the tibia.
  This moves against the
  inner side of the upper
  surface of the talus
- Trochlea of the talus.
  Named for its pulley shape,
  this upper part of the talus
  fits into the ankle joint, and
  articulates with the lower
  ends of the tibia and fibula.

Anterior view, left ankle

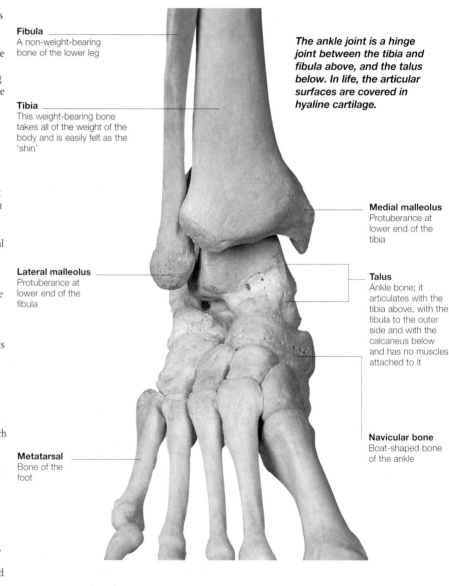

Fibula
A non-weight-bearing
bone of the lower leg

Tibia
This weight-bearing bone
takes all of the weight of the
body and is easily felt as the
'shin'

*The ankle joint is a hinge
joint between the tibia and
fibula above, and the talus
below. In life, the articular
surfaces are covered in
hyaline cartilage.*

Medial malleolus
Protuberance at
lower end of the
tibia

Lateral malleolus
Protuberance at
lower end of the
fibula

Talus
Ankle bone; it
articulates with the
tibia above, with the
fibula to the outer
side and with the
calcaneus below
and has no muscles
attached to it

Navicular bone
Boat-shaped bone
of the ankle

Metatarsal
Bone of the
foot

Ligaments of the ankle

The ankle is supported by strong ligaments which help to stabilize this important weight-bearing joint.

The ankle joint needs to be stable as it bears the weight of the body. The presence of a variety of strong ligaments around the ankle helps to maintain this stability, while still allowing the necessary freedom of movement.

Like most joints, the ankle is enclosed within a tough fibrous capsule. Although the capsule is quite thin in front and behind, it is reinforced on each side by the strong medial (inner) and lateral (outer) ankle ligaments.

MEDIAL LIGAMENT
Also known as the deltoid ligament, the medial ligament is a very strong structure which fans out from the tip of the medial malleolus of the tibia. It is usually described in three parts, each named for the bones that they connect:
• Anterior and posterior tibiotalar ligaments. Lying close against the bones, these parts of the medial ligament connect the tibia to the medial sides of the talus beneath
• Tibionavicular ligament. More superficially, this part of the ligament runs between the tibia and the navicular, one of the bones of the foot
• Tibiocalcaneal ligament. This strong ligament runs just under the skin from the tibia to the sustentaculum tali, a projection of the calcaneus (large heel bone).
Together, these parts of the medial ligament support the ankle joint during the movement of eversion (where the foot is turned out to the side).

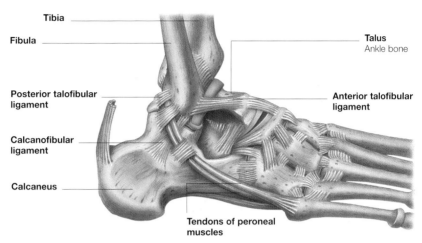

Lateral view (right foot)

Tibia

Fibula

Talus
Ankle bone

Posterior talofibular ligament

Anterior talofibular ligament

Calcanofibular ligament

Calcaneus

Tendons of peroneal muscles

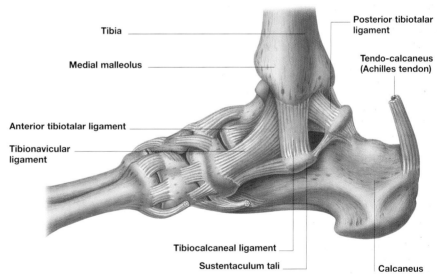

Medial view (right foot)

Tibia

Medial malleolus

Posterior tibiotalar ligament

Tendo-calcaneus (Achilles tendon)

Anterior tibiotalar ligament

Tibionavicular ligament

Tibiocalcaneal ligament

Sustentaculum tali

Calcaneus

LATERAL LIGAMENT
The lateral ligament is weaker than the medial ligament, and is made up of three distinct bands:
• Anterior talofibular ligament. This runs forward from the lateral malleolus of the fibula to the talus
• Calcanofibular ligament. This passes down from the tip of the lateral malleolus to the side of the talus
• Posterior talofibular ligament. This is a thick, stronger band which passes back from the lateral malleolus to the talus behind.

Bones of the foot

The human foot has 26 bones in total: seven larger, irregular tarsal bones; five metatarsals running the length of the foot; and 14 phalanges forming the skeleton of the toes.

The tarsal bones in the foot are equivalent to the carpal bones in the wrist, but there are seven tarsals as opposed to eight wrist bones. The tarsal bones differ somewhat from the wrist in terms of their arrangement, reflecting the different functions of the hand and the foot.

TARSAL BONES

The tarsal bones consist of:

- The talus – articulates with the tibia and fibula at the ankle joint. It bears the full weight of the body, transferred down from the tibia. Its shape is such that it can then spread this weight by passing this force backwards and downwards, and forwards to the front of the foot
- The calcaneus – the large heel bone
- The navicular – a relatively small bone, named for its boat-like appearance. It has a projection, the navicular tuberosity which, if too large, may cause foot pain as it rubs against the shoe
- The cuboid – a bone roughly the shape of a cube. It lies on the outer side of the foot, and has a groove on its under surface to allow passage of a muscle tendon
- The three cuneiforms – bones named according to their positions: medial, intermediate and lateral. The medial cuneiform is the largest of these three wedge-shaped bones.

Tarsal bones

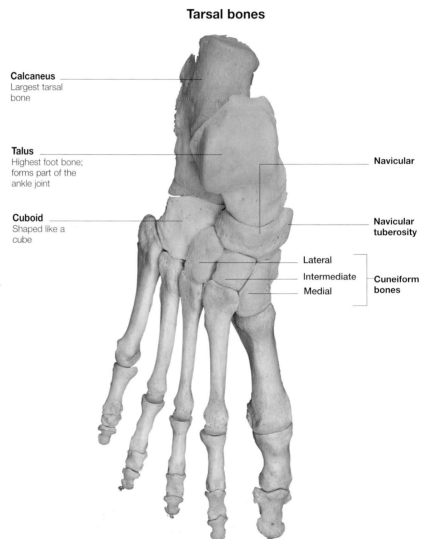

Calcaneus
Largest tarsal bone

Talus
Highest foot bone; forms part of the ankle joint

Cuboid
Shaped like a cube

Navicular

Navicular tuberosity

Lateral
Intermediate **Cuneiform**
Medial **bones**

The tarsal bones of the foot are classified as short bones. Their function is to bear the weight of the body and assist in locomotion (walking, running).

Metatarsals and phalanges

The metatarsals and phalanges in the foot are miniature long bones, consisting of a base, shaft and head.

Like the metacarpals in the hand, there are five metatarsals in the foot. While the individual bones tend to resemble the metacarpals in structure, their arrangement is slightly different. This is mainly due to the fact that the big toe lies in the same plane as the other toes and is not opposable like the thumb.

METATARSALS

Each metatarsal has a long shaft with two expanded ends, the base and the head. The bases of the metatarsals articulate with the tarsal bones in the middle of the foot. The heads articulate with the phalanges of the corresponding toes.

The metatarsals are numbered from 1 to 5 starting with the most medial, which lies behind the big toe. The first metatarsal is shorter and more sturdy than the rest. It articulates with the first phalanx of the big toe.

PHALANGES

The phalanges of the toe resemble the phalanges of the fingers. There are 14 phalanges in the foot, the big toe (hallux) having just two while the other four toes have three each.

The base of the first phalanx of each toe articulates with the head of the corresponding metatarsal. The phalanges of the big toe are thicker than those of the other toes.

Lateral view of the foot

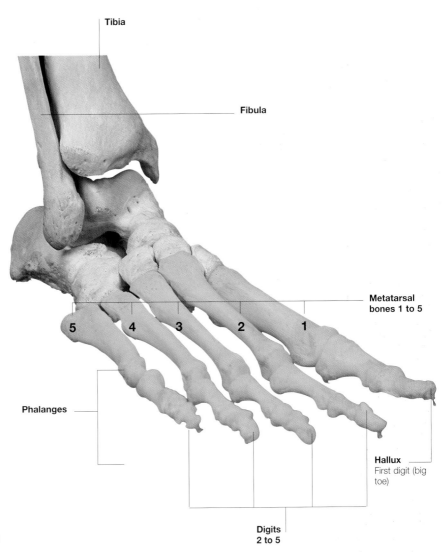

Tibia

Fibula

Metatarsal bones 1 to 5

5 4 3 2 1

Phalanges

Hallux
First digit (big toe)

Digits 2 to 5

The metatarsals of the foot provide stability while standing. The phalanges of the toes are a stable support during movement.

Ligaments and arches of the foot

The bones of the foot are arranged in such a way that they form bridge-like arches. These bones are supported by the presence of a number of strong ligaments.

The main supportive ligaments of the foot lie on the plantar (under) surface of the bones. The three most prominent ligaments are:
• The plantar calcaneonavicular, or spring ligament – stretches forward from the sustentaculum tali, a projection of the calcaneus (heel bone), to the back of the navicular (boat-shaped) bone. This ligament is important in helping to maintain the longitudinal arch of the foot
• The long plantar ligament – runs forward from the underside of the calcaneus to the cuboid (outer) bone and to the bases of the metatarsals (foot bones). It helps to maintain the arches of the foot
• The plantar calcaneocuboid, or short plantar, ligament – lies under the long plantar ligament and runs from the front of the undersurface of the calcaneus forward to the cuboid.

OTHER LIGAMENTS
Many other ligaments support and bind together the long metatarsals and the phalanges (toe bones). The metatarsals are bound to the tarsals and to each other by ligaments running across the foot on both their dorsal and plantar surfaces.

Ligaments of foot (plantar view)

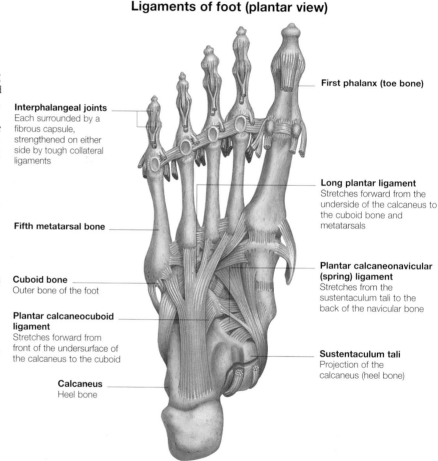

Interphalangeal joints
Each surrounded by a fibrous capsule, strengthened on either side by tough collateral ligaments

Fifth metatarsal bone

Cuboid bone
Outer bone of the foot

Plantar calcaneocuboid ligament
Stretches forward from front of the undersurface of the calcaneus to the cuboid

Calcaneus
Heel bone

First phalanx (toe bone)

Long plantar ligament
Stretches forward from the underside of the calcaneus to the cuboid bone and metatarsals

Plantar calcaneonavicular (spring) ligament
Stretches from the sustentaculum tali to the back of the navicular bone

Sustentaculum tali
Projection of the calcaneus (heel bone)

Strengthening ligaments ensure that the foot provides a firm but flexible base to bear the weight of the body. The ligaments also facilitate locomotion.

Arches of the foot

A distinctive feature of the human foot is that the bones within it are arranged in bridge-like arches. This allows the foot to be flexible enough to cope with uneven ground, while still being able to bear the weight of the body.

Bones forming medial longitudinal arch of foot

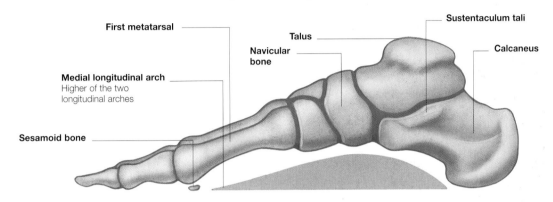

First metatarsal

Talus

Sustentaculum tali

Navicular bone

Calcaneus

Medial longitudinal arch
Higher of the two longitudinal arches

Sesamoid bone

Bones forming lateral longitudinal arch of foot

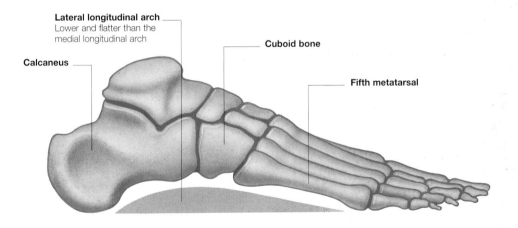

Lateral longitudinal arch
Lower and flatter than the medial longitudinal arch

Cuboid bone

Calcaneus

Fifth metatarsal

The arched shape of the foot can be illustrated by looking at a footprint. Only the heel, the outer edge of the foot, the pads under the metatarsal heads and the tips of the toes leave an impression. The rest of the foot is lifted away from the ground.

THREE ARCHES
The foot has two longitudinal arches (medial and lateral) running along its length, and a transverse arch lying across it:
• Medial longitudinal arch. This is the higher and more important of the two longitudinal arches. The bones involved are the calcaneus, the talus, the

navicular bone, the three cuneiform bones and the first three metatarsals. The head of the talus supports this arch
• Lateral longitudinal arch. This arch is much lower and flatter, the bones resting on the ground when standing. The lateral arch is formed by the calcaneus, the cuboid and

the fourth and fifth metatarsal bones
• Transverse arch. This arch runs across the foot, supported on either side by the longitudinal arches, and is made up of the bases of the metatarsal bones, the cuboid and the three cuneiform bones.

Muscles of the upper foot

Many of the muscles which move the foot lie in the lower leg, rather than in the foot itself. This allows them to be more powerful than if they were contained within the small space of the foot.

To have an effect upon the bones and joints of the foot, the leg muscles have long tendons. To reach the bones of the foot these tendons must first cross the ankle joint, where they are held in place by a series of retaining bands, or retinacula. If these bands were not present, the tendons would run straight to their attachments like a bow-string rather than following the contours of the ankle joint.

RETINACULA OF THE FOOT
There are four main retinacula in this area:
• Superior extensor retinaculum. Lies just above the ankle joint and retains the long tendons of the extensor muscles
• Inferior extensor retinaculum. Lies beneath the ankle joint. It also retains extensor muscles
• Peroneal retinaculum. Lies on the outer side of the ankle. It is in two parts, upper and lower, and retains the long peroneal muscle tendons
• Flexor retinaculum. Lies on the inner side of the ankle and retains the long flexor tendons as they pass under the medial malleolus to reach the sole of the foot.

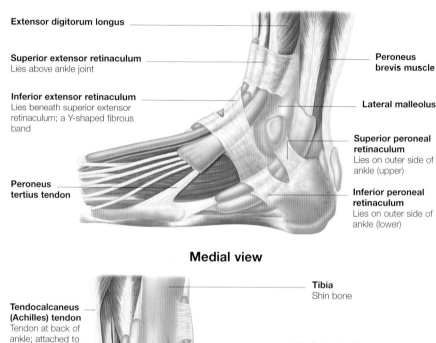

Lateral view

Extensor digitorum longus

Superior extensor retinaculum
Lies above ankle joint

Inferior extensor retinaculum
Lies beneath superior extensor retinaculum; a Y-shaped fibrous band

Peroneus tertius tendon

Peroneus brevis muscle

Lateral malleolus

Superior peroneal retinaculum
Lies on outer side of ankle (upper)

Inferior peroneal retinaculum
Lies on outer side of ankle (lower)

Medial view

Tendocalcaneus (Achilles) tendon
Tendon at back of ankle; attached to heel bone

Flexor hallucis longus

Tibialis posterior tendon

Posterior tibial artery and nerve

Tibia
Shin bone

Medial malleolus

Flexor retinaculum

Sheath of extensor hallucis longus

Tibialis anterior tendon

The leg muscles have long tendons which connect to the foot bones like puppet strings. Retinacula are fibrous bands that hold tendons in place.

Muscles of the top of the foot

Although they are not particularly powerful, the muscles that lie over the top of the foot play an important part in helping to extend the toes. The extensor digitorum brevis muscle tends to be used when the foot is already pointing upwards.

Extensor digitorum longus muscle

Peroneus tertius tendon
This lies over the extensor digitorum brevis

Extensor digitorum brevis

Superior extensor retinaculum

Inferior extensor retinaculum

Extensor hallucis brevis

Extensor hallucis longus tendon

Most of the muscles which lie within the foot, the intrinsic muscles, are in the sole. The top, or dorsal surface, of the foot has just two muscles: the extensor digitorum brevis and the extensor hallucis brevis.

MUSCLES OF THE DORSAL SURFACE
- Extensor digitorum brevis. As its name suggests, this is a short muscle which extends (straightens or pulls

upwards) the toes. It arises from the upper surface of the calcaneus, or heel bone, and the inferior extensor retinaculum. This muscle divides into three parts, each with a tendon that joins the corresponding long extensor tendon to insert into the second, third and fourth toes
- Extensor hallucis brevis. This short muscle is really part of the extensor digitorum brevis. It runs

down to insert into the big toe, or 'hallux', from which its name derives

ACTION OF THE MUSCLES
Together these two muscles assist the long extensor tendons in extending the first four toes. Although they do not have a particularly powerful action, they are useful in extending the toes when the foot itself is already pointing up, or dorsiflexed, as in this

position the long extensors are unable to act further.

CLINICAL RELEVANCE
The top of the foot is one of the sites in the body where excess tissue fluid (oedema) may accumulate and be visible to a doctor. The position of the muscle bellies of these two short extensor muscles must be known to prevent them being mistaken for such oedema.

Muscles of the sole of the foot

Many of the movements of the bones and joints of the feet are brought about by muscles in the lower leg. However, there are also many small 'intrinsic' muscles which lie entirely within the foot.

The sole of the foot has four layers of intrinsic muscles, which work with the extrinsic muscles to meet the varying demands placed upon the foot during standing, walking, running and jumping. They also help to support the bony arches of the foot.

FIRST MUSCLE LAYER

The first layer of sole muscles is the most superficial, lying just under the thick plantar aponeurosis. The muscles of this layer include:

- Abductor hallucis – this muscle lies along the medial (inner) border of the sole. It acts to abduct the big toe, or 'hallux', which means moving it away from the mid-line. It also flexes, or bends down, the big toe
- Flexor digitorum brevis - this fleshy muscle lies down the centre of the sole and inserts into each of the lateral four toes. Contraction of this muscle causes those toes to flex
- Abductor digiti minimi - lying along the lateral (outer) border of the sole within this first layer, this muscle acts to abduct and flex the little toe. These muscles are similar to the corresponding muscles in the hand, but their individual function is less important as the toes do not have such a wide range of movement as the fingers.

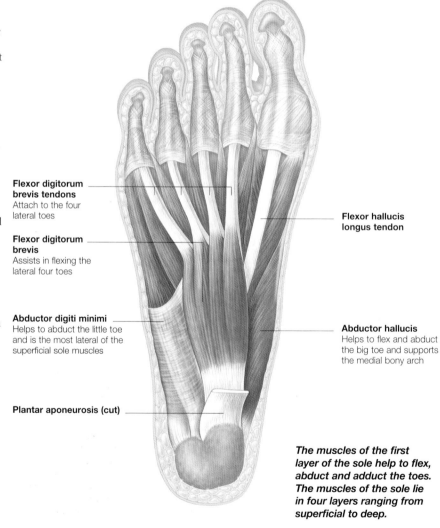

Flexor digitorum brevis tendons
Attach to the four lateral toes

Flexor digitorum brevis
Assists in flexing the lateral four toes

Abductor digiti minimi
Helps to abduct the little toe and is the most lateral of the superficial sole muscles

Plantar aponeurosis (cut)

Flexor hallucis longus tendon

Abductor hallucis
Helps to flex and abduct the big toe and supports the medial bony arch

The muscles of the first layer of the sole help to flex, abduct and adduct the toes. The muscles of the sole lie in four layers ranging from superficial to deep.

Deeper muscle layers of the sole

The muscles of the sole of the foot are made up of four different layers; three of these layers lie under the top layer of the sole of the foot. All of these muscles act together, to help to keep the bony arches of the feet stable.

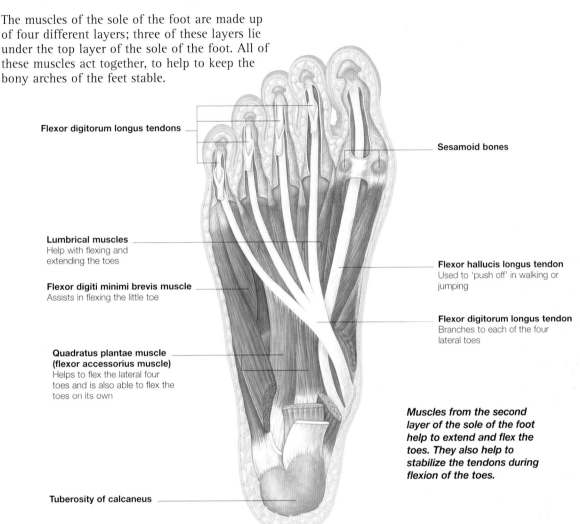

Flexor digitorum longus tendons

Sesamoid bones

Lumbrical muscles
Help with flexing and extending the toes

Flexor hallucis longus tendon
Used to 'push off' in walking or jumping

Flexor digiti minimi brevis muscle
Assists in flexing the little toe

Flexor digitorum longus tendon
Branches to each of the four lateral toes

Quadratus plantae muscle (flexor accessorius muscle)
Helps to flex the lateral four toes and is also able to flex the toes on its own

Muscles from the second layer of the sole of the foot help to extend and flex the toes. They also help to stabilize the tendons during flexion of the toes.

Tuberosity of calcaneus

Beneath the superficial layer of intrinsic muscles of the sole lie three further layers. These all have a contribution to make to the stability and flexibility of the foot, both at rest and in motion.

Although the deep muscles each have individual actions, their main role is to act together to maintain the stability of the bony arches of the feet.

SECOND MUSCLE LAYER OF THE SOLE
The second muscle layer of the sole of the foot includes some tendons from the extrinsic muscles, as well as some smaller intrinsic muscles.

Muscles and tendons that are included within this second layer of the sole are:
• Quadratus plantae (or flexor accessorius) muscle – this wide, rectangular muscle arises from two heads on either side of the heel.

It inserts into the edge of the tendon of flexor digitorum longus where it acts by pulling backwards on this tendon and so stabilizing it while it flexes the toes
• Tendons of flexor hallucis longus and flexor digitorum longus – these tendons enter the second muscular layer of the sole after winding around the medial malleolus (inner 'ankle bone')
• The four lumbrical muscles – named for their worm-

like appearance, these four muscles arise from the tendons of flexor digitorum longus. They are similar to the lumbrical muscles in the hand. These muscles act to extend (straighten) the toes while the long tendons are flexing them, which helps to prevent the toes 'buckling under' when walking or running.

The skeleton

The skeleton is made up of bone and cartilage, and it accounts for one-fifth of the body's weight. Over 200 bones form a living structure, superbly designed to support and protect the body.

The human skeleton provides a stable yet flexible framework for the other tissues of the body. Cartilage is more flexible than bone and is found in the places where movement occurs.

FUNCTIONS OF BONE
The bones of the skeleton have a number of vital functions:

- Support – bones support the body when standing, and hold soft internal organs in place
- Protection – the brain and spinal cord are protected by the skull and vertebral column, while the rib cage protects the heart and lungs
- Movement – throughout the body, muscles attach to bones to give them the leverage to bring about movement
- Storage of minerals – calcium and phosphate ions are stored in bone to be drawn upon when necessary
- Blood cell formation – the marrow cavity of some bones, such as the sternum, is a site of production of red blood cells.

Bone is actually living tissue – the skeleton has its own blood vessels and nerve supply. Up to 5 per cent of bone is recycled every week by its cells.

Anterior view of the human skeleton

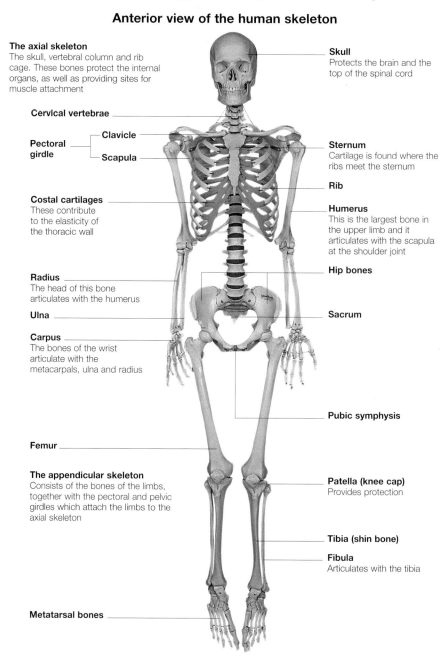

The axial skeleton
The skull, vertebral column and rib cage. These bones protect the internal organs, as well as providing sites for muscle attachment

Cervical vertebrae

Pectoral girdle

Clavicle

Scapula

Costal cartilages
These contribute to the elasticity of the thoracic wall

Radius
The head of this bone articulates with the humerus

Ulna

Carpus
The bones of the wrist articulate with the metacarpals, ulna and radius

Femur

The appendicular skeleton
Consists of the bones of the limbs, together with the pectoral and pelvic girdles which attach the limbs to the axial skeleton

Metatarsal bones

Skull
Protects the brain and the top of the spinal cord

Sternum
Cartilage is found where the ribs meet the sternum

Rib

Humerus
This is the largest bone in the upper limb and it articulates with the scapula at the shoulder joint

Hip bones

Sacrum

Pubic symphysis

Patella (knee cap)
Provides protection

Tibia (shin bone)

Fibula
Articulates with the tibia

Bone markings and features

Each bone of the skeleton is shaped to fulfil its own functions. Bones bear marks, ridges and notches which relate to other structures with which they come into contact.

Over the years, anatomists have given names to the various types of feature which can be found on bones. Using these names, a bone can be described quite clearly and accurately, something which can be of importance clinically.

PROJECTIONS
Projections on the surface of a bone often occur where muscles, tendons or ligaments are attached or where a joint is formed. Examples include:
- Condyle – rounded projection at a joint (such as the femoral condyle at the knee)
- Epicondyle – the raised area above a condyle (such as on the lower humerus, at the elbow)
- Crest – prominent ridge of bone (such as the iliac crest of the pelvic bone)
- Tubercle – small raised area (such as the greater tubercle at the top of the humerus)
- Line – long, narrow raised ridge (such as the soleal line at the back of the tibia).

The bones of the human skeleton are rarely smooth. Markings are often found on bones where tendons, ligaments, and fasciae attach.

Posterior view of the human skeleton

External occipital protuberance
This projection is usually easily palpable

Greater tubercle of humerus
A raised area at the lateral margin of the humerus

Spinous processes of vertebrae

Greater sciatic notch
A deep indentation of the ischium

Ischial tuberosity
A protuberance of the ischium

Lateral femoral condyle

Soleal line of tibia
A rough diagonal ridge of the shin bone

Spine of scapula
A thick ridge of bone which continues as the acromion of the shoulder

Infraspinous fossa of scapula

Lateral epicondyle of humerus
When the elbow joint is partially flexed, this can be felt

Iliac crest
A ridge that forms the rim of the fan-shaped ilium

Greater trochanter of femur
This protuberance occurs above the femur

Obturator foramen
A large aperture in the hip bone

Lateral malleolus of fibula
A projection at the lower end of the outer bone of the fibula

Types of joints

A joint is formed where two or more bones meet. Some allow movement and so give mobility to the body while others protect and support the body by holding the bones rigid against each another.

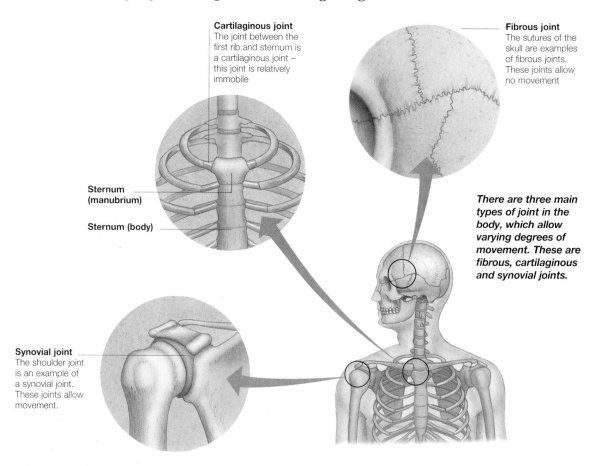

Cartilaginous joint
The joint between the first rib and sternum is a cartilaginous joint – this joint is relatively immobile

Fibrous joint
The sutures of the skull are examples of fibrous joints. These joints allow no movement

Sternum (manubrium)

Sternum (body)

There are three main types of joint in the body, which allow varying degrees of movement. These are fibrous, cartilaginous and synovial joints.

Synovial joint
The shoulder joint is an example of a synovial joint. These joints allow movement.

The joints of the body can be divided into three main structural groups, according to the tissues that lie between the bones. These groups are fibrous, cartilaginous and synovial.

FIBROUS JOINTS
Where two bones are connected by a fibrous joint, they are held together with collagen (a protein).

Collagen fibres allow little, if any, movement. Fibrous joints are located in the body where the movement of one bone upon the other should be prevented, such as in the skull.

CARTILAGINOUS JOINTS
The ends of the bones in a cartilaginous joint are covered with a thin layer of hyaline (glass-like)

cartilage, with the bones being connected by tough fibrocartilage. The whole joint is covered by a fibrous capsule.

Cartilaginous joints do not allow much movement but they can 'relax' under pressure, so giving flexibility to structures such as the spinal column.

SYNOVIAL JOINTS
Most joints of the body are synovial, and allow easy movement between the bones. In a synovial joint, the bones are covered by hyaline cartilage and separated by fluid. The joint cavity is lined by a synovial membrane and the whole joint is enclosed by a fibrous capsule.

Fibrous and cartilaginous joints

Fibrous and cartilaginous joints have an important role to play in the human skeleton. Unlike the more widespread synovial joints, which are designed to allow mobility, fibrous and cartilaginous joints help to maintain stability of the body's frame.

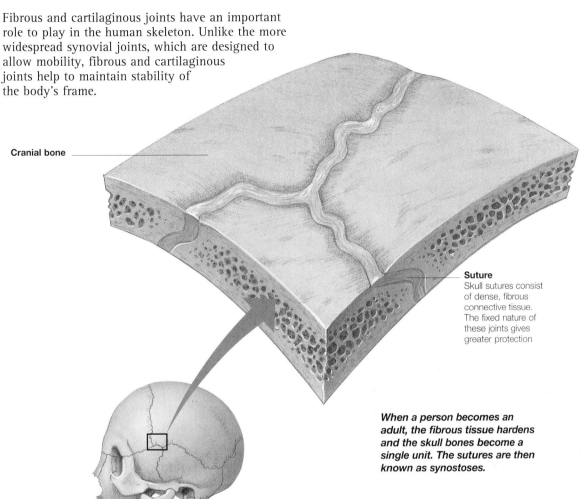

Cranial bone

Suture
Skull sutures consist of dense, fibrous connective tissue. The fixed nature of these joints gives greater protection

When a person becomes an adult, the fibrous tissue hardens and the skull bones become a single unit. The sutures are then known as synostoses.

The bones of a fibrous joint are connected solely by long collagen fibres; there is no cartilage, and no fluid-filled joint cavity. Because of its structure, a fibrous joint does not allow much real movement of the bones against each other. What little movement there is, is determined by the length of the collagen fibres.

GROUPS OF FIBROUS JOINTS
Fibrous joints can be further subdivided into three groups:
• Sutures – literally meaning 'seams', sutures are the tough fibrous joints between the interlocking bones of the skull. Short collagen fibres allow no side-to-side movement of these bones upon each other although there may be some slight 'springing'

of the bones if pressure is applied. The presence of fixed fibrous joints in the skull gives great protection to the vulnerable brain tissue that lies beneath
• Syndesmoses – here, the bones are connected by a sheet of fibrous connective tissue, and the length of the fibres varies from joint to joint. These may also be known as interosseous membranes and are a feature of the forearm and the lower leg, where two bones lie side by side, acting as a unit. Syndesmoses tend to have

longer fibres than sutures and so allow a little more movement
• Gomphoses – this is a very specialized type of fibrous joint with only one example in the human body, the tooth socket. In a gomphosis, a peg-like process sits in a depression, or socket, and is held in place by fibrous tissue, in this case the periodontal ligament. Movement is generally abnormal but micromovement is essential to eating to allow adjustment of the pressure of the bite.

Types of muscle

There are three main types of muscle in the body – skeletal muscle is used for voluntary movement, smooth muscle controls internal organs and cardiac muscle keeps the heart beating.

The most familiar muscles in the body are the skeletal muscles, (also known as striated, or voluntary, muscles), many of which are visible under the skin. Voluntary muscles can be under conscious control, and can also contract in a reflex action such as when the knee straightens when the patellar tendon is tapped (the knee jerk reflex).

STRUCTURE
The muscle fibres of each skeletal muscle are bound together by connective tissue (epimysium), and divided into groups or fascicles by a sheath (perimysium). Within these fascicles, each muscle fibre is surrounded by an endomysium. The whole muscle is attached to bone by a tough fibrous band, the muscle tendon.

FUNCTION
Skeletal muscle can be very adaptable. These muscles can contract powerfully, exerting a great deal of force, such as in lifting a heavy object. Alternatively, they can exert a small force to perform a delicate action such as picking up a feather. Another feature of skeletal muscle, which becomes obvious after performing exercise, is that it tires easily. Whereas the heart can beat all day, every day, without ceasing, skeletal muscle needs a period of rest after a contraction.

Each skeletal muscle is made up of muscle fibres running along its length, together with connective tissues, nerves and blood vessels.

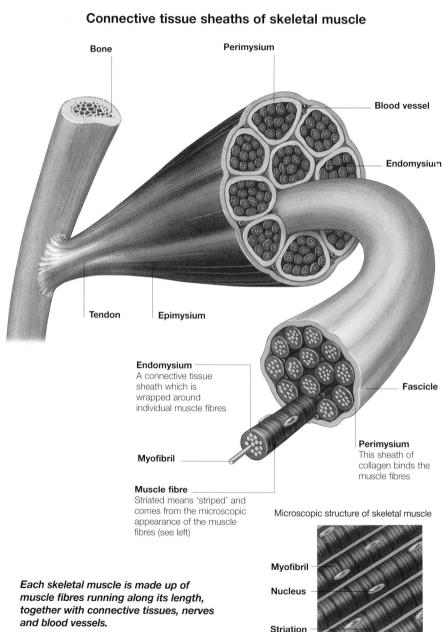

Connective tissue sheaths of skeletal muscle

Bone

Perimysium

Blood vessel

Endomysium

Tendon

Epimysium

Endomysium
A connective tissue sheath which is wrapped around individual muscle fibres

Fascicle

Myofibril

Perimysium
This sheath of collagen binds the muscle fibres

Muscle fibre
Striated means 'striped' and comes from the microscopic appearance of the muscle fibres (see left)

Microscopic structure of skeletal muscle

Myofibril

Nucleus

Striation

Shapes of skeletal muscle

Although all skeletal muscles are made up of fascicles, or groups of muscle fibres, the arrangement of these fascicles may vary. This variation leads to a number of different muscle shapes throughout the body.

There are several ways of describing the various shapes of muscle, including:
- Flat – muscles, such as the external oblique in the abdominal wall, may be flat, yet fairly broad. They may cover a wide area and sometimes insert into an aponeurosis (a broad sheet of connective tissue)
- Fusiform – many muscles are of this 'spindle-shaped' form, where the rounded belly tapers at each end. Examples include the biceps and triceps muscles of the upper arm which have more than one head
- Pennate – these muscles are named for their similarity to a feather (the word 'penna' means feather). They may be described as being unipennate (for example, extensor digitorum longus), bipennate (such as rectus femoris) or multipennate (for example, the deltoid). Multipennate muscles resemble a number of feathers placed next to one another
- Circular – these muscles, also known as sphincteral muscles, surround body openings. Contraction of these muscles, where the fibres are arranged in concentric rings, closes the opening. Circular muscles within the face include the orbicularis oculi, which closes the eye

Fascicle arrangement in relation to muscle structure

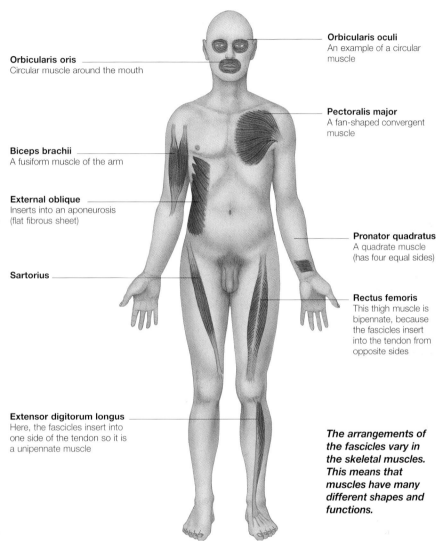

Orbicularis oris
Circular muscle around the mouth

Biceps brachii
A fusiform muscle of the arm

External oblique
Inserts into an aponeurosis
(flat fibrous sheet)

Sartorius

Extensor digitorum longus
Here, the fascicles insert into
one side of the tendon so it is
a unipennate muscle

Orbicularis oculi
An example of a circular
muscle

Pectoralis major
A fan-shaped convergent
muscle

Pronator quadratus
A quadrate muscle
(has four equal sides)

Rectus femoris
This thigh muscle is
bipennate, because
the fascicles insert
into the tendon from
opposite sides

*The arrangements of
the fascicles vary in
the skeletal muscles.
This means that
muscles have many
different shapes and
functions.*

- Convergent – these muscles are fan-shaped and the muscle fibres arise from a wide origin and converge on a narrow tendon. In some cases, these muscles take on a triangular shape. Examples include the large pectoral muscles.

FUNCTION
The arrangement of the fascicles within a muscle influences that muscle's action and power. When muscle fibres contract, they shorten to about 70 per cent of their relaxed length. If the muscle is long with parallel fibres, such as the sartorius muscle in the leg, it can shorten a great deal but has little strength.

If the degree of shortening is not as important as the power it can produce, the muscle may have numerous fibres packed tightly together and converging on a single point. This is the arrangement in multipennate muscles such as the deltoid in the shoulder.

237

Overview of blood circulation

There are two blood vessel networks in the body. The pulmonary circulation transports blood between the heart and lungs; the systemic circulation supplies blood to all parts except the lungs.

The blood circulatory system can be divided into two parts:
- Systemic circulation – those vessels that carry blood to and from all the tissues of the body
- Pulmonary circulation – the vessels that carry blood through the lungs to take up oxygen and release carbon dioxide.

SYSTEMIC ARTERIAL SYSTEM

The systemic arterial system carries blood away from the heart to nourish the tissues. Oxygenated blood from the lungs is first pumped into the aorta via the heart. Branches from the aorta pass to the upper limbs, head, trunk and the lower limbs in turn. These large branches give off smaller branches, which then divide again and again. The tiniest arteries (arterioles) feed blood into capillaries.

PULMONARY CIRCULATION

With each beat of the heart, blood is pumped from the right ventricle into the lungs through the pulmonary artery (this carries deoxygenated blood). After many arterial divisions, the blood flows through the capillaries of the alveoli (air sacs) of the lung to be reoxygenated. The blood eventually enters one of the four pulmonary veins. These pass to the left atrium, from where the blood is pumped through the heart to the systemic circulation.

Major arteries of the body

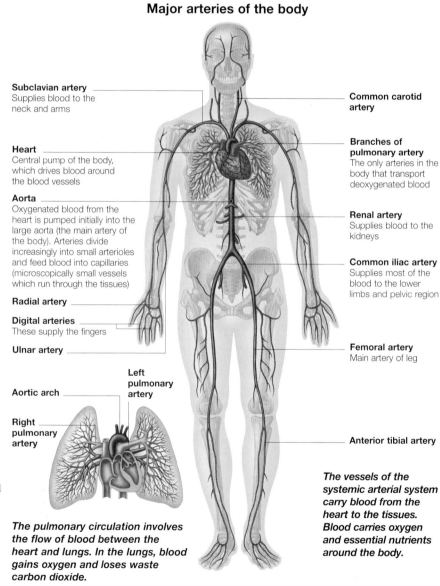

Subclavian artery
Supplies blood to the neck and arms

Heart
Central pump of the body, which drives blood around the blood vessels

Aorta
Oxygenated blood from the heart is pumped initially into the large aorta (the main artery of the body). Arteries divide increasingly into small arterioles and feed blood into capillaries (microscopically small vessels which run through the tissues)

Radial artery

Digital arteries
These supply the fingers

Ulnar artery

Common carotid artery

Branches of pulmonary artery
The only arteries in the body that transport deoxygenated blood

Renal artery
Supplies blood to the kidneys

Common iliac artery
Supplies most of the blood to the lower limbs and pelvic region

Femoral artery
Main artery of leg

Anterior tibial artery

The vessels of the systemic arterial system carry blood from the heart to the tissues. Blood carries oxygen and essential nutrients around the body.

Aortic arch

Left pulmonary artery

Right pulmonary artery

The pulmonary circulation involves the flow of blood between the heart and lungs. In the lungs, blood gains oxygen and loses waste carbon dioxide.

The venous system

The systemic venous system carries blood back to the heart from the tissues. This blood is then pumped through the pulmonary circulation to be reoxygenated before entering the systemic circulation again.

Veins originate in tiny venules that receive blood from the capillaries. The veins converge upon one another, forming increasingly large vessels until the two main collecting veins of the body, the superior and inferior vena cavae, are formed. These then drain into the heart. At any one time, about 65 per cent of the total blood volume is contained in the venous system.

DIFFERENCES

The systemic venous system is similar in many ways to the arterial system. But, there are some important differences:

- Vessel walls – arteries tend to have thicker walls than veins to cope with the greater pressure exerted by arterial blood.
- Depth – most arteries lie deep within the body to protect them from injury, but many veins lie super-ficially, just under the skin.
- Portal venous system – the blood that leaves the gut in the veins of the stomach and intestine does not pass directly back to the heart. It first passes into the hepatic portal venous system, which carries the blood through the liver tissues before it can return to the systemic circulation.
- Variations – the pattern of systemic arteries tends to be the same from person to person, but there is greater variability in the layout of the systemic veins.

Major veins of the body

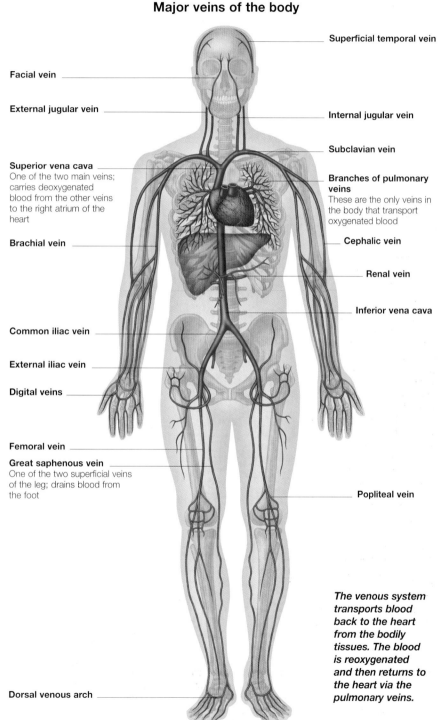

Superficial temporal vein

Facial vein

External jugular vein

Internal jugular vein

Subclavian vein

Superior vena cava
One of the two main veins; carries deoxygenated blood from the other veins to the right atrium of the heart

Branches of pulmonary veins
These are the only veins in the body that transport oxygenated blood

Brachial vein

Cephalic vein

Renal vein

Inferior vena cava

Common iliac vein

External iliac vein

Digital veins

Femoral vein

Great saphenous vein
One of the two superficial veins of the leg; drains blood from the foot

Popliteal vein

Dorsal venous arch

The venous system transports blood back to the heart from the bodily tissues. The blood is reoxygenated and then returns to the heart via the pulmonary veins.

239

Peripheral nervous system

The peripheral nervous system includes all the body's nerve tissue that is not in the brain and spinal cord. Its principal anatomical components are the cranial and spinal nerves.

The nervous system of the human body is divided into two parts: the central nervous system (CNS) and the peripheral nervous system (PNS).

The major components of the PNS are:
• Sensory receptors – specialized nerve endings which receive information about temperature, touch, pain, muscle stretching, and taste
• Peripheral nerves – bundles of nerve fibres which carry information to and from the CNS
• Motor nerve endings – specialized nerve endings which cause the muscle on which they lie to contract in response to a signal from the CNS.

ARRANGEMENT
Peripheral nerves are of two types:
• Cranial nerves
These emerge from the brain and are concerned with receiving information from, and allowing control of, the head and neck. There are 12 pairs of cranial nerves
• Spinal nerves
These arise from the spinal cord, each containing thousands of nerve fibres, to supply the rest of the body. Many of the 31 pairs of spinal nerves enter one of the complex networks, such as the brachial plexus which serves the upper limb, before becoming part of a large peripheral nerve.

Major nerves of the peripheral nervous system

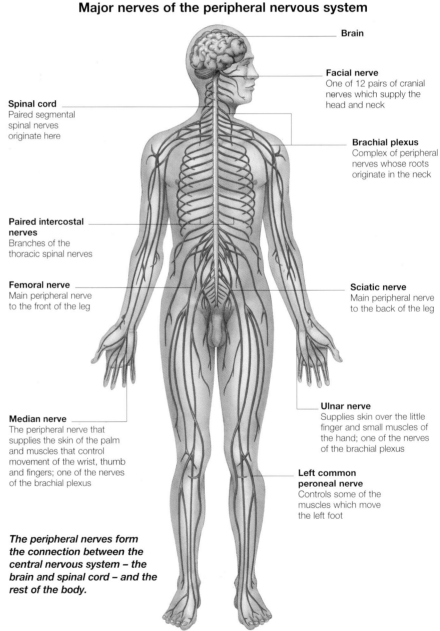

Brain

Facial nerve
One of 12 pairs of cranial nerves which supply the head and neck

Spinal cord
Paired segmental spinal nerves originate here

Brachial plexus
Complex of peripheral nerves whose roots originate in the neck

Paired intercostal nerves
Branches of the thoracic spinal nerves

Femoral nerve
Main peripheral nerve to the front of the leg

Sciatic nerve
Main peripheral nerve to the back of the leg

Median nerve
The peripheral nerve that supplies the skin of the palm and muscles that control movement of the wrist, thumb and fingers; one of the nerves of the brachial plexus

Ulnar nerve
Supplies skin over the little finger and small muscles of the hand; one of the nerves of the brachial plexus

Left common peroneal nerve
Controls some of the muscles which move the left foot

The peripheral nerves form the connection between the central nervous system – the brain and spinal cord – and the rest of the body.

Structure of a peripheral nerve

Each peripheral nerve consists of separate nerve fibres, some with an insulating layer of myelin, enclosed within connective tissue.

The greater part of the final bulk of a peripheral nerve is made up of three protective connective tissue coverings, without which the fragile nerve fibres would be vulnerable to injury.

• Endoneurium
The endoneurium is a layer of delicate connective tissue that surrounds the smallest unit of the peripheral nerve, the axon. This layer may also enclose an axon's myelin sheath.

• Perineurium
The perineurium is a layer of connective tissue that encloses a group of protected nerve fibres, called fascicles, that are tied together in bundles.

• Epineurium
Nerve fascicles are bound together by a tough connective tissue coat, the epineurium, into a peripheral nerve. The epineurium also encloses blood vessels which help to nourish the nerve fibres and their connective tissue coverings.

NERVE FUNCTION
Most peripheral nerves carry information to and from the central nervous system (sensory and motor functions respectively), and thus are known as 'mixed' nerves.
Nerves that are either purely sensory or purely motor are very rare within the body.

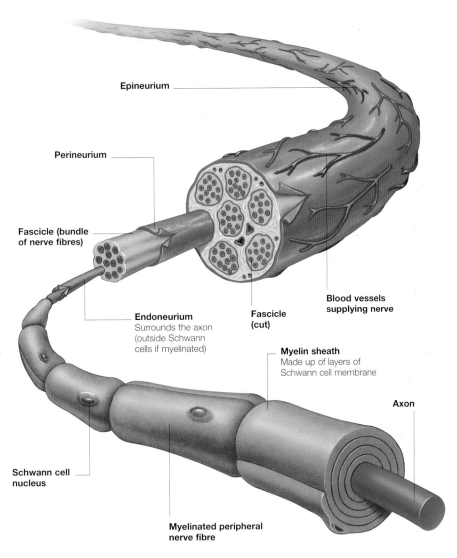

Epineurium

Perineurium

Fascicle (bundle of nerve fibres)

Endoneurium
Surrounds the axon (outside Schwann cells if myelinated)

Fascicle (cut)

Blood vessels supplying nerve

Myelin sheath
Made up of layers of Schwann cell membrane

Axon

Schwann cell nucleus

Myelinated peripheral nerve fibre

Peripheral nerve fibres are grouped together in bundles, called fascicles. These carry both sensory (afferent) and motor (efferent) fibres.

Autonomic nervous system

The autonomic nervous system provides the nerve supply to those parts of the body which are not consciously directed. It can be subdivided into the sympathetic and parasympathetic systems.

The autonomic nervous system is divided into two parts: the sympathetic system and the parasympathetic system. Both systems supply the same organs, but with opposing effects. In each system two neurones (nerve cells) make up the pathway from the central nervous system (CNS) to the organ which is being supplied.

SYMPATHETIC NERVOUS SYSTEM

The effects upon the body of stimulation by the sympathetic nervous system are often referred to as the 'fight or flight' response. In exciting or dangerous situations, the sympathetic nervous system becomes more active, causing the heart rate to increase and the skin to become pale and sweaty as blood is diverted to muscle.

STRUCTURE

The cell bodies of the neurones of the sympathetic nervous system lie within a section of the spinal cord. Fibres from these cell bodies exit the spinal column at the ventral root and pass through the white rami communicantes to the paravertebral sympathetic chain.

Some of the fibres which enter the sympathetic chain connect there with the second cell of their pathway. Fibres then exit through the grey rami communicantes to join the ventral spinal nerve.

Anatomy of a sympathetic trunk

Spinal cord

Vertebral body

Intervertebral disc

Thoracic splanchnic nerves
Pass downwards and forwards from the chain of sympathetic ganglia near the spinal cord

Paravertebral/ sympathetic (chain) ganglion
Comprises grouped nerve cell bodies; chains of ganglia run down each side of the spinal cord

Intercostal nerve
(Ventral ramus of spinal nerve)

Intercostal muscle

Sympathetic trunk

Grey ramus communicantes
Branch through which sympathetic fibres pass on their way to the ventral spinal nerve

White ramus communicantes
Branch through which sympathetic fibres pass on their way to the paravertebral sympathetic chain

The sympathetic nervous system prepares the body for action. Fibres leave the CNS via a chain of ganglia close to the spinal cord.

242

Parasympathetic nervous system

The parasympathetic system is the part of the autonomic nervous system which is most active during periods of rest.

The structure of the parasympathetic nervous system is simpler than the sympathetic nervous system.

LOCATION OF CELL BODIES

The cell bodies of the first of the two neurones in the pathway are located in only two places:

- The brainstem – fibres from the parasympathetic cell bodies in the grey matter of the brainstem leave the skull as part of a number of cranial nerves. Together, these fibres make up what is known as the cranial parasympathetic outflow
- The sacral region of the spinal cord – the sacral outflow arises from parasympathetic cell bodies which lie within part of the spinal cord. Fibres leave through the ventral root.

Because of the locations of the origins of parasympathetic fibres, the parasympathetic system is sometimes known as the craniosacral division of the autonomic nervous system; the sympathetic system is known as the thoracolumbar division.

DISTRIBUTION

The cranial outflow provides parasympathetic innervation for the head, and the sacral outflow supplies the pelvis. The area between (the majority of the abdominal and thoracic internal organs) is supplied by part of the cranial outflow which is carried within the vagus (tenth cranial nerve).

Organs controlled by the parasympathetic nervous system

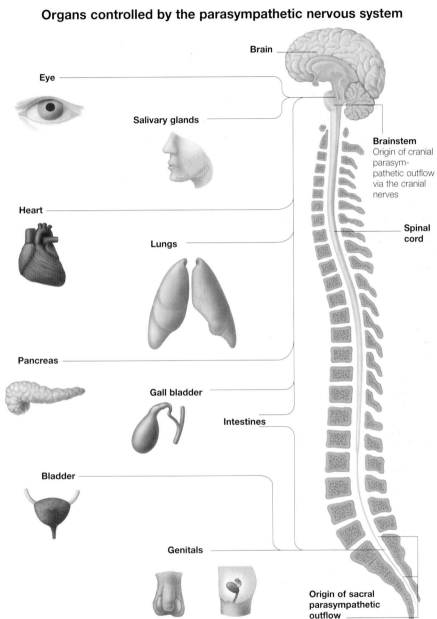

Brain

Eye

Salivary glands

Heart

Lungs

Pancreas

Gall bladder

Intestines

Bladder

Genitals

Brainstem
Origin of cranial parasympathetic outflow via the cranial nerves

Spinal cord

Origin of sacral parasympathetic outflow

The parasympathetic nervous system is most active when the body is at rest. Fibres leave the CNS from the brain and sacral region of the spinal cord.

Lymphatic system

The lymphatic system consists of a network of lymph vessels and organs and specialized cells throughout the body. It is an essential part of the body's defence against invading micro-organisms.

The lymphatic system works with the cardiovascular system to transport a fluid called lymph around the body. The lymphatic system plays a vital role in the defence of the body against disease.

LYMPH FLUID

Lymph is a clear, watery fluid containing electrolytes and proteins which is derived from blood and bathes the body's tissues. Lymphocytes – specialized white blood cells involved in the body's immune system – are found in lymph. They attack and destroy foreign micro-organisms. This is known as an immune response.

The vessels of the lymphatic system carry lymph, but the fluid is not pumped around the body as blood is; instead, contractions of muscles surrounding the lymph vessels move the fluid along.

CONSTITUENT PARTS OF THE LYMPHATIC SYSTEM

The lymphatic system is made up of these interrelated parts:
- Lymph nodes – lie along the routes of the lymphatic vessels and filter lymph
- Lymphatic vessels – small capillaries leading to larger vessels that eventually drain lymph into the veins
- Lymphoid cells (lymphocytes) – cells through which the body's immune response is mounted
- Lymphoid tissues and organs – scattered throughout the body, act as reservoirs for lymphoid cells

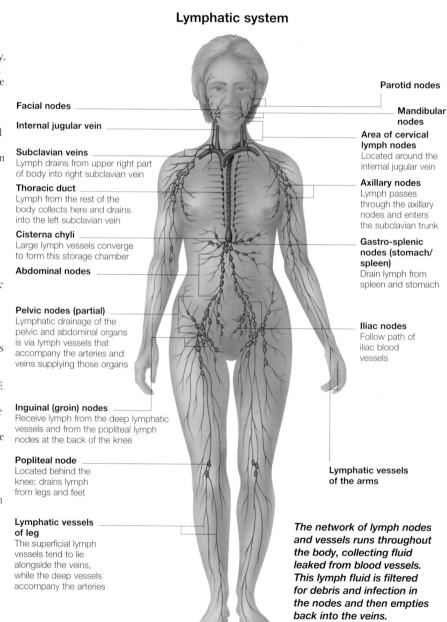

Lymphatic system

Parotid nodes

Facial nodes

Internal jugular vein

Mandibular nodes

Area of cervical lymph nodes
Located around the internal jugular vein

Subclavian veins
Lymph drains from upper right part of body into right subclavian vein

Axillary nodes
Lymph passes through the axillary nodes and enters the subclavian trunk

Thoracic duct
Lymph from the rest of the body collects here and drains into the left subclavian vein

Cisterna chyli
Large lymph vessels converge to form this storage chamber

Gastro-splenic nodes (stomach/ spleen)
Drain lymph from spleen and stomach

Abdominal nodes

Pelvic nodes (partial)
Lymphatic drainage of the pelvic and abdominal organs is via lymph vessels that accompany the arteries and veins supplying those organs

Iliac nodes
Follow path of iliac blood vessels

Inguinal (groin) nodes
Receive lymph from the deep lymphatic vessels and from the popliteal lymph nodes at the back of the knee

Popliteal node
Located behind the knee; drains lymph from legs and feet

Lymphatic vessels of the arms

Lymphatic vessels of leg
The superficial lymph vessels tend to lie alongside the veins, while the deep vessels accompany the arteries

The network of lymph nodes and vessels runs throughout the body, collecting fluid leaked from blood vessels. This lymph fluid is filtered for debris and infection in the nodes and then empties back into the veins.

Lymph nodes

Lymph nodes lie along the route of the lymphatic vessels. They filter the lymph for invading micro-organisms, infected cells and also other foreign particles.

Lymph nodes are small, rounded organs that lie along the course of the lymphatic vessels and act as filters of the lymph. Lymph nodes vary in size, but they are mostly bean-shaped, 1–25 mm in long, surrounded by a fibrous capsule and usually embedded in connective tissue.

LYMPH NODE FUNCTION

As well as fluid, the tiny lymphatic vessels in the tissues may pick up other items, such as parts of broken cells, bacteria and viruses. Within the lymph node, fluid slows and comes into contact with lymphoid cells which ingest any solid particles and recognize foreign micro-organisms. To prevent these particles from entering the bloodstream – and to allow the body to mount a defence against invading organisms – lymph is filtered through a number of lymph nodes before draining into the veins.

Some lymph nodes are grouped together in regions and given names according to their position, the region in which they are found (for example, the axillary nodes in the axilla, or armpit), the blood vessels they surround (such as the aortic nodes around the large central artery of the body, the aorta), or the organ they receive lymph from (pulmonary nodes in the lungs).

Structure of a lymph node

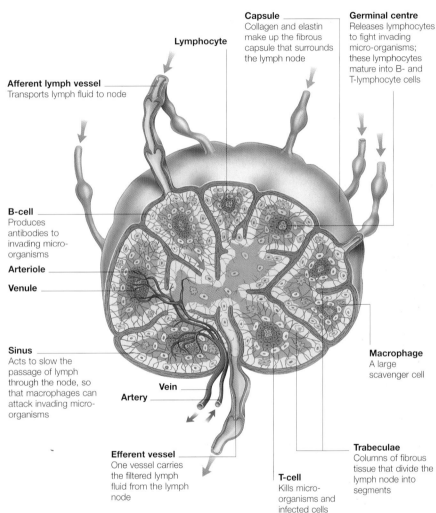

Capsule
Collagen and elastin make up the fibrous capsule that surrounds the lymph node

Germinal centre
Releases lymphocytes to fight invading micro-organisms; these lymphocytes mature into B- and T-lymphocyte cells

Lymphocyte

Afferent lymph vessel
Transports lymph fluid to node

B-cell
Produces antibodies to invading micro-organisms

Arteriole

Venule

Sinus
Acts to slow the passage of lymph through the node, so that macrophages can attack invading micro-organisms

Vein

Artery

Macrophage
A large scavenger cell

Efferent vessel
One vessel carries the filtered lymph fluid from the lymph node

T-cell
Kills micro-organisms and infected cells

Trabeculae
Columns of fibrous tissue that divide the lymph node into segments

The internal structure of a lymph node slows the passage of lymph fluid, so that specialized lymphocyte cells can filter out any micro-organisms.

Skin and nails

The skin, together with the hair and nails, makes up
the integumentary system. Functions of the skin include heat regulation
and defence against microbial attack.

Cross section of skin

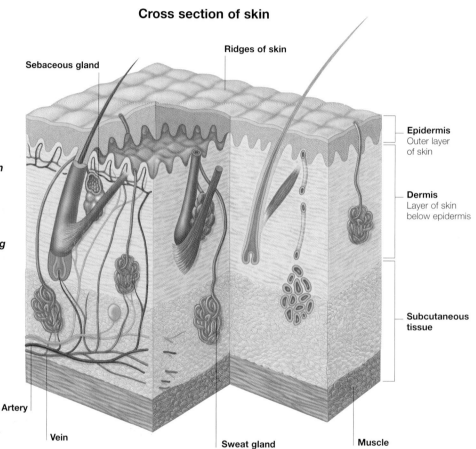

Sebaceous gland

Ridges of skin

Epidermis
Outer layer
of skin

Dermis
Layer of skin
below epidermis

*The skin has been
described as the
largest organ in
the body. It helps
to regulate
temperature
through narrowing
and widening of
blood vessels in
the dermis.*

Subcutaneous
tissue

Artery

Vein

Sweat gland

Muscle

The skin covers the entire human body and has a surface area of about 1.5–2 m2. It accounts for about 7 per cent of the weight of the body and weighs around 4 kg.

TWO LAYERS
Skin is composed of two layers – the epidermis and dermis.
• Epidermis – this is the thinner of the two layers of skin and serves as a tough protective covering for the underlying dermis. It is made up of numerous layers of cells, the innermost of which consist of living cube-shaped cells that divide rapidly, providing cells for the outer layers.
By the time these cells reach the outer layers, they have died and become flattened, before being 'sloughed off' by abrasion. The epidermis has no blood supply of its own, and depends upon diffusion of nutrients from the plentiful supply of blood to the dermis below
• Dermis – this is the thicker layer of skin, which lies protected under the epidermis. It is composed of connective tissue which has elastic fibres to keep it stretchy, and collagen fibres for strength. The dermis contains a rich supply of blood vessels as well as numerous sensory nerve endings. Lying within this layer are the other important structures of the integumentary system, including hair follicles and oil (sebaceous) and sweat glands.

246

Nails

Human nails are the equivalent of the hooves or claws of other animals. They form a hard protective covering for the vulnerable fingers and toes, and they provide a useful tool for scratching or scraping when this is required.

Cross section of nail

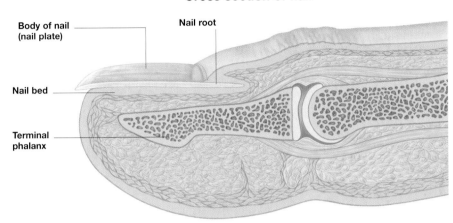

Body of nail (nail plate)

Nail root

Nail bed

Terminal phalanx

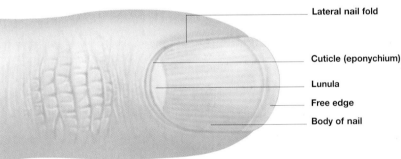

Nails provide protection for the delicate fingertips and toes, and in some cases act as a tool. They are formed from a durable substance known as keratin.

Lateral nail fold

Cuticle (eponychium)

Lunula

Free edge

Body of nail

Nails lie on the dorsal (back) surfaces of the ends of the fingers and toes, overlying the terminal phalanx, or final bone, of each.

CONSTITUENT PARTS
The parts of the nail include:
- Nail plate – each nail is composed of a plate of hard keratin (the same substance as is found in hair) which is continuously produced at its root
- Nail folds – except for the free edge of the nail, at its furthermost end, the nail is surrounded and overlapped by folds of skin (nail folds)

- Free edge – the nail separates from the underlying surface at its furthermost point to form a free edge. The extent of this nail at the free edge depends upon preference and wear and tear
- Root, or matrix – this lies at the base of the nail beneath the nail itself and the nailfold. This part of the nail is closest to the skin, and it is here that the hard keratin of the nail is produced by cell division. If the root of the nail is destroyed the nail cannot grow back

- Lunula – the paler, crescent-shaped area located at the base of the nail where the matrix is visible through the nail
- Cuticle (eponychium). This covers the proximal (near) end of the nail and extends over the nail plate to help protect the matrix from infection by invading micro-organisms.

GROWTH
Fingernails grow much more quickly than toenails. A mark made over the lunula of a fingernail will take three months to reach the free

edge, whereas the corresponding time for a toenail may be up to two years. For a normal rate of growth, and to produce normal, pink, healthy nails, there needs to be a good blood supply to the root of the nail; nails look pink because of the large number of blood vessels in the dermis. Nails grow at a rate of about 0.1mm a day, but when there is injury to a nail, the growth speeds up.

Index

248